Advances in Civil Engineering

Advances in Civil Engineering

Edited by **Armando Ruiz**

Willford Press

New York

Published by Willford Press,
118-35 Queens Blvd., Suite 400,
Forest Hills, NY 11375, USA
www.willfordpress.com

Advances in Civil Engineering
Edited by Armando Ruiz

International Standard Book Number: 978-1-68285-242-2 (Hardback)

Printed in China.

Contents

Preface

The main aim of this book is to educate learners and enhance their research focus by presenting diverse topics covering this vast field. This is an advanced book which compiles significant studies by distinguished experts. This book addresses successive solutions to the challenges arising in the area of application, along with it; the book provides scope for future developments.

Civil engineering is one of the oldest branches of engineering. The concepts and applications of this field can be found in every structure that surrounds us, from buildings to bridges to roads. This book elucidates the diverse principles and innovative models around prospective developments with respect to civil engineering. It strives to provide a fair idea about this discipline and to help develop a better understanding of the latest advances within this field through the various topics such as surveying, geotechnical engineering, etc. which have been discussed with an analytical perspective within this book. This book has been compiled by internationally renowned panel of experts and is highly recommended for students and researchers engaged in the field of civil engineering, architecture, and associated disciplines.

It was a great honour to edit this book, though there were challenges, as it involved a lot of communication and networking between me and the editorial team. However, the end result was this all-inclusive book covering diverse themes in the field.

Finally, it is important to acknowledge the efforts of the contributors for their excellent chapters, through which a wide variety of issues have been addressed. I would also like to thank my colleagues for their valuable feedback during the making of this book.

Editor

A New Approach for Modelling Variability in Residential Construction Projects

1

Mehrdad Arashpour, (RMIT University, Australia)
Ron Wakefield, (RMIT University, Australia)
Nick Blismas, (RMIT University, Australia)
Eric Wai Ming Lee, (City University, Hong Kong)

Abstract
The construction industry is plagued by long cycle times caused by variability in the supply chain. Variations or undesirable situations are the result of factors such as non-standard practices, work site accidents, inclement weather conditions and faults in design. This paper uses a new approach for modelling variability in construction by linking relative variability indicators to processes. The mass homebuilding sector was chosen as the scope of the analysis because data is readily available. Numerous simulation experiments were designed by varying size of capacity buffers in front of trade contractors, availability of trade contractors, and level of variability in homebuilding processes. The measurements were shown to lead to an accurate determination of relationships between these factors and production parameters. The variability indicator was found to dramatically affect the tangible performance measures such as home completion rates. This study provides for future analysis of the production homebuilding sector, which may lead to improvements in performance and a faster product delivery to homebuyers.

Keywords: Computer simulaion, Production, Project management, Queuing, Residential construction, Variability in supply chain, Optimim level of work-in-process inventory

Introduction
Simulation of construction processes has received much attention in recent years due to its ability to estimate the behaviour of systems in the presence of variability. Variations or undesirable situations that arise are the result of delays or interruptions in the workflow. Performance measures such as project completion time or resource utilization rates are very sensitive to changes in production variables.

Attention should be paid to address present variability in production systems otherwise the cost will be paid later on in forms of lost output (throughput) rate, wasted capacity, inflated completion (cycle) times, and poor customer service (Arashpour and Arashpour 2012).

Construction processes are different in nature with unequal levels of variability. In residential construction, for instance, an outdoor process such as roofing is more prone to inclement weather conditions comparing with an indoor process such as plumbing. Also, other factors such as accident risks differ from one process to another. In the construction management literature, some researchers have modelled the variability by means of longer mean process times (Walker and Shen 2002, Walsh, et al. 2007, Arashpour, et al. 2013) and some others by assuming a larger variance in process times (Peña-Mora and Dwivedi 2002, Sawhney, et al. 2009, Ghoddousi, et al. 2013). However, the negative influence of variability has been more precisely modelled in other sectors such as manufacturing. Using relative measures of variability have led to a more accurate measurement of system performance in the manufacturing sector (Hopp and Spearman 2008, Jeong, et al. 2011).

Evidence such as lengthened completion times and poor client service particularly during boom periods calls for new approaches for variability modelling in construction projects (Dalton, et al. 2011). On this basis, the present paper uses an innovative approach for modelling variability in residential projects by linking variability indicators to processes. Volume homebuilding was chosen as the scope of this analysis because data is readily available. A two-level hierarchical model was developed to represent the typical production of detached suburban houses in Melbourne, Australia. Several simulation experiments were then conducted by varying: 1. Size of the buffers (queue of jobs to be processed) in front of trade contractors; 2. Level of resource availability; 3. Variability level in the production homebuilding network. In this paper, the effects of variability on the key performance measures such as project completion times and resource utilisation rates are explored.

Review of the Existing Approaches to Model and Address Variability in the Construction Industry

Data obtained in previous studies indicate that variability, which is non-uniformity in building processes, always degrades the performance and productivity measures in construction projects (Moyal 2010, Chia, et al. 2012). Existing strategies are discussed in this section:

A. Using Capacity Buffers against Production Variability

Construction processes are usually defined by the trade contractors who are responsible for them. Buffers between processes can prevent downstream trade contractors to become idle when upstream contractors experience a delay (González, et al. 2011, Koskela and Ballard 2012). Disadvantages of large buffers between interacting trade contractors include a large work-in-process (WIP) inventory and higher costs. In order to investigate this approach to model and address variability, different capacity buffer sizes are modelled and compared in the first and second simulation experiments in the next section.

B. Increasing Resource Availability

Availability of trade contractors can directly affect the completion time of construction processes. During boom periods, homebuilders often use more trade contractors or overtime as buffers against undesirable situations in the work sites (Arashpour, et al. 2012). By authorization of over-time the work capacity increases temporarily and overtakes the demand rate. However, the cycle of over-time would start again by any future randomness in demand or production rate (Hopp and Spearman 2004). Any change in availability of resources has an impact on costs, similar to the crashing concept in project planning. The third simulation experiment in this paper focuses on resource availability.

C. Variability Reduction Approaches

Different approaches are available to reduce the variability level in the mass production homebuilding sector. For example, complications at work sites can be decreased by using modular designs. Furthermore, using prefabrication, modularization and preassembly can dramatically leverage the constructability (Blismas, et al. 2010). Another initiative is to use advanced design and marketing methods, which enables the construction firms to schedule the production in advance (Bouchlaghem, et al. 2005, Veryzer 2005).

Flow-smoothing is another way of reducing variability in the construction environment. Different techniques can be used for this aim such as standardizing construction practices (Carlos, et al. 2002), quality management and reducing rework (Henry 2000), and applications of lean principles in industrialized housing production (Höök and Stehn 2008). Furthermore, variability caused by project-based subcontractors can be decreased by developing long-term business relationships with them. In this way, much of the capacity buffer against variability is carried by subcontractors (Kumaraswamy and Matthews 2000, Greenwood 2001). The variability reduction approach has been modelled and analysed in the fourth simulation experiment.

Research Design

Interconnected work processes are main building blocks of construction projects. They are performed either serially or in parallel until the project is completed. In the first step of this study, process times were plotted for main processes in volume homebuilding projects (see Figure 1). Then, statistical parameters of the data were calculated to perform chi-square check (Halpin and Woodhead 1976). Care was taken to match the process times to the optimum statistical distribution.

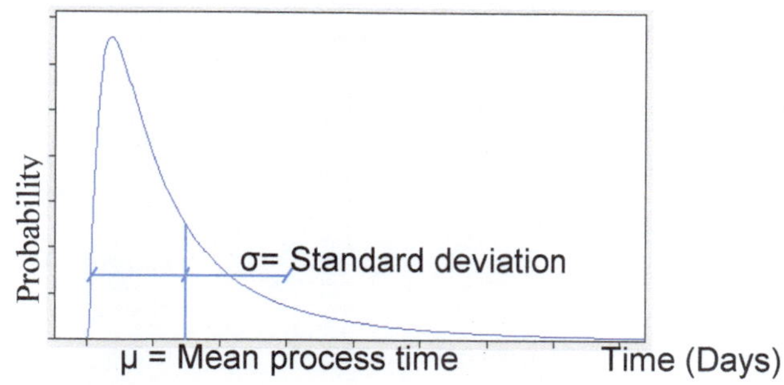

Figure 1 Probability density of cycle times for the construction production system

In the next step, the house building model was developed using similar method to Bashford, et al. (2003). The ARENA simulation software was selected for modelling due to its flexibility in using both ready-to- use constructs and user-written codes by the general-purpose procedural language SIMAN. User-written codes enable precise modelling of unique situations in the production homebuilding sector such as several hand-offs (workflows) among trade contractors. Numerous experiments were designed by varying the size of the capacity buffer between trade contractors, availability level of trade contractors, and level of variability in homebuilding processes.

A new indicator to measure the relative (not absolute) variability were introduced and used in simulation experiments. Then, in order to check the validity of results, they were verified against Little's law, which is a basic equation used in manufacturing management. Finally, conclusions were drawn based on the comparison of results.

Case Study

The typical process of building detached suburban homes in Melbourne, Australia was modelled. Extensive production data are usually kept in volume homebuilding, which makes this sector an ideal subject for investigation. Allocating an ID code to each trade contractor enabled us to trace upstream and downstream processes and analyse the effects of resource availability. In the Australian mass homebuilding sector, all the main building processes are subcontracted to trade contractors. Table 1 shows the list of operations for 20 selected subcontractors.

In production homebuilding, the builder is solely focused on sales and construction management. Subcontractors are in charge of performing construction operations (Walsh, et al. 2004). Due to congestion in work sites, subcontractors are required to finish their job quickly and vacate the workface for the next contractor. The transfers of work among trade contractors is sometimes called 'hand-offs' and becomes complicated by increasing the number of involved trade contractors.

Simulation is a suitable approach to empirical work as it is very costly to play with real systems and examine (pre-test and post-test) their behaviour upon changes in input variables (Fellows and Liu 2008, Martinez 2010). Despite the fact that performance measures in simulated systems involving variability might be subject to error, long simulation runs allow production systems to stabilize and achieve reliable outputs (Hopp and Spearman 2008)[1].

Selected homebuilding elements			
Process	**Subcontractor ID**	**Process**	**Subcontractor ID**
Site preparation	1	Drywall	11
Foundation	2	Trim carpentry	12
Framing	3	Plumbing fit-out	13
Brickworks	4	Electrical fit-out	14
Roofing	5	Painting	15
HVAC rough in	6	Tiling	16
Plumbing rough-in	7	Flooring	17
Electrical rough in	8	External paving	18
Cladding	9	Cleaning	19
Insulation	10	Finishing and handover	20

Table 1 Construction processes and related subcontractors

Variability in Process Times

The mean construction process time (μ) is not fixed and there is always variability around each process. The variability can be caused by several factors such as queuing time to use resources, rework, inclement weather conditions, and accidents on the work site. Both commonly used parameters of mean (μ) and standard deviation (σ) of construction process times reflect absolute variability. However, relative variability is more important in the production process (Hopp, Iravani et al. 2011). As an example, consider a 2 mm dimension error that is not critical in the thickness of foundation slabs. The same error, however, can affect the stability and internal tensions of structural elements if it is a deviation from the vertical access of columns. Therefore, a relative Variability Indicator (VI) can be a very robust parameter in analysing construction processes. We propose VI to be the standard deviation of a given construction process time divided by the mean process time:

$$VI = \frac{\sigma}{\mu}$$

Variability indicator is similar to the coefficient of variation in manufacturing proposed by Hopp and Spearman (2008).The key contribution of the proposed approach is to enable

[1] Some might suggest that the assumptions used to build the model are not supported by the state of the building industry. However, it is worth mentioning that all models are abstractions of reality. While there is a considerable debate about how realistic the assumptions of a model need to be, there is a general agreement on accurate prediction as the major aim of any model. In this way, the validity of assumptions is of the second importance. Friedman (1953) argued that a "useful" theory should be judged not by its descriptive realism but by its simplicity and fruitfulness as an engine of prediction. In other words, the value of a model is an empirical question – how useful it is, and how well it predicts. Therefore, the validity of a model cannot be settled by theoretical arguments but only by empirical investigations.

homebuilders to evaluate the long term performance of trade contractors and consider both mean and standard deviation of process times.

Trade contractors closely interact in the interconnected network of projects. In this way, the departure rate (r_d) of a predecessor is the arrival rate (r_a) for the successor:

r_d (subcontractor #1) = r_a (subcontractor #2)

Figure 2 illustrates how two arbitrary trade contractors are interconnected.

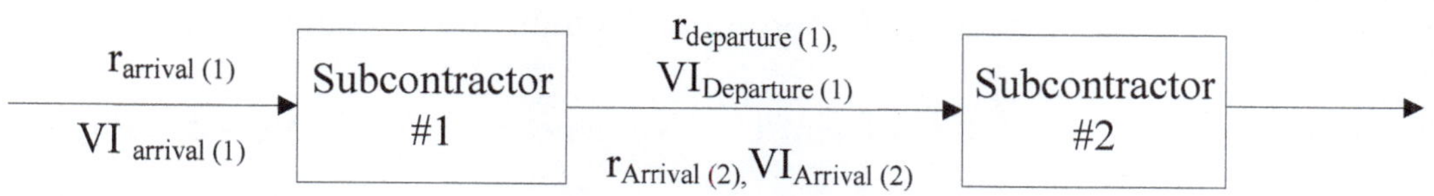

Figure 2 Illustration of the flows of work and variability among trade contractors

Since several interacting contractors are involved in the complex operations of house building, it is logical to consider the maximum randomness for completion times and also job arrival rates. That is, the mean and standard deviation of construction process times can be represented by exponential distribution (VI =1). In this way, once a trade contractor undergoes a very long process time due to bad weather conditions or accidents in the work site, the following trade contractor becomes idle.

Size of the Capacity Buffers in front of Each Trade Contractor

In the absence of variability, the optimum number of houses under construction is equal to the number of trade contractors. This minimises the completion time and keeps every trade contractor busy at all times. This special level of work-in-process (WIP) inventory is called critical WIP (W_0). Upon the presence of variability, average completion time of each house will inflate. To improve the situation, the first two experiments were conducted to find the optimum size of the capacity buffers in order to optimize tangible performance measures: system throughput rate (TH), house completion time (CT), and the number of houses under construction (WIP).In the first experiment, the size of the capacity buffers in front of each trade contractor is quite large and up to 3 houses can stand in a queue to be processed. Exponentially distributed process times introduce the maximum randomness to construction operations.

The second experiment decreased the size of the capacity buffers to only one house. It is worth mentioning that the policy used here is very similar to Kanban squares that are used in manufacturing production lines. At the end of each experiment WIP/W_0 was calculated in order to quantitatively determine how efficient the homebuilding network is working.

Number of Trade Contractors (Resource Availability)

In the third experiment construction processes were accelerated by increasing the resource availability level. Using two dedicated (available for 100 per cent of time) trade contractors for each process resulted the mean process time decreasing to almost half. Similar to the second simulation experiment, a small capacity buffer of one job in front of each trade contractor was used.

Level of Variability in Construction Process Times

In the previous experiments we assumed maximum randomness in the homebuilding network (VI = 1). Variability can be decreased by smoothing the work flow, upgrading the quality of operations in order to minimise the amount of rework, avoid delaying successors, and reducing accidents by means of improved safety measures (Arashpour and Arashpour

2010). The Variability Indicator (VI) of the processes was decreased from 1 to half in the fourth simulation experiment. Trade contractors can be promoted to a new class of variability by reducing the ratio of mean process time to standard deviation over the long term.

Experiments and Output Analysis

Care was taken to model experiments as close as possible to a typical production homebuilding scenario in Melbourne, Australia. For this reason, deterministic distributions for process times and job arrivals were not used. Each experiment was replicated 20 times in order to achieve statistical accuracy in the results. The desired confidence interval was 95% in our study. The simulation experiments were run for 1000 working days. Table 2 illustrates a quantitative comparison of average performance metrics in production runs.

Parameters	Experiment 1	Experiment 2	Experiment 3	Experiment 4
Size of the capacity buffer	3	1	1	1 (houses)
Variability indicator (VI)	1	1	1	0.5
Average throughput rate (TH)	0.55	0.54	0.56	0.58
Resource utilization rate	91.8%	89%	93%	96%
Averagecycle time (CT)	195	135	115	120 (days)
WIP inventory (houses)	36	24	22	23
Per cent of the optimum WIP	180%	122%	180%	116%

Table 2 Performance measures of the volume homebuilding network in the four simulation experiments

In order to cross-check the precision of results and for the sake of verification, the outputs of simulation modelling were compared with an analytical model. Due to the long simulation period, which let the production network reach its steady state, Little's law was selected for verification. Little's law is a queuing formula, which is widely used in manufacturing, in order to predict the performance measures of steady state systems over the long run (Little 2011). It correlates the work-in-process inventory (WIP) to the throughput (TH) rate and completion time (CT):

$$WIP = TH \times CT$$

Experiment	WIP inventory (simulation results)	WIP inventory (analytical results)
1	36	35.7
2	24	24.3
3	22	21.4
4	23	23.2

Table 3 Verification of the outputs (Quantitative comparison of simulation and analytical results)

A two-sample t-test was conducted to verify the results of simulation experiments. No statistically significant difference was found between the performance measures computed by simulation and queuing theory. Table 3 illustrates the results for one of the measures.

As can be seen, running the simulation over the long term caused the system to stabilize and consequently our results complied with Little's law.

Relationship between Capacity Buffer and Production Parameters

In the first experiment and by using a capacity buffer of three houses, the number of homes under construction reached a peak of 36. Consequently, the average cycle time for a single house was inflated to 195 days, as in table 2. This indicates that although capacity buffers prevent downstream contractors from work starvations and idleness, increasing the number of houses under construction results in lengthened completion times. There is a similar situation during construction boom periods, when demand exceeds the capacity of trade networks and houses have to stand in long queues before being processed. Large capacity buffers create big WIP inventories resulting in late completions and decreased service level.

By decreasing the size of capacity buffers in front of trade contractors to 1 house, average completion times decreased dramatically in our second experiment. It is worth mentioning that no extra resources and investment are needed. Improvements in this scenario are the results of changing the control and management policies by limiting the size of capacity buffers. In the second experiment, the number of houses under construction declined to 24, which reduced the average completion time to 135 days (see table 2). The construction output rate (TH) is slightly less than TH in the first production scenario. This is because of occasional job starvations that downstream trade contractors undergo.

Relationship between Resource Availability and Production Parameters

In the third simulation experiment, by increasing the level of resource availability, the average house completion time decreased to 115 days. Furthermore, resource utilization level stood at 93 per cent. This is more than utilization level of 89 per cent in the second experiment where trade contractors occasionally were idle. Although the third experiment achieved the shortest average CT, trade-offs need to be made as reducing the mean value of construction process times here is linked to employing more trade contractors and costs might offset the revenue (unlike the second scenario with no extra costs).

Relationship between VI and Production Parameters: Applications of the New Variability Modelling Approach

The relative variability indicator introduced in this study can differentiate construction processes. Different policies can be used by trade contractors to reduce mean process times and consequently VI. These include avoiding rework by improving quality controls and preventing workflow interruption by improving safety measures.

In the fourth simulation experiment, by reducing the variability indicator to half, a completion time of 120 days was achieved, which is almost identical to the third experiment with its necessary investments. The number of houses under construction is 23, which is very close to the number of involved trade contractors (there is almost no capacity buffer in front of the trades contractors). The work-in-process inventory in this experiment is only 16% more than the optimum critical WIP. In other words, the homebuilding network worked more efficiently comparing with other production experiments (116% of the optimum WIP was the minimum amongst all experiments).

Overall, although there are several opportunities on construction sites to buffer against variability, there are several advantages in attempting to reduce variability. Successful variability reduction strategies, through custom-designed policies, could be implemented in a firm's future projects (Hopp and Spearman 2008). Additionally, improving a specific construction process by finding the source of excess variability would create the mind-set of variability reduction and a culture of continual improvement within the homebuilding networks (Arashpour and Farzanehfar 2011).

Conclusion

Data obtained in previous studies indicate that variability is not accurately modelled and addressed in construction projects. This fact in the mass homebuilding sector results in inflated house completion times, reduction in outputs, and higher capital costs for homebuyers (Bashford, et al. 2005, Arashpour, et al. 2013). In the construction management literature, variability has been mostly modelled by assuming longer process times (pessimistic durations) and/or a larger variance in process times.

To bridge this gap, the present paper has modelled the variability in the production homebuilding sector using an innovative approach. Several experiments were designed by varying the size of the capacity buffers between trade contractors, the availability of trade contractors, and the intensity of the variability indicator in homebuilding processes. We found that there are solid relationships between these factors and production parameters. The findings extend those of Kamat and Martinez (2008) and Li, et al. (2009), confirming that tracing, modelling and addressing sources of variability in construction can lead to achieving optimum performance measures.

The key contribution of the proposed approach is to enable homebuilders to evaluate the long term performance of their trade contractors and decide on the best size of the capacity buffers (queue length of houses to be processed) in front of each trade. Although we investigated the production homebuilding sector, results may be generalized to other sectors within the construction industry.

Future research could include works designed to model variability and investigate its effects on production parameters. The variables within construction projects are numerous and the underlying logic for many system behaviours in the construction sector is still unknown.

Acknowledgements

Authors would like to thank anonymous reviewers of the Australasian Journal of Construction Economics and Building whose constructive comments and recommendations brought more depth to this investigation. Furthermore, the kind support of Ms Jennifer Anderson in RMIT Study and Learning Centre is highly appreciated.

References

Arashpour, M. and M. Arashpour, (2010) 'Gaining the best value from hr in construction companies', *Proceedings of the 6th European Conference on Management Leadership and Governance*, 23-33

Arashpour, M. and M. Arashpour (2012) 'A collaborative perspective in green construction risk management', *37th Annual Conference of the Australasian Universities Building Educators Association (AUBEA)*, University of New South Wales

Arashpour, M. and P. Farzanehfar (2011) Project Management and Control, vol. 1, Arkan

Arashpour, M., M. Shabanikia and M. Arashpour (2012) 'Valuing the contribution of knowledge-oriented workers to projects: a merit based approach in the construction industry', *Australasian Journal of Construction Economics and Building,* **12** (4), 1- 12

Arashpour, M., R. Wakefield and N. Blismas (2013) 'Improving construction productivity: implications of even flow production principles', *CIB World Building Congress 2013: Construction and Society*, Queensland University of Technology

Arashpour, M., R. Wakefield and N. Blismas (2013) Role of simulation in construction processes-harmony in capturing resources, Research, Development and Practice in Structural Engineering and Construction (ASEA-SEC), Research Publishing Services (RPS)

Bashford, H. H., A. Sawhney, K. D. Walsh and K. Kot (2003) 'Implications of even flow production methodology for U.S. housing industry', *Journal of Construction Engineering and Management,* **129** (3), 330-337

Bashford, H. H., K. D. Walsh and A. Sawhney (2005) 'Production system loading-cycle time relationship in residential construction', *Journal of Construction Engineering and Management,* **131** (1), 15-22

Blismas, N., R. Wakefield and B. Hauser (2010) 'Concrete prefabricated housing via advances in systems technologies - Development of a technology roadmap', *Engineering, Construction and Architectural Management*, **17** (1), 99-110

Bouchlaghem, D., H. Shang, J. Whyte and A. Ganah (2005) 'Visualisation in architecture, engineering and construction (AEC)', *Automation in Construction*, **14** (3), 287-295

Carlos, T. F., A. Dos Santos and A. P. James (2002) 'An exploratory study on the applicability of process transparency in construction sites', *Journal of Construction Research,* **3** (01), 35-54

Chia, F. C., M. Skitmore, G. Runeson and A. Bridge (2012) 'An analysis of construction productivity in Malaysia', *Construction Management and Economics*, **30** (12), 1055-1069

Dalton, T., R. Wakefield and R. Horne (2011) 'Australian suburban house building: industry organisation, practices and constraints', *Australian Housing and Urban Research Institute Positioning Paper Series*, 1-56

Fellows, R. F. and A. M. M. Liu (2008) *Research Methods for Construction*, Wiley-Blackwell

Friedman, M. (1953) *The Methodology of Positive Economics*

Ghoddousi, P., E. Eshtehardian, S. Jooybanpour and A. Javanmardi (2013) 'Multi-mode resource-constrained discrete time-cost-resource optimization in project scheduling using non-dominated sorting genetic algorithm', *Automation in Construction,* **30,** 216-227

González, V., L. Alarcón, S. Maturana and J. Bustamante (2011) 'Site Management of Work-in-Process Buffers to Enhance Project Performance Using the Reliable Commitment Model: Case Study', *Journal of Construction Engineering and Management,* **137** (9), 707-715

Greenwood, D. (2001) 'Subcontract procurement: are relationships changing?', *Construction Management & Economics,* **19** (1), 5-7

Halpin, D. W. and R. W. Woodhead (1976) *Design of Construction and Process Operations*, John Wiley & Sons, Inc

Henry, E. (2000) 'Quality management standardization in the French construction industry: singularities and internationalization prospects', *Construction Management & Economics,* **18** (6), 667-677

Höök, M. and L. Stehn (2008) 'Applicability of lean principles and practices in industrialized housing production', *Construction Management and Economics,* **26** (10), 1091-1100

Hopp, W. J., S. M. R. Iravani and Z. Liu (2011) 'Mitigating the Impact of Disruptions in Supply Chains', *Supply Chain Disruptions: Theory and Practice of Managing Risk*, 21

Hopp, W. J. and M. L. Spearman (2004) 'To pull or not to pull- what is the question?', *Manufacturing and Service Operations Management,* **6** (2),133-148

Hopp, W. J. and M. L. Spearman (2008) *Factory Physics*, McGraw-Hill Irwin Irwin

Jeong, J. G., M. Hastak, M. Syal and T. Hong (2011) 'Internal relationship modeling and production planning optimization for the manufactured housing', *Automation in Construction,* **20** (7), 864-873

Kamat, V. R. and J. C. Martinez (2008) 'Software mechanisms for extensible and scalable 3D visualization of construction operations', *Advances in Engineering Software,* **39** (8), 659-675

Koskela, L. and G. Ballard (2012) 'Is production outside management?', *Building Research & Information,* **40** (6), 724-737

Kumaraswamy, M. M. and J. D. Matthews (2000), 'Improved subcontractor selection employing partnering principles', *Journal of Management in Engineering,* **16** (3), 47-57

Li, H., N. Chan, T. Huang, H. L. Guo, W. Lu and M. Skitmore (2009) 'Optimizing construction planning schedules by virtual prototyping enabled resource analysis', *Automation in Construction,* **18** (7) 912-918

Little, J. D. C. (2011) 'Little's Law as Viewed on Its 50th Anniversary', *Operations Research,* **59** (3), 536-549

Martinez, J. C. (2010) 'Methodology for Conducting Discrete-Event Simulation Studies in Construction Engineering and Management', *Journal of Construction Engineering and Management-Asce*, **136** (1), 3-16

Moyal, P. (2010) 'The queue with impatience: construction of the stationary workload under FIFO', *Journal of Applied Probability,* **47** (2), 498-512

Peña-Mora, F. and G. H. Dwivedi (2002) 'Multiple device collaborative and real time analysis system for project management in civil engineering', *Journal of Computing in Civil Engineering,* **16** (1), 23-38

Sawhney, A., K. D. Walsh, H. H. Bashford and S. Palaniappan (2009) 'Impact of inspected buffers on production parameters of construction processes', *Journal of Construction Engineering and Management***, 135** (4), 319-329

Veryzer, R. W. (2005) 'The Roles of Marketing and Industrial Design in Discontinuous New Product Development', *Journal of Product Innovation Management,* **22** (1), 22-41

Walker, D. H. T. and Y. J. Shen (2002) 'Project understanding, planning, flexibility of management action and construction time performance: two Australian case studies', *Construction Management & Economics*, **20** (1) 31-44

Walsh, K. D., H. H. Bashford, A. Sawhney and A. Witjakso (2004) 'Cost of risk transfer: Pricing agreements in residential supply chains', *Journal of Architectural Engineering,* **10** (3), 112-118

Walsh, K. D., A. Sawhney and H. H. Bashford (2007) 'Production equations for unsteady-state construction processes', *Journal of Construction Engineering and Management,* **133** (3), 254-261

Management Strategies to Harness Cultural Diversity on Australian Construction Sites – a Social Identity Perspective

2

Martin Loosemore, (University of New South Wales, Australia)
Florence Phua, (University of Reading, UK)
Melissa Teo, (Queensland University of Technology, Australia)
Kevin M. Dunn, (University of Western Sydney, Australia)

Abstract

Construction sites around the world employ large numbers of people from diverse cultural backgrounds. The effective management of this cultural diversity has important implications for the productivity, safety, health and welfare of construction workers and for the performance and reputation of firms which employ them. The findings of a three year, multi-staged study of cultural diversity management practices on construction sites are critiqued using social identity theory. This reveals that so called "best-practice" diversity management strategies may have an opposite effect to that intended. It is concluded that the management of diversity on construction projects would benefit from being informed by social identity research.

Keywords: Cultural diversity, Racism, Discrimination, Social identity theory.

Introduction

Major construction projects in most countries employ large numbers of ethnic minorities (Ofori 1994; Hay 1996, Loosemore and Lee 2003, Gale and Davidson 2006, Dainty et al 2007). For example, in Australia, Loosemore et al's (2008) survey of 1150 site operatives found that 57% spoke a language other than English at home, 54% had been born in another country and 70% of parents had been born in another country. Loosemore and Lee's (2003) analysis of Australian construction sites found that large construction sites segregate into ethnically-based occupational groups and cultural ghettos characterised by language and cultural barriers which exclude outsiders (especially those from different ethnic groups). According to Loosemore and Andonakis (2006) and Loosemore et al (2010), these divisions present project managers with a complex dynamic of linguistic and cultural interfaces – some posing more challenges than others – which can affect performance in areas such as productivity, waste, quality and safety. Certain groups are more likely to be seen as problematic by others and that intolerance towards them was more likely to be expressed by those who did not speak a language other than English. In contrast, multilingual speakers were more likely to be tolerant towards other groups. Furthermore, cultural differences and stereotypes play a major role in shaping workers' perceptions of other groups. For example, there is a widespread belief among workers that Australian-Lebanese have a different work culture reflected in their aggressive behaviours. In contrast, intolerance towards Asian-Australians mostly occurs on the grounds that Asian-Australians are perceived to dramatically reduce safety standards, wages and professional quality on site. Other research by Dunn et al (2009) indicate that these intergroup attitudes and perceptions might be associated with and influenced by tensions between the two groups within broader society, where similar attitudes are often expressed.

These types of attributes and challenges make the construction industry an especially important setting to study the management of cultural diversity. While other industries such as apparel, pharmaceuticals, entertainment media, information technology, retailing and fast moving consumer goods manage diversity as an asset to improve productivity, innovation and profitability (Cox and Blake 1991, De Cieri and Kramar 2003), management strategies used in the construction industry apparently fail to harness its productive potential and this appears to be a common characteristic of the industry globally (Steele and Sodhi 2006, Dainty et al 2007). Within this context, the aim of this paper is to explore and critique current strategies used in the industry to manage cultural diversity and to critique, using social identity theory, the strategies employed by what are perceived to be "best practice" firms in other industries.

A Social Identity Perspective on the Management of Diversity

As Loosemore et al (2003) and Dainty et al (2007) point out, the construction sector represents one of the most complex and problematic arenas within which to manage people. The challenges inherent in managing a highly diverse workforce distributed across temporary coalitions of loosely coupled firms are significant. And these challenges are exacerbated by the operational realities faced by construction organizations which mean that all too often the needs of those who work for construction firms are subjugated by performance concerns. According to Dainty et al (2007:4) this approach to human resource management reflects rationalist management paradigms which are "strangely disconnected' from theoretical developments in other disciplines". Such approaches underplay the importance of people as a key resource and have likely ingrained many negative assumptions into operational and management practices relating to the value of diversity in the construction sector. Social identity theory offers a new and potentially powerful paradigm to better understand these issues and reframe management practices. This theory offers a theoretical framework for understanding the cognitive mechanisms by which corporative intergroup relations are presumed to work. In doing so it has laid the conceptual foundations for understanding the factors and conditions that encourage in-group favouritism and out-group discrimination to occur. Social identity theory rests of two basic premises. The first premise is that people understand the social world through categorisation of continuous social variables into discrete classes which has the effect of minimising perceived differences but accentuating inter-category differences (Tajfel 1970). The second premise is that this social categorisation carries with it implicit in-group and out-group (us and them) distinctions. Van Oudenhoven et al (1996) point out in their investigation of interethnic attitude formation, in order to engage member identification and loyalty, groups must develop clear rules for inclusion and exclusion which in turn gives rise to "clique selfishness", ethnocentrism and other forms of in-group ethnic bias which encourage we-they distinctions and further demark boundaries of loyalty and cooperation between ethnic groups. In turn, the attitudes and values of group members become shaped by these in-group and out-group distinctions in that people from other groups are viewed as inferior from an in-group perspective. The obvious conclusion from this theory was that one should eliminate difference, or at least play it down, and pursue the "melting pot" assimilationist approach to diversity management. However, as Brewer (1997) points out, while this would seem to a well founded strategy, implementation of this disaggregation approach in real world settings has proved less than successful and has met with deep resistance to attempts at interethnic assimilation, integration and mergers. Consequently, many social psychologists reconceptualised the importance of in-group differentiation and identification to social relations and advocated an alternative "pluralistic" model for managing inter group relations which recognises the importance of difference (Brewer 2009). This approach argues that distinctive group identities can be complementary and mutually respected leading to positive interdependencies and attributes which can be capitalised on through effective management. These pluralistic notions of "positive distinctiveness" underpin what is now widely known as multiculturalism. However, notions of multiculturalism have also been problematic when it has been used to preserve unequal structures in societies by legally locking-in differential group access to power and resources. The creation of privileged group status has the effect of encouraging monolithic perceptions of cultural groups, reducing opportunities for inequalities to be eliminated and for a common identity to arise (conditions which have long been associated with reduced discrimination in society). In this way, differences become institutionalised. The fact that both assimilationist and multicultural strategies appear to be equally problematic poses a dilemma for managers who have to operate in a multicultural setting. The answer according to Brewer (1997) is to capitalise more effectively on people's capacity for multiple social identities. In other words, people are capable of being members of and contributing to more than one social group and these cross-cutting identities are the secret to reducing the social comparisons and stereotypes that motivate intercultural conflict. Brewer's research shows that most people have a multicultural heritage and are capable of juggling multiple identities over their lifetime. This approach has very different implications for the development of cultural diversity management strategies in the construction industry. For example, it would question the common practice of making all signage and documentation available in many languages and promote the alternative idea of encouraging bilingualism by teaching everyone to speak two languages (at least in the restricted vocabulary needed to work safety on site etc). The first strategy reinforces difference and teaches people to think of each other as members of distinct groups whereas the second strategy eliminates difference and encourages "collective multiculturalism" as a shared value.

Method

Using the above ideas, a three-staged method was employed to explore the attitudes that underpin cultural diversity in the Australian construction industry, the management challenges that these pose and the range of strategies that can be employed to manage them more effectively.

		per cent (%)
Gender	Male	100
	Female	0
Age	under 18	1
	18-34	42
	35-54	47
	55-64	10
	65 & over	1
Education	No schooling	2
	Primary	7
	Secondary	83
	Bachelors degree	8
	Postgraduate	1
Birthplace	Australia	47
	Non-English speaking countries	40
	English speaking countries	13
Language other than English	Yes	57
	No	43
Religion	Christian	71
	no religion	15
	Islam	6
	All other beliefs	3
	inadequately described	5

Table 1 Respondent demographics, Operatives' survey, Sydney construction sites, 2008

In stage one, we undertook a survey of operatives to identify attitudes about cultural diversity and towards other cultural groups; the positive and negative functions of cultural groupings; the impact of discrimination and racism on constructing inter-communal relations and work practices; and finally existing policies and strategies in dealing with cultural diversity on Australian construction sites. The survey was administered on twenty-nine construction sites in the Sydney metropolitan area. The sites were chosen as typically representative of large construction sites with culturally diverse workforces and most of the questions were closed-ended in yes-no format since many operatives were from a lower educational background and had poor English language skills (particularly Chinese and Koreans). Where necessary we translated questions using an interpreter (most subcontractors have a "gatekeeper" who translates for the rest of the group). During the survey sessions which were administered in tea and smoking breaks on site, respondents were also permitted to openly discuss the questions which provided further insights into the issues being investigated.

Participation in the survey was entirely voluntary and a useable sample of 1155 questionnaires were collected (See Table 1 for sample structure).

The second stage survey aimed to explore supervisors' perceptions of cultural diversity on construction sites and to explore the strategies they used to manage diversity. It was distributed on sixteen construction sites and administered by hand during weekly subcontractors meetings where all managers, supervisors, foremen and engineers could be found at the same time. Participation in the survey was entirely voluntary a total of 204 questionnaires were collected (See Table 2 for sample structure).

		per cent (%)
Gender	Male	98
	Female	2
Age	18-34	43
	35-54	47
	55-64	10
Educational background	Technical or trade certificate	40
	School certificate (Year 10)	5
	Higher secondary school HSC/VCE	9
	College certificate or diploma	16
	Undergraduate degree	23
	Postgraduate degree	7

Table 2 Respondent demographics, Supervisors' survey, Sydney construction sites, 2009

The third stage was a desk-top study of "best practices" in managing cultural diversity within the workplace. The desk-top study consisted of a review of a large number of online and published sources including specialised diversity reports, journal papers, books, company annual reports, company websites and press releases. In total, 156 companies were investigated across 15 industry sectors including hospitality, apparel, pharmaceuticals, entertainment media, information technology, internet services and retailing, megabanks, construction and engineering, health care, aviation, petroleum refining, and fast moving consumer goods. These companies were sourced from lists of the largest and most successful companies around the world:

- Engineering News Record's Top 400 Contractor List 2009;
- Fortune Magazine's World's Most Admired Company List for 2010 (People Management; and Management Quality) ;

- Fortune Magazine's 100 Best Companies to Work For 2009; and
- Human Rights Campaign Foundation's Corporate Equality Index 2009.

A number of companies profiled were also acknowledged in the field as leaders in diversity management and some have won awards and accolades for their commitment and innovative solutions to managing cultural diversity and have been featured on award winning lists such as:

- Diversity Inc.'s Top 50 Companies List 2010;
- Diversity Inc.'s Top 10 Companies for Supplier Diversity List 2010; and
- Reader's Choice Best Diversity Companies 2009

Findings

The stage one survey of construction workers found that the majority of the respondents (85 per cent) interact with other ethnic groups during social situations and work-based activities suggesting that there is a good deal of cross-cultural interaction on construction sites. Half of the respondents indicated that most of their friends at work belonged to their own ethnic group and almost 31 per cent of respondents reported that they did not make an effort to talk with workers of different ethnic backgrounds. The majority of the respondents (76 per cent) believed "their own ethnic group understood them better" and 45 per cent of respondents reported that members of their own ethnic group needed to stick together to 'survive'. Our findings suggested that ethnic groupings can have both negative and positive functions. On the one hand, such ethnic groupings play a role in perpetuating the boundaries between different groups; potentially exacerbating a lack of cross-cultural understanding on construction sites. On the other hand, ethnic groupings (or in-group favouritism) have some positive functions such as maintaining positive bonds among group members, group support and providing safe-havens in conflict situations. For example, 45 per cent of respondents reported that members of their own ethnic group need to stick together to "survive" on construction sites. This is supported by the following statement made by a worker, which is typical of many made:

> *Greeks stick together definitely... [Points out his Greek friend and says] For example, we don't talk too much, we're not close friends. But if something happens on the site or something that concerns all of us, then we stick together, definitely. We support each other (Site diary).*

Maintaining in-group identification as a response to a threat is the strongest among respondents from Asian background. Almost 76 per cent of Asian respondents thought that they needed to stick together to survive on construction sites. These views suggest that threats to one's own ethnic group (or identity) provide stronger in-group identification. In this regard one may argue that in-group identification as a response to a threat is strongest among the members of major out-groups. The major out-groups on sites were perceived to be Australian-Lebanese, Asian-Australians, and those from the former Yugoslavia (mostly Australian-Serbians and Croatians). Having said this, the vast majority of workers were comfortable with cultural diversity and thought that it worked well. A vast majority of respondents (88 per cent) were of the view that cultural diversity is good for society and the construction industry. The majority of respondents are comfortable with working with people of different ethnic groups (88 per cent) and 77 per cent thought that different ethnic groups worked well on construction sites. A considerable percentage of respondents (32 per cent) thought that different ethnic groups should stay away from each other, thus implying that support for cultural diversity is not consistent and depends on the nature of the relationship between workers. There was a wide-spread belief that lack of interaction between certain groups is normal in the construction work environment. Despite this negative picture, 64 per cent of the respondents indicated that they would like to see more opportunities to mix with people from other ethnic groups while at work. So the picture which emerged was of high levels of cross-cultural interaction, and a general desire for more opportunities for meaningful cross-cultural contacts. At the same time, there is a comfort and confidence with in-group affiliation and association which indicates a co-existent politics of difference and a politics of identity, for which most workers can see merit in each.

Communication problems caused by language barriers were one of the major challenges affecting work and social relations between different ethnic groups on sites. Communication problems caused by language barriers are one of the major challenges affecting work and social relations between different ethnic groups on construction sites. A large number of respondents made comments about the low English proficiency of some groups (particularly Asian-Australian workers). The association of certain trades with certain ethnic groups (Italians in concreting, Koreans in tiling, Chinese in plaster boarding etc) deepens the complexity of language problems. Since each ethnic group tends to specialise in a particular trade, the workers of each trade tend to use their first language (if other than English) at the workplace. The intensity of communication problems becomes more challenging when workers of different subcontractors have to work together. One worker mentioned that different ethnic groups do not work well on site:

> *"due to language barriers and people not wanting to mix e.g all companies, subbies are made up of one ethnic group" (Comments in questionnaires).*

The safety implications of language barriers were also raised by many workers.

> *Cultural diversity is a good thing for construction sites, yes but need to speak and understand English for safety aspects (Comments in questionnaires).*
>
> *People should be able to read English when working on sites to read safety notices (Comments in questionnaires).*
>
> *Cultural diversity is not a good thing "no, not when safety is compromised cause they are unable to understand, comprehend or speak English (Comments in questionnaires).*

The majority (71 per cent) of respondents believed that there is equality of opportunity in the construction industry. However, 29 per cent of respondents reported differential treatment of ethnic groups. Derogative name calling was identified as one of the common experiences of workers with 32 per cent of respondents indicating that they have been called names or insulted because of their ethnic background. Offensive graffiti and joke telling were the most common forms of racial prejudice on construction sites.

The stage two survey of managers found that the majority of respondents believed that responsibility for managing ethnic diversity belonged to site managers, supervisors and project managers. Safety and equal opportunities policies were perceived to have more importance and be more widely implemented within the industry than affirmative action, Aboriginal employment and managing ethnic diversity policies. There was also a low level of awareness about ethnic diversity policies. 57 per cent of managers reported that they did not receive any training that aimed to reduce stereotyping and ways of managing ethnic diversity effectively. 54 per cent of the managers reported that they embraced a personal ethnic diversity management strategy to compensate for the lack of centralised policy. The other 46 per cent ignore the issue and rely on others to manage it for them. Indeed, most managers saw ethnic diversity strategies and affirmative action plans as discriminatory and unfair since such strategies may favour some groups over others. Managers of subcontractors were more actively involved in developing their own strategies than managers of main contractors. 13 per cent of the managers who did not have a personal strategy reported that having a strategy to manage ethnic diversity was not a priority. Organizing social events (such as barbeques and drinks in the pub) were listed as the most common form of ethnic diversity management strategy followed by multilingual induction processes. However, such events are recognised as inappropriate for some cultural groups and there was little imagination in managing this issue, probably because it was seen as such a low priority. A comparison of managers' and operatives' surveys suggests that operatives were slightly but not significantly less supportive of ethnic diversity (87 per cent), perhaps because they are directly exposed to its consequences. However, the overall level of support for diversity was very high in both groups. Significantly, around 23 per cent of managers thought that it would be better if different ethnic groups were separated. The majority of managers and operatives believed that there were sufficient opportunities for operatives to interact with workers from different ethnic backgrounds. More than half of the managers did not identify any problems with ethnic diversity and did not see any need to manage it proactively. In summary, the cultural diversity of the industry appears to be taken for granted by project managers and it appears as if the problems associated with it are accepted as an inevitable part of daily life on construction sites.

The stage three desk-top study findings highlight the types of management strategies which could be employed by construction firms to address the above problems. These are critically appraised below from a social identity perspective and are categorised under four main headings: leadership and commitment; education; networking and; subcontractor and supplier diversity programs.

Leadership and Commitment

Creating a diversity portfolio

The appointment of a senior executive to champion and lead diversity initiatives is a common strategy used by the firms we reviewed. Senior management positions such as 'Chief Diversity Officer' or 'Global Chief of Diversity' are common with direct reporting responsibilities to the CEO to ensure diversity remains a "strategic" priority. While on the surface this approach would seem positive in creating a focus on the management of cultural diversity which our stage one and two results show is missing from the construction industry, this approach can also be criticised from a social identity perspective. In particular, the "separating out" diversity as an issue could be counter-productive by potentially reinforce workplace differences as a management "problem". From a social identity perspective, a more positive approach would be to "normalise" diversity and multiculturalism by making it an integral part of "everyone's" job rather than separated which everyone is expected to give attention to on a day-to-day basis.

Measuring and reporting diversity performance

Companies committed to diversity initiatives often establish measurable key performance indicators (KPIs) that enable them to monitor and report the return on investment and the effectiveness of diversity initiatives implemented. Diversity KPIs may include: number of clients from different minority groups; number of senior managers from minority groups; levels of acceptance and satisfaction among employees; grievances and complaints; workplace conflict and disputes; retention; recruitment; community and public image etc. These companies also actively participate in specialist rankings as a way to build brand recognition in the marketplace. Once again, this might seem a useful way to "measure" and thereby track and adjust diversity performance and address the "taken for granted" nature of the problem discovered in stages one and two of our research. However, the effectiveness of this strategy from a social identity perspective lies in whether the metrics reinforce or erode perceptions of difference. For example, by highlighting a certain group as one which needs a proportion to be represented as certain levels is to accentuate difference. In social identity terms, it would be better to measure these indicators for every cultural group.

Education

Diversity training

Leading companies integrate diversity training into their overall training program and this would appear to be an effective strategy for addressing the general ignorance of cultural diversity management strategies which we discovered in our stage one and two research. Typically, these courses cover a wide range of topics such as the rights and obligations of employees, anti-discrimination legislation and strategies for dealing with workplace discrimination. Other companies emphasise more practical day-to-day managerial approaches for dealing with diversity, such as how to overcome negative behaviours, attitudes and stereotypes. Some organisations utilise more centralised and specialised methods of disseminating information on cultural diversity. For example, one company uses a 'Cultural Dictionary' to provide cultural insights and quick access to a variety of information about different groups including language, religion, general attitudes, greetings, gestures, eating habits, lifestyle and family etc. From a social identity perspective, education (for all – not just for one group) is a positive way to erode difference in the workplace and encourage a sense of collective multiculturalism.

Language proficiency

The ability to communicate in a language other than English is a quality that companies increasingly value as a resource to help bridge cultural divides. For example, some companies proactively encourage native speakers from culturally and linguistically diverse backgrounds (CALD) to undertake a language competency test so that they can become qualified interpreters. Other companies provide language courses to encourage multilingualism. Certainly, given the language problems uncovered in stage one and two of this research, this would appear to be an important strategy for the construction sector to consider. However, from a social identity perspective, these initiatives are positive as long as they are not imposing a dominant language on a minority group. Social identity theory would advocate "meeting in the middle" by encouraging all groups to learn each other's language rather than expecting one language to be the "nominated and accepted" language that everyone speaks.

Cultural integration opportunities

Our stage one and stage two results showed that people working on construction sites desire more opportunities to interact with people from other cultural groups. Our stage three research showed that leading companies employ numerous strategies to bring different cultural groups together. For example, some companies organise a diversity week where members of different cultural groups are invited informally to share their cultural food, drinks, cooking lessons and cultural knowledge. Other companies employ a far more structured approach with a diversity calendar with a detailed listing of cultural events. Once again, a social identity perspective would see this as "setting aside" special resources and events outside the work context. According to social identity theory, if such events are to have any effect, they should be played out in a work context and not be separated out. They should happen naturally and spontaneously all the time and they should not be targeted at specific groups.

Cultural holidays

Different cultural and ethnic groups celebrate events and holidays that have specific cultural value and significance to them. In acknowledging this, some companies implement a 'Floating Cultural Holiday Policy' that gives employees the option of 'trading' official public holidays for a day that is of cultural significance to them. From a social identity perspective this strategy is negative in that it requires someone to "trade" their identity for another. A far better approach would be to make these holidays an additional entitlement on top of shared holidays that everyone receives.

Role modelling

Many companies believe that positive role modelling of successful and inspirational employees from minority groups can have a significant motivational impact within minority groups. For example, some companies have started role modelling successful ethnic minorities in senior positions. From their "stories", it is argued that other employees of similar ethnicity can learn important lessons about the challenges that they may experience in achieving and maintaining senior positions. However, from a social network perspective, role models simply highlight the fact that it is more difficult for someone from this group to progress through a company. In other words, this person is not "normal" and has a separate identity to the rest of the organisation. To be effective, role models should target roles not groups.

Networking

Employee groups

Employee groups are highly valued by some companies which often contribute resources and expertise to help them build capacity and which frequently consult them on areas relating to strategy, diversity policy and professional development. However, from a social identity perspective, the idea of creating a Chinese group to advise on Chinese strategy is an anathema. It is difficult to think of a management strategy that could be more delineating or divisive.

Mentoring

Mentoring programs are increasingly used to enable employees from minority backgrounds to access support, identify pathways and source knowledge they need to reach their full potential at work. Some programs match people from different backgrounds and provides training to mentors to help them manage this process, regularly monitoring mentor and mentee performance in achieving pre-determined diversity and inclusion goals. Other companies go one step further by implementing a "reverse mentoring" programme that pairs very senior and successful managers up with junior mentors of culturally diverse backgrounds. The program aims to provide very senior managers with a practical insight into diversity issues from an alternative viewpoint. From a social identity perspective this is excellent since it blurs and mixes the strata's of identity creating a collective corporate identity which is devoid of difference.

Supplier or subcontractor diversity

Supplier or subcontractor guidelines and codes of practice

To ensure a high standard of behaviour and integrity of the suppliers and subcontractors they use, many companies expect suppliers and subcontractors to adhere to rules and guidelines governing the way they operate their business, including codes of conduct and ethics and requirements for maintaining a culturally diverse workforce. From a social identity perspective, the problem with this approach is the underlying assumption that parent company policies and procedures are superior. Given the critical analysis of the above strategies, this clearly cannot be assumed.

Strategic allocation of contracts

A large number of companies strategically allocate a proportion of contracts to minority-owned businesses. For example, some companies allocate between 12% to 15% of total annual procurement dollars to minority-owned businesses in the belief that it contributes towards the development of historically under-represented entrepreneurs. Other companies sponsor procurement fairs, to inform potential minority-owned businesses about the opportunities of working on their projects and to introduce them to prime contractors. And several companies have online supplier registry to nurture partnerships with qualified diverse suppliers and subcontractors. From the perspective of social identity theory, this is equivalent to an affirmative action program which once again reinforces difference by separating out particular groups for special attention. A more effective approach would be to support companies so that they can compete equally.

Conclusion

The aim of this paper was to use social identity theory to explore the attitudes that underpin cultural diversity in the Australian construction industry, the management challenges that these pose and the range of strategies that can be employed to manage them effectively. While the research reported here has been undertaken in Australian construction industry, the findings are of relevance to construction industries elsewhere given the workforce demographic and institutional similarities between them. Our findings indicate that construction workers take the cultural diversity of the industry for granted, are optimistic and positive about diversity and have cross-ethnic experiences which generally leave them happy for further contact. Yet, they exist in an environment of ambient racism and retain a set of assumptions regarding the desirability of ethnic uniformity (sameness) over diversity. While there was some acceptance that management should take responsibility for ethnic diversity management, it was not seen as a priority and for some there was no perceived problem to manage. The latter views are inconsistent with the findings on the perceptions of uneven opportunities and of experiences of everyday racism. They also conflict with the findings of cross cultural social identity research which indicates that the best way to manage diversity is to diffuse responsibility throughout the workforce rather than assume it is the responsibility of one group (managers). Our research also shows that most of the extant strategies for managing diversity are ad-hoc and informal, and principally include social functions arranged by workers or unions. This is hardly surprising given that there is an absence of formal training on diversity management for both workers and managers. Not only is diversity being poorly managed, exposing the industry to the costs of poor inter-ethnic relations, but the wider social benefits from improving attitudes among this section of society are not being reaped. No construction or engineering companies reside in the recognised global leaders of best practice which in terms of the attitudes discovered towards diversity is not surprising. In stark contrast to the finely detailed information we found in the companies we reviewed, the limited number of initiatives we could find in the construction sector were typically poorly documented and ad hoc and there was little evidence of them fitting into strategic, detailed and fully integrated diversity programs which are systematically measured and reported in terms of their success and return of investment. However, while more research is needed to explore, analyse and test the effectiveness of the full variety of diversity strategies higlighted in this paper, our critical analysis using social identity theory suggest that firms should be wary of many of the so called best practice strategies which are promoted in the literature. We have shown that many of these could reinforce rather than eliminate cultural difference and can actively undermine attempts to build a cooperative workplace culture where different cultural groups can work in the harmony that most of them appear to desire. In conclusion, we would argue that any research into effective workplace strategies to manage diversity in the construction and engineering sectors be informed by relevant social psychological research and not be uncritically adopted without any thought to their application and unintended consequences. We would also point out that further research needs to be undertaken into differences in attitude towards cultural diversity between different ethnic groups. One of the limitations of this research was that attitudes could not be analyzed and compared across these groups. Such research would enable more targeted strategies than the generic strategies highlighted in this paper.

References

Brewer, M. B. (1997) 'The social psychology of intergroup relations: can research inform practice?', *Journal of social issues*, **53** (1) 197-211

Brewer, M. B. (2009), 'The importance of being *we*: Human nature and intergroup relations', *American Psychologist*, 62, 728-738

Cox, T.H. and Blake, S. (1991) 'Managing cultural diversity: implications for organizational competitiveness', *Academy of Management Executive*, **5** (3), 45-56

Dainty, A., Green, S., Bagilhole, B. (eds.) (2007) *People and Culture in Construction: A Reader*, Taylor & Francis, Oxon

De Cieri, H. and Kramar, R. (2003) *Human Resource Management in Australia: Strategy, People, Performance,* McGraw Hill, Sydney

Dunn, Kevin M., Forrest J., Pe-Pua, R., Hynes M. and Maeder-Han, K. (2009) 'Cities of race hatred? The spheres of racism and anti-racism in contemporary Australian cities', *Cosmopolitan Civil Societies: An Interdisciplinary Journal*, **1** (1), 1-14

Gale, A. W. and Davidson, M. J. (eds.) (2006) *Managing Diversity and Equality in Construction Initiatives and practice*, Taylor &Francis, Oxon

Hay, C. (1996) *Managing cultural diversity: Opportunities for enhancing the competitive advantage of Australian Business*, AGS, Canberra

Loosemore, M. and Lee, P. (2003) 'An investigation into communication problems with ethnic minorities in the construction industry', *International Journal of project Management,* **20** (3), 517 – 524

Loosemore , M., Dainty, A. and Lingard, H. (2003) *Human Resource Management in the Construction Industry: Strategic and Operational Approaches*, Taylor and Francis, London, UK

Loosemore, M. and Andonakis, N.(2006) 'Subcontractor barriers to effective OHS compliance in the Australian Construction Industry', in Fang D. P., Choudry, R.M. and Hinze, J. W.(eds.), *Global Unity for safety and health in construction proceedings of the CIB W99 International conference, Tsinghua University*, Beijing, 61-68

Loosemore, M., Phua, F., Dunn, K. and Ozguc, U. (2010) Key findings of survey with site operatives: Managing Cultural Diversity on Australian Construction Sites, First report of Australian Research Council Grant, UNSW, Sydney, Australia

Loosemore, M., Phua, F., Dunn, K. and Ozguc, U. (2010), 'Managing cultural diversity in Australia construction sites', *Construction Management and Economics*, **28** (2), 177-188

Ofori, G. (1994) *Foreign construction workers in Singapore*, International labour Organization, International labour Office, Geneva

Steele, A. and Sodhi, D. (2006) Race, ethnic minorities and construction industry in Gale, A.W. and Davidson, M.J.(eds.), *Managing Diversity and Equality in construction: Initiatives and practice*, Taylor & Francis, Oxon

Tajfel, H. (1970) Experiments in intergroup discrimination, Scientific American, **223** (2), 96-102

Van Oudenhoven, J. P., Groenewoud, J. T. and Hewstone, M. (1996) 'Cooperation, ethnic salience and generalization of interethnic attitudes', *European Journal of Social Psychology*, **26**, 649-661

A Comparative Approach to Determining the Growth of Productivity of the New Zealand Construction Industry

3

Malcolm Abbott, (Swinburn University of Technology, Australia)
Chris Carson, (Unitech Institute of Technology, New Zealand)

Abstract
In recent years there has been increasing interest in the productivity and efficiency of the construction industry in New Zealand. In part this interest has manifested itself in the increased use of numerous statistical techniques to determine the productivity and efficiency of the industry. These efforts have, however, some degree of controversy. The purpose of this paper is, therefore, threefold. First it summarises the key structural findings that have been determined from past research into the construction industry in New Zealand. Secondly it makes some comparisons between the construction industry's productivity in New Zealand with that of the six states of Australia. Finally it also considers potential areas for potential future research.

Introduction
Over the years attempts have been made to determine the reasons behind the differences in productivity and income between New Zealand and Australia. One area where this gap in productivity is regarded as being most marked is in the case of the construction industry. In the case of New Zealand this perceived gap has been recognised by official statistics, which seems to show that the construction industry has a productivity level at much the same level that it was in the late 1970s (Statistics New Zealand 2012, *Industry productivity statistics*). A number of possible reasons have been put forward for this gap between the Australian and New Zealand construction industries, including such things in the latter country as low economies of scale in the industry, a lack of competition, regulatory impediments, faulty innovation and management practice, poor investment quality and a low levels of skills (New Zealand, Department of Building and Housing, Productivity Taskforce 2009; Van Dai Tran 2010; New Zealand Productivity Commission 2011). One aspect that has received less attention has been the validity of the productivity measures themselves, which in a number of cases overseas such as in the United States have been the subject of some controversy.

The aim of this paper is to discuss some of the issues surrounding the estimation of productivity levels in the construction industry, to review past productivity analysis that has been conducted on the New Zealand construction industry and then to undertake productivity estimates of the New Zealand industry using a data envelopment approach (DEA), which involves a comparison of New Zealand figures with those of the various state based jurisdictions of Australia.

The paper is structured as follows. In the first section the general background to the issue of productivity in New Zealand and Australia is given. A section is then provided that examines past studies of productivity in the construction industry in New Zealand (and references the major Australian work). The method used in this paper is then provided. A section which gives estimates of the productivity level of the industry in Australia and New Zealand follows, and in the final section some conclusions are made.

Background
New Zealand and Australia are similar countries in that they are both geographically isolated from the world's major markets, are sparsely populated, have populations of

predominately British background, as well as similar common-law based legal systems. The structure of the two economies is also similar in that they both specialise in the export of mainly resource based commodities to the countries of the Asia-Pacific region. Despite these similarities, Australia has a considerably higher per capita GDP than New Zealand of US$40,800, (2011) compared to only US$28,000 (Central Intelligence Agency 2012. The large per capita output gap between Australia and New Zealand is a relatively new phenomenon. Up until the 1970s average output and income of the two countries were of similar size, but during that decade a substantial gap between the two countries opened up. Both countries went through a process of economic reform in the 1980s and 1990s and despite these reforms being similar in character (although undertaken at a faster rate in New Zealand); the gap in living standards that had opened up in the 1970s has persisted.

Given the similarity of the two countries it would be expected that over the longer term some sort of convergence would have occurred between the two countries. The economic theory of 'convergence' suggests that over time, wealthy countries will end up at around the same level of income, as technology and ideas spill over between them. In the case of Australia and New Zealand it would normally be expected that given the similarities of the two countries these spillovers from the former to the latter would be considerable, and would bring about an eventual convergence. This expected convergence, did not occur during the 1990s and 2000s (Conway and Hunt 1998).

In recent years a number of researchers and policy groups have attempted to put forward reasons for the persistence in the gap in average output and income levels between the two countries. These reasons include the following.

1. Smaller population in New Zealand than Australia. New Zealand's market of 4.2 million people does not allow for as much depth, development and economies of scale as does Australia's population of around 20.9 million. New Zealand's small population means that firms in that country will reach their limit in terms of domestic growth fairly quickly, and will then have to export or merge with a larger company to expand further. Smaller markets might also mean less competition with the result being less pressure on companies to operate at high levels of efficiency. The isolation of New Zealand from world markets intensifies this problem and reduces the gains from economies of scale and externalities between businesses, which occurs in large, globalised cities (McCann 2009).

2. A greater abundance of natural resources in Australia. Another common explanation for Australia's higher income has been luck, in the form of its abundant mineral resources. In recent times, however, New Zealand has also experienced a boom in commodity prices, with rising demand for agricultural products.

3. Low household savings. Some have argued that the New Zealand's low level of saving is a weakness. This forces firms to borrow money from overseas lenders, who charge higher interest rates, and thereby encourage less investment (Skilling 2004). Some doubts have been expressed about the importance of this explanation in recent times New Zealand's long-term real interest rates and real cost of capital are, as it happens, similar to Australia's (Cameron et. al. 2007; Lally 2000).

4. Policy. Improving the policy environment has been a goal of economic reform in New Zealand since 1984, with deregulation, privatisation, trade liberalisation, and tax cuts. In the case of Australia, however, similar reforms have occurred and the level of regulation across the two countries is of a similar level. Both countries tend to rank fairy highly in international rankings of freedom and competitiveness.[1]

[1] World Bank (2011); The Heritage Foundation (2012); World Economic Forum (2011).

5. Lower levels of labour productivity. Lower levels of labour productivity in New Zealand, which in turn are driven by lower levels of investment in physical capital and human capital. This last reason seems the most likely to be true but begs the question of why there may be less investment in physical and human capital in New Zealand than Australia. It could be that there is a lack of investment opportunities and that returns are too low. For New Zealand firms and investors, the returns are either too low compared to investments that could be made overseas, or not worth the risk. Another explanation could be the lower relative cost of labour. Australia has a more rigid, unionised system of industrial relations than New Zealand, with a high minimum wage and national awards that set pay levels. This makes it expensive to hire staff, and therefore more rewarding to invest in capital.

Whatever the reasons there might be, there is considerable evidence that labour productivity and productivity levels in general in New Zealand in many sectors are simply lower than they are in Australia, which in turn explains the lower levels of income in New Zealand. No sector is perceived to illustrate this problem more than that of the construction industry, where there appears to be a considerable gap between productivity levels in Australia compared to New Zealand. In terms of the construction industry it is possible to gain some appreciation of some of these issues by simply comparing education, profit levels and average business size across the two countries (Tables 1 to 3).

	Post-graduate Degree	Graduate Diploma & Graduate Certificate	Bachelor Degree	Advanced Diploma & Diploma	Certificate	Inadequately described/ Not stated	Total qualifications	Total employed
Australia	4,619	2,441	35,237	34,796	321,946	33,557	432,596	709,840
%	0.7	0.3	5.0	4.9	45.4	4.7	60.9	
New Zealand	717	681	5,394	10,908	87,363	5,733	110,796	149,370
%	0.5	0.5	3.6	7.3	58.5	3.8	74.2	

Table 1 Education levels of the construction workforce in Australia and New Zealand, Construction 2006

Source: Australian Bureau of Statistics, Census of population and housing, cat. no. 2001.0. Statistics New Zealand, 2006 Census

	New Zealand				Australia			
	Income $m NZ	Expenditure $m NZ	EBIT $m NZ	Profit margin %	Income $m Au	Expenses $m Au	EBIT $m Au	Profit margin %
2002	16,570	15,232	1,952	11.8	123,328	112,596	11,232	9.1
2003	18,730	17,271	2,096	11.2	140,984	128,960	13,078	9.3
2004	20,872	19,383	2,333	11.2	159,928	145,434	15,935	10.0
2005	23,391	25,902	3,394	14.5	174,162	157,655	16,930	9.7
2006	32,734	29,517	4,275	13.1	198,802	179,329	19,485	9.8
2007	34,507	30,640	5,043	14.6	233,348	202,348	33,896	14.5
2008	38,104	34,749	3,893	10.2	253,820	234,279	28,227	11.1
2009	36,965	34,991	2,231	6.0	260,154	240,499	29,010	11.2
2010	33,451	31,263	1,972	5.9	275,290	254,002	31,308	11.4

Table 2 Returns to business in the New Zealand and Australian construction industry

Source: Statistics New Zealand, Annual enterprise survey. Australian Bureau of Statistics, Australian industry, cat. no. 8155.0. EBIT- Earnings before interest and tax. Profit margin-EBIT/Income

In the case of education there does not appear to be any lower level of education overall in the New Zealand industry (Table 1) compared to Australia, except to say the mix of qualifications is somewhat different (i.e. degrees versus diplomas). In the New Zealand case the latter type of qualification seems more popular. Likewise levels of profitability in the industry seem to be of comparable level (Table 2), so it would be expected that investment levels would be similar. Finally the average size of businesses in the construction industry in New Zealand is smaller than that of Australia (Table 3), probably a product of the greater dominance of the Australian economy by large sized state capital cities such as Melbourne, Sydney and Brisbane. In light of these factors it is difficult to determine why there should be estimated in the case of the New Zealand industry a significant stagnation of productivity compared to the Australian case.

	Australia			New Zealand		
	Enterprises no.	Employed no.	Employed/ enterprises	Enterprises no.	Employed no.	Employed/ enterprises
2000	309,691	688,000	2.2	41,389	75,970	1.8
2001	341,457	670,500	2.0	40,181	76,560	1.9
2002	344,561	693,300	2.0	39,485	80,370	2.0
2003	339,982	717,800	2.1	41,168	87,410	2.1
2004	365,780	773,700	2.1	44,715	97,630	2.2
2005	383,515	832,800	2.2	47,971	110,870	2.3
2006	na	877,400	na	50,813	119,200	2.3
2007	351,612	973,000	2.8	52,263	125,710	2.4
2008	349,118	983,000	2.8	54,139	131,140	2.4
2009	342,436	983,000	2.9	53,191	122,110	2.3

Table 3 Average size of construction enterprises in New Zealand and Australia

Source: Statistics New Zealand, Business demography statistics. Australian Bureau of Statistics, Australian industry, cat. no. 8155.0

Past Measures of Productivity

Over the years a number of researchers in New Zealand have made estimates of New Zealand's productivity change (for a more detailed overview of studies of this sort see Abbott and Carson 2012). In some of these cases researchers broke down their estimates into industry sectors, including that of the construction industry. Industry breakdowns that include the construction industry occurred in the following studies: Orr (1989); Chapple (1994); Philpott (1991, 1993, 1995); Fare, Grosskopf and Margaritis (1996), Diewert and Lawrence (1999); Black, Guy & McLellan; Mason & Osborne (2007) and Janssen & McLoughlin (2008). In addition Davis (2007) undertook a study that concentrated just on construction productivity.

Figures 1 and 2 provide a summary of some of these works on the productivity of the construction industry. As can be seen from Figure 1 the estimates of productivity in the construction industry tended to rise through the 1960s and early 1970s, before peaking around 1976 and then declining. A recovery in productivity appears to have taken place in the early 1980s. Figure 2 provides more pessimistic estimates. In each case the estimates of productivity appear to show a stagnation of growth in productivity. In the Australian case some studies have also shown that there has been a slowdown of productivity growth (see for instance: Parham and Zheng 2006; Li and Liu 2010; KPMG Econtech 2010).

Regardless of the periods selected it appears from the various studies that although construction productivity increased to some degree up until the mid-1970s, since then the results have been mixed. This appears to be the case even though there have certainly been a range of technological changes that have occurred in the industry since then. New tools and equipment have been introduced that have tended to be labour savings. This has mean the introduction of hand held powered tools (nail drivers, sanders, saws and drills), and improved lifting and moving machinery (cranes, loaders, earth movers, graders, forklifts). New materials and processes have also been introduced along with a greater use of pre-fabricated materials.

The introduction and greater use of new equipment, materials and processes has meant that it is hard to explain why growth in the productivity of the New Zealand construction industry has been sluggish. One possible explanation is that the various studies of productivity growth may have underestimated the growth in output of the industry, and therefore productivity. This explanation has been forward for a variety of reasons in the case of the United States, where a similar slowdown in productivity growth was estimated in the 1970s and 1980s. In the American context some thought that this decline could be explained by the use of inappropriate price indexes (Dacy 1965; Stokes 1981). Rosefielde and Mills (1979), in contrast, argued that official data did not fully account for changes in quality and so tried to adjust for changes in the scale and structure of the quality of output. They argued that as construction is locationally immobile and designed to achieve some particular function, at some specific site, a diverse and changing type off output is produced by the industry. This coupled with a changing composition of materials makes it difficult to measure quality improvements over time, which in turn leads to an underestimation of output levels, and therefore productivity improvements. Schriver and Bowlby (1985) also found that much of the fall in productivity could be attributed to a change in the output mix from high to low productivity building projects.

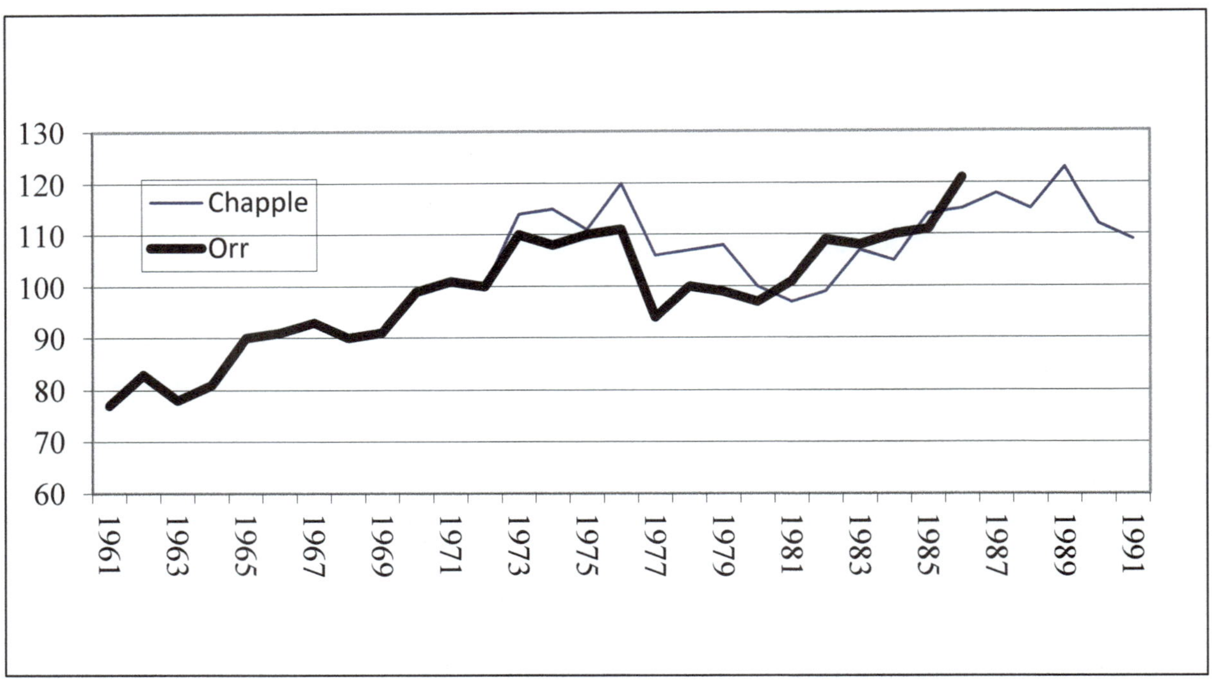

Figure 1 Total factor productivity in New Zealand construction, 1972=100
Source: Orr 1989; Chapple 1994

In using quality adjusted figures for output Rosefielde and Mills relied upon the work of Cassimatis who used the number of rooms per dwelling and square footage of buildings as an indicator of the quality of output. In the New Zealand case past studies have all used some constant dollar valuation of the industry's production as an indication of output, which assumes that this value reflects any changes that are occurring in quality and size, whether it be in terms of area or other form of quality. This might, perhaps, be a fair assumption but nowhere has it been justified, and if it is not a fair assumption may have led to some distortion of results.

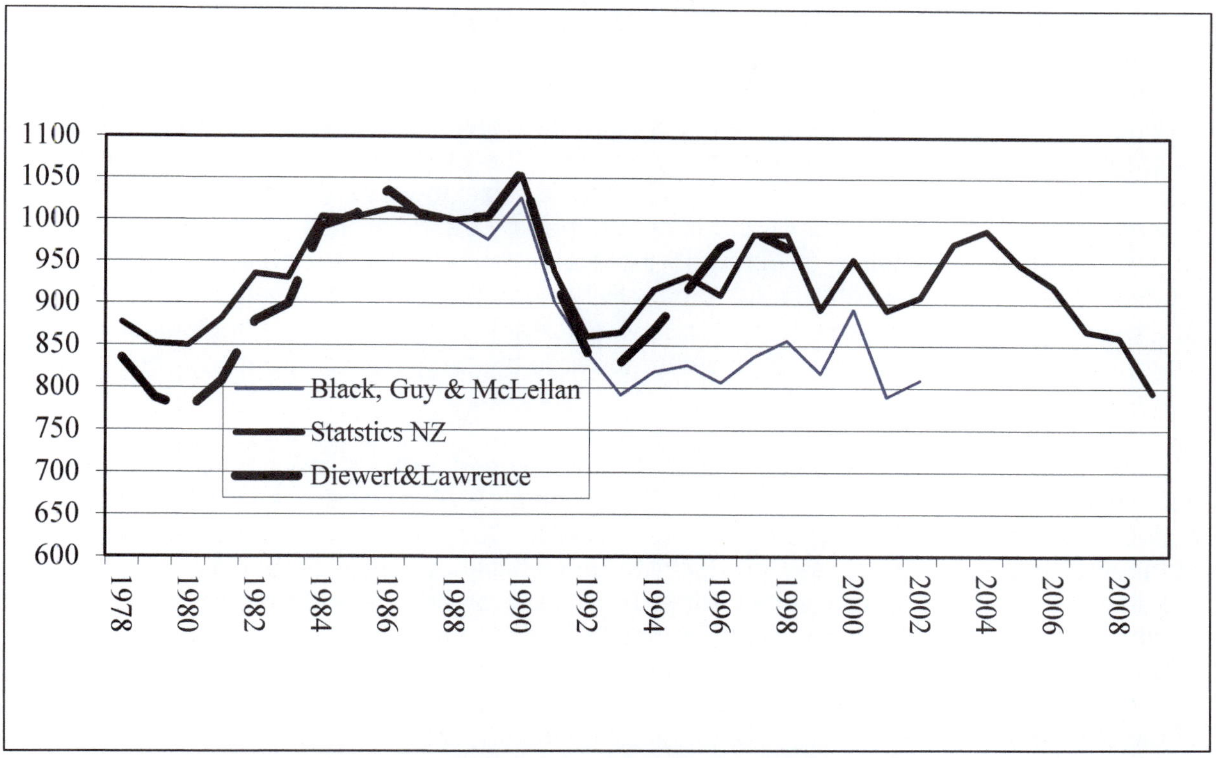

Figure 2 Total factor productivity of the New Zealand, 1978=1000 construction industry

Source: Statistics New Zealand 2012, Industry productivity statistics. Diewert & Lawrence 1999, Black, Guy & McLellan 2003

One way of accounting for this issue in the New Zealand context would be simply to make use of the square metres of area constructed as an indicator of output, rather than some monetary valuation of property. In addition to avoiding the issue of quality it would also avoid the issue of defective price indexes mentioned earlier. In the New Zealand case Statistics New Zealand has data on building consents for area to be constructed for residential housing from 1974 onwards and for non-residential construction from 1991. As construction also involves work on altering existing building as well as the construction of major engineering projects some indicator of output for this sector would also need to be used. In order to achieve this, a productivity analysis method would need to be used, which enables the combination of different measures of output (i.e. floor area and value added). Luckily the DEA approach is ideal for this purpose.

Methodology

In order to incorporate the physical indicators of output, such as square metres, with financial indicators, such as the dollar of value added, the DEA method can be used. DEA was pioneered by Charnes et al (1978) and based on the earlier work by Farrell (1957). There are many texts offering a detailed discussion on DEA, including the algorithms used (see, for example, Lovell and Schmidt 1988; Fare, Grosskopf and Lovell 1985; and Coelli, et. al. 2005). DEA is a linear programming technique which estimates organisational efficiency by measuring the ratio of total inputs employed to total output produced for each organisation (or country or industry sector). This ratio is then compared to others in the sample group to derive an estimate of relative efficiency.

DEA identifies the most efficient providers of a good or service by their ability to produce a given level of output using the least number of inputs. Other organisations in the sample group receive an efficiency score determined by the variance in their ratio of inputs employed to outputs produced relative to the most efficient producer in the sample group. DEA is therefore a measure of relative efficiency against the sample group's benchmark best practice. If input and output prices are used then allocative, and technical efficiency levels can be determined. If prices are not available, or distorted, then DEA can be used to estimate levels of technical efficiency. Technical efficiency under constant returns to scale can be decomposed into scale efficiency and pure technical efficiency under variable returns to scale. Scale efficiency can be defined as being the extent to which an organisation can take advantage of returns to scale by altering its size towards the optimal scale. Pure technical efficiency on the other hand is determined by the difference between the observed ratio of combined quantities of output to input and the ratio achieved by best practice. In studies taken over time total factor productivity change can be estimated using the DEA approach, which is comprised of technical efficiency change and technological change. Technical efficiency change is a movement towards the best practice frontier. Technological change is the movement of the frontier.

The advantage of this approach is that it can be used without input or output prices to determine levels of technical efficiency (pure and scale), which when used in the case of the construction industry might lead to distorted results. DEA has been used to extensively in a number of industries to assess productivity and efficiency levels (especially public sector providers and utilities where prices either don't exist or are distorted by abuses of market power). In the case of the construction industry only a small number of studies have been undertaken using this approach. This is regrettable as studies of the productivity of the construction industry often run into difficulties in finding adequate price indices. Examples of the DEA approach in the construction industry include studies by Chau and Wang (2005), Wang (1998); Wang and Chau (1997, 2001) on the Hong Kong industry and in the case of New Zealand a single study by Fare, Grosskopf and Margaritis (1996).

It is possible, using the DEA approach, to combine two indicators of output (square metres residential constructed, and the value of new alterations undertaken plus engineering and commercial construction) with two indicators of input (labour employed and net capital stock used) over the period 1991 to 2010. Furthermore it is possible to combine the estimates of the New Zealand results with those of the state based jurisdictions in Australia. Data for the years 1991 to 2010 can be obtained from the Australian Bureau of Statistics and Statistics New Zealand.[2] Monetary figures (i.e. for non-residential construction and net capital stock) are converted into constant $US, using the purchasing parity in order to make the Australian and New Zealand figures comparable.

New Zealand and Australian Construction Productivity Figures

To begin with it is possible to see if the use of a physical indicator of output (i.e. floor area) has any impact on the productivity results. Tables 4 and 5 benchmark the Australian state based construction industries with that of New Zealand using data from between the years 1991 and 2010. The difference between the two tables is that in the first, the value added of the construction industry is used as the output, whereas in the second table to determine the level of new residential building activity the figures for square metres of floor area is used rather than value added (with alterations commercial and engineering construction continuing to use value added). In both cases the inputs used are the numbers employed in industry and the value of the net capital stock.

[2] Australian Bureau of Statistics: Residential floor area – *Building approvals, Australia*, cat no. 8731.0; Value added – *Australian national accounts*, cat. no. 5220.0; Labour - *Labour force, Australia*, cat. no. 6291.0; Capital stock - *Australian national accounts*, cat. no. 5220.0.
Statistics New Zealand: Residential floor area – *Building consents issued*; Value added: National accounts; Labour – *Household labour force survey*; Capital stock – *National accounts.*

In Table 4 there does appear to be a considerable gap in the level of efficiency of the New Zealand industry compared to that of the Australian state based industries. The level of technical efficiency is only 0.505 compared to a mean of 0.816 for all seven jurisdictions. In the second case, however, this gap has largely disappeared. In Table 5 the New Zealand level of technical efficiency is 0.846 compared to a mean of 0.897 for all seven jurisdictions. What is notable is that the New Zealand level is really only lower than the two main states of Western Australia and Queensland, states that have a disproportionately high level of engineering construction, rather than residential construction.

What this means is that the residential construction industry in New Zealand is not especially inefficient compared to that of the Australian industry in building square metres of floor area. This is reflected in the higher level of efficiency when this measure is used. Instead as the New Zealand residential industry tends to use lower value materials (timber versus brick), the low productivity measures in Table 4 are probably largely caused by prices indices creating a lower level of value added as an indication of output. What does help to create a gap between the two countries in productivity is probably caused by the larger proportion of large scale engineering projects in states like Western Australia and Queensland, as well as differences in productivity levels in this sector of the construction industry. What this means is that the composition of the industry in different jurisdictions has important implications for the evaluation of relative productivity.

	Technical efficiency	Pure technical efficiency	Scale efficiency
New South Wales	0.778	1.000	0.778
Victoria	0.803	0.915	0.878
Queensland	0.940	1.000	0.940
South Australia	0.687	0.699	0.984
Western Australia	1.000	1.000	1.000
Tasmania	1.000	1.000	1.000
New Zealand	0.505	0.533	0.948
Mean (for all states & New Zealand)	0.816	0.878	0.932

Table 4 Efficiency levels 1991-2010 Australian and New Zealand jurisdictions (output=value added; inputs=labour & capital stock)

	Technical efficiency	Pure technical efficiency	Scale Efficiency
New South Wales	0.704	1.000	0.704
Victoria	0.917	1.000	0.917
Queensland	1.000	1.000	1.000
South Australia	0.810	0.838	0.966
Western Australia	1.000	1.000	1.000
Tasmania	1.000	1.000	1.000
New Zealand	0.846	0.852	0.993
Mean (for all states & New Zealand)	0.897	0956	0.940

Table 5 Efficiency levels 1991-2010 Australian and New Zealand jurisdictions (output=residential floor area & value added; inputs=labour & capital stock)

If the floor area/value added as output approach is used in determining productivity change over time more optimistic results are obtained. Table 6 provides data on the total factor productivity change over the period 1990 to 2010. What is found in these estimations is that the Australian industry is relatively stagnant in terms of productivity improvements over the period. The average productivity change over the period is -1.1 per annum for the seven jurisdictions together (as indicated by the figure 0.989).

	Technological change	Technical Efficiency change	Pure efficiency change	Scale efficiency change	Total Factor Productivity change
New South Wales	0.940	1.000	1.000	1.000	0.940
Victoria	0.983	0.999	1.000	0.999	0.983
Queensland	0.985	0.992	0.992	0.999	0.977
South Australia	0.996	0.995	1.000	0.005	0.991
Western Australia	0.996	1.000	1.000	1.000	0.996
Tasmania	1.007	1.004	1.005	0.999	1.010
New Zealand	1.034	0.992	0.994	0.998	1.026
Mean (for all states & New Zealand)	0.991	0.997	0.999	0.999	0.989

Table 6 Total Factor Productivity change, Australia and New Zealand, 1990 to 2010

In the New Zealand case, however, there does appear to be some improvement over the period (growth of productivity of 2.6 per cent per annum). This growth in productivity in New Zealand seems to indicate that there is some convergence throughout the period. Figure 3 provides estimates of the annual growth in productivity of the New Zealand construction industry and estimates for the whole economy produced by Statistics New Zealand. Although the growth of productivity of the New Zealand construction industry appears to be cyclical (strong in periods of economic growth such as the early to id 1990s and mid-2000s, it does seem to fluctuate to a greater degree than that of the economy as a whole.

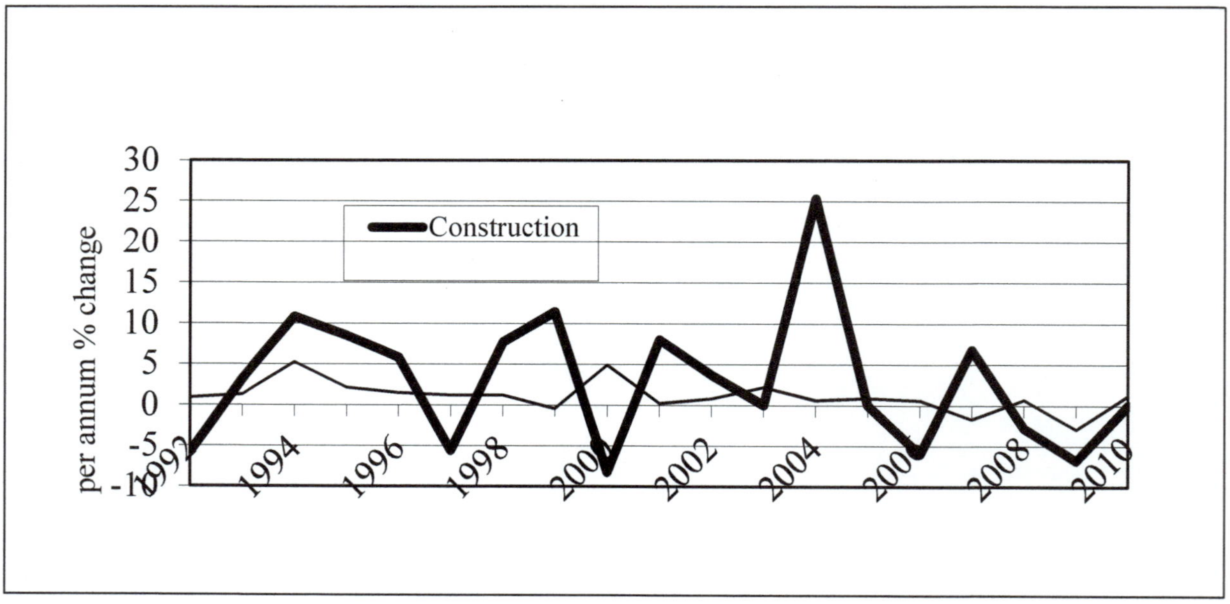

Figure 3 Productivity growth, construction and whole economy, New Zealand 1991 to 2010
Source: Statistics New Zealand 2012, Industry productivity statistics

Growth of productivity in New Zealand's construction industry, therefore, appears from these figures to be stronger than growth overall in the economy, but perhaps this is because the industry is coming off a low base after stagnant periods of productivity growth from the mid-1970s, until the late 1980s. These results are in line with those studies that show an uptake in productivity from the early-1990s, and work that shows similar levels of productivity in the Auckland building industry to those in major centres in Australia. In the latter case Best (2008), for instance, found that in the construction of standard sized hotel complexes the productivity of the industry in Brisbane, Melbourne, Sydney and Auckland was of comparable levels.

Conclusion

In light of the work presented in this paper a number of conclusions can be made. The first is that in the case of the estimation of the productivity change in the New Zealand construction industry the manner in which the outputs are defined can have a crucial impact on the results; just as was the case in earlier American studies. The use of prices indices in determined output levels and changes in quality and location can affect the nature of the outputs and therefore productivity levels.

Bearing this in mind some hesitant conclusions can be made. First of all productivity in the New Zealand construction probably has grown through the period 1991 to 2010. Even if it can be shown that the industry's productivity growth was reduced in the period 1975 to 1990 there is reason to believe that the performance has been better in more recent years.

The second conclusion that can be made is that the New Zealand industry does have productivity levels similar to those in some Australian jurisdictions although there does appear to be a gap between that of New Zealand and some jurisdictions (i.e. Queensland and Western Australian). In all likelihood this is due to the composition of the construction industry as those Australians states with the highest level of productivity tend to be those states with the highest proportion of construction in heavy engineering projects rather than residential and commercial construction. Attempts by policy makers to encourage improvements of construction industry productivity should be mindful of the differences in the composition of the industry across different jurisdictions.

This conclusion does not mean that the New Zealand industry by any means is close to world's best practice levels of productivity, but instead are just on a similar level to its near neighbour.

Finally it is clear that the notion of the level of productivity is a complex one in the case of the construction industry. A need for further experimentation with methods of approach needs to be undertaken before any really firm conclusions can be made.

References

Abbott, M. & Carson, C. 2012, 'A review of productivity analysis of the New Zealand construction industry', *Australasian Journal of Construction and Building Economics*, **12** (3), 1-15

Australian Bureau of Statistics, 2012, *Australian Industry*, cat no. 8155.0, ABS, Canberra

Australian Bureau of Statistics, 2012, *Australian National Accounts*, cat. no. 5220.0, ABS, Canberra

Australian Bureau of Statistics, 2012, *Building Approvals*, Australia, cat. No. 8731, ABS, Canberra

Australian Bureau of Statistics, 2006, *Census of Population and Housing 2006*, cat. no. 2001.0, ABS, Canberra

Australian Bureau of Statistics, 2012, *Labour Force: Australia*, cat. no. 6291.0, ABS, Canberra

Best, R. 2008, 'The Development and Testing of Purchasing Power Parity Method for Comparing Construction Costs Internationally', PhD Thesis, University of Technology Sydney, Sydney

Black, M., Guy, M. & McLellan, N. 2003, *Productivity in New Zealand 1988 to 2002*, Treasury Working paper 03/06, The Treasury, Wellington

Cameron, L., Chapple, B., Davis, N., Kousis, A., & Lewis, G. 2007, *New Zealand Financial Markets, Saving, and Investment*, Policy Perspectives Paper 07/01, New Zealand Treasury, Wellington

Central Intelligence Agency, 2012, *CIA Factbook*, CIA, Washington

Chapple, S. 1994, 'Searching for the Heffalump?', Anexploration into sectoral productivity and growth in New Zealand', New Zealand Institute of Economic Research Working paper 99/10, Wellington

Charnes, A. Cooper, W.W & Rhodes, E. 1978, 'Measuring the efficiency of decision making units, *European journal of operational research*, **2** (6), 429–444

Chau, K.W. & Wang, Y.S. 2005, 'An analysis of productivity growth in the construction industry: a non-parametric approach', in Khosrowshahi, F. ed. *21st Annual ARCOM conference, 7-9 September 2005, SOAS, University of London,* Association of Researchers in Construction Management, vol. 1, 159-69

Coelli, T., Prasada Rao, D.S. & Battese, G.E. 1998, *An introduction to efficiency and productivity Analysis*, Boston, Kluwer

Conway, P. & Hunt, B. 1998, 'Productivity growth in New Zealand: economic reform and the convergence hypothesis', Reserve Bank of New Zealand, Internal Paper, Economics Department, February, Wellington

Dacy, D.C. 1965, 'Productivity and price trends in construction since 1947', *Review of Economics and Statistics,* **47** (4), 406-411

Davis, N. 2007, *Construction Sector Productivity Scoping Report*, MartinJenkins, Wellington

Diewert, E. & Lawrence, D. 1999, *Measuring New Zealand's Productivity*, Treasury Working Paper 99/5, Treasury, Wellington

Fare, R., Grosskopf, S. & Knox Lovell, C.A. 1985, *The Measurement of Efficiency of Production*, Kluwer-Nijhoff, Boston.

Fare, R. Grosskopf, S. & Margaritis, D. 1996, 'Productivity growth' in A study of Economic Reform: the Case of New Zealand, B. Silverstone, A. Bollard, & R. Lattimore, eds., Elsevier Science, New York

Farrell, M. 1957, 'The measurement of productive efficiency', *Journal of the Royal Statistical Society*, Series A, CXX, 253-81

Janssen, J. & McLoughlin, S. 2008, *New Zealand's Productivity Performance*, New Zealand Treasury Productivity Paper 08/02, The Treasury, Wellington

KPMG Econtech, 2010, *Economic Analysis of Building and Construction Industry Productivity*, Report prepared for the Master Builders Australia, Melbourne

Lally, M. 2000, The Real Cost of Capital in New Zealand: is it too High?, New Zealand Business Roundtable, Wellington

Li, Y. & Liu, C. 2010, 'Malmquist indicies of total factor productivity in the Australian construction industry', *Construction Management and Economics*, **28**, 933-45

Lovell, C.A. Knox & Schmidt, P. 1988, 'A comparison of alternative approaches to the measurement of productive efficiency', in *Applications of Modern Production Theory: Efficiency and Productivity*. eds. A. Dogramaci & R. Färe, Kluwer, Boston

McCann, P. 2009, 'Economic geography, globalization and New Zealand's productivity paradox', *New Zealand Economic Papers*, **43** (3), 279-314

Mason, G. & Osborne, M. 2007, *Productivity, Capital-Intensity and Labour Quality at Sector Level in New Zealand and the UK*, New Zealand Treasury Working Paper 07/01, Wellington.

New Zealand, Department of Building and Housing, *Productivity Taskforce, 2009, Report of the Building and Construction Sector Productivity Taskforce: a Modern Efficient and Productive New Zealand Built Infrastructure Industry*, N.Z.: Dept. of Building and Housing, Wellington

New Zealand, Productivity Commission, 2011, *Housing Affordability Inquiry: Draft Report*, Productivity Commission, Wellington

New Zealand Institute of Economic Research, 2006, *New Zealand-Australian Income Differential*, NZIER, Wellington

Organisation of Economic Cooperation and Development, 2011, *GDP PPPs and Derived Indices for OECD Countries*, OECD, Paris

Parham, D., & Zheng, S., 2006, 'Aggregate and industry productivity estimates for Australia', *Australian Economic Review*, **39** (2), 216-26

Philpott, B. 1991, 'Economic growth in New Zealand models and Experience', in Long Run Perspectives on the New Zealand economy. *Proceedings of the Sesqicentennial Conference of the New Zealand Association of Economists,* New Zealand Association of Economists, Wellington

Philpott, B. 1993, 'Data for sectoral productivity analysis and some preliminary results for 1978-93', RPEP Paper 256

Philpott, B. 1995, *New Zealand's Aggregate and Sectoral Productivity Growth 1960-1995*, Research Project on Economic Planning Paper 274, Victoria University of Wellington, October, Wellington

Rosenfielde, S. & MillsD.Q. 1979, 'Is construction technologically stagnant?', in J.E. Lange & D.Q. Mills, eds., *The Construction Industry*, D.C. Heath and Company, Lexington MA

Schriver, W.R. & Bowlby, R.L. 1985, 'Changes in productivity and composition of output in building construction, 1972-1982', *Review of Economics and Statistics*, **67** (2), 318-322

Skilling, D. 2004, *Investment, Savings, and Growth*, New Zealand Institute, Auckland

Statistics New Zealand, 2006, *2006 Census of Population and Dwellings*, Statistics New Zealand, Wellington

Statistics New Zealand, 2012, *Annual Enterprise Survey*, Statistics New Zealand, Wellington

Statistics New Zealand, 2012, *Building Consents Issued*, Statistics New Zealand, Wellington

Statistics New Zealand, 2012, *Business Demography Statistics*, Statistics New Zealand, Wellington

Statistics New Zealand 2012, *Household Labour Force Survey*, Statistics New Zealand, Wellington

Statistics New Zealand 2012, *Industry Productivity Statistics*, 1978-2009, Statistics New Zealand, Wellington

Statistics New Zealand 2012, *National Accounts*, Statistics New Zealand, Wellington

Stokes, H.K. 1981, 'An examination of the productivity decline in the construction industry', *Review of Economics and Statistics*, **63** (4), 495-502

The Heritage Foundation, 2012, 'Index of economic freedom 2012-top 10 countries', http://www.heritage.org/ accessed 1 January 2013

Van Dai Tran, 2010, 'Exploring Construction Productivity Statistics in New Zealand', Master of Engineering Thesis, Auckland University of Technology, Auckland

Wang, Y.S. 1998, 'An Analysis of the Technical Efficiency in Hong Kong's Construction Industry', Unpublished PhD Thesis, University of Hong Kong, Hong Kong

Wang, Y.S. & Chau, K.W. 1997, 'An evaluation of the technical efficiency of construction industry in Hong Kong using the DEA approach', *Proceedings of ARCOM97 Conference,* 690-701, Cambridge

Wang, Y.S. & Chau, K.W. 2001, 'An assessment of the technical efficiency of construction firms in Hong Kong', *International Journal of Construction Management*, **29**, 105-122

World Bank, 2011, 'Economy rankings - doing business'. http://www.doingbusiness.org/rankings accessed 1 January 2013

World Economic Forum, 2011, The Global Competitiveness Report 2011-2012, World Economic Forum, Geneva

Issues of Construction Health and Safety in Developing Countries: A Case of Jordan

4

Suhair Zaid Alkilani, (University of Technology Sydney, Australia)

Julie Jupp, (University of Technology Sydney, Australia)

Anil Sawhney, (Indian Institute of Technology, India)

Abstract

The construction industry is widely regarded as one of the most significant in terms of its impact on health and safety (H&S). Recent studies suggest that in developing countries H&S awareness and performance is low. In this paper, the current state of Jordan's on site construction H&S is explored using a two-part investigation. The first part introduces the area of research in a literature based study of on-site safety. The second part presents the findings of a case study on the Jordanian construction industry and its H&S practices. Primary data was collected from field visits, expert interviews and semi-structured questionnaires. Supporting secondary data was collected from archival studies and related research literature. The findings highlight a lack of government commitment exemplified by regulation, policy and legal constraints that limit the operational efficiency of government departments responsible for H&S management, thus hindering the development of good H&S practice. Research results also highlight the key constraints of good H&S practice from the perspective of construction contractors.The study concludes with a discussion of potential solutions to improve H&S performance on construction sites in Jordan.

Keywords: Health and safety, Construction industry performance, Management processes, Procurement methods, Jordan

Introduction

The construction industry is widely recognized as being both economically and socially important (Ofori 2011). Yet the industry is also considered to be the most hazardous (Larcher and Sohail 1999, Jannadi and Bu-Khamsin 2002, Konkolewsky 2004, Charles 2007, Hamalainen *et al.* 2007, Farooqui *et al.* 2008). Global estimates by the International Labor Organization (ILO) show that occupational health and safety (OHS) problems in the construction industry are more extensive than previously recorded (Larcher and Sohail 1999, Hamalainen *et al.* 2007, Murie 2007). Globally, poor performance in the industry accounts for more than 100,000 fatalities annually, equating to approximately 30-40% of the world's work related fatal injuries (Jill and John 2010, ILO. 2010).

Whilst the governments and industry bodies of developed countries acknowledge that the construction industry's undesirable accident record must be improved (Chileshe and Dzisi 2012), with many advocating the adoption of a zero accident policy (Jill and John 2010), poor safety records continue to hinder performance improvement. This is particularly evident in developing countries where a dearth of reliable information on occupational accidents exists due to the absence of proper recording and notification systems, coupled with low construction H&S awareness (Fang *et al.* 2004, Paivi *et al.* 2006, Hamalainen *et al.* 2007, Chiocha *et al.* 2011). In a comparative study of a developing country, South Africa, with a developed country, Singapore, Toe *et al.* (2008) illustrate the disparities in construction H&S performance.The construction sectors of developing countries are typically dominated by contractors that can be classified as small and medium sized enterprises (SMEs) which operate mainly within domestic markets (Kheni *et al.* 2006). As previous studies show these contractors give little attention to H&S issues (Weil 2001, Petrovoc-Lazarevic and Perry 2004), which results in a significant number of accidents and health problems (Weil 2001).

Numerous factors have been identified that could influence H&S management of construction sites in developing countries (Toole 2002, Gibb and Bust 2006, Chiocha *et al.* 2011, Alkilani *et al.* 2012b); they include: poor infrastructure, lack of appropriate H&S regulation, legislation and laws, low literacy levels among laborers, lack of proper recording and notification systems, competitive tendering, extreme weather conditions, and corruption (Kartam *et al.* 2000, Gibb and Bust 2006, Alkilani *et al.* 2012b).

As in many other developing economies, Jordan's society and economy have suffered human and financial losses as a result of poor H&S management in the construction industry (El-Mashaleh *et al.* 2010). The construction sector is an integral part of Jordan's infrastructure development and significantly boosts its economy (Alkilani *et al.* 2012b). The sector accounts for between 5 and 8.3% of GDP (during the period of 1998-2012). The structure of the sector is tied to the Ministry of Public Works and Housing (MPWH) and together with the Jordanian Construction Contractors Association (JCCA) contractors are registered and classified according to five grade categories. Grading is based on a company's registered capital, financial status and qualifications (JDOS 2010). The classification system ranges from 'Grade 1', which designates the largest and most experienced contractors, to 'Grade 5', designating the smallest (JCCA 2012). The sector is also highly dependent upon manual labor, of which a high proportion is comprised of unskilled foreign workers (mostly Egyptian), most with minimal education who work for low wages (MOL 2010). Annual reports published by the Ministry of Labor and Social Security Department (between 2006-2010) show that the sector is associated with a disproportionately high number of job-related accidents and diseases compared to other industries, accounting for 10.5% of all workplace accidents (El-Mashaleh *et al.* 2010).

This study presents a literature review and findings from investigative interviews with government institutions and local contractors on the issues affecting H&S performance in developing countries and specifically in Jordan. The study aims to:

1- Review H&S initiatives undertaken globally to improve performance in the construction industry.
2- Investigate existing H&S procedures, regulations, polices, and practices in Jordan's construction industry.
3- Identify the constraints and challenges for good H&S practices in Jordan's construction industry.
4- Suggest possible solutions for government and industry bodies to improve H&S performance in Jordan.

Methodology

Data for the study was collected in three stages. Stage one comprised a review of H&S literature in the construction industry with a focus on internationally recognized research and initiatives that address H&S via performance improvement measures, management processes and procurement methods. The research publications surveyed included studies of construction H&S from both developed and developing economies. The literature was then analyzed to create a framework that informed the design of a case study of Jordan and helped form the semi-structured interview questionnaires used to gather primary data in Stage two.

Stage two of the research design therefore encompassed data collection. The case study focused on conducting semi-structured interviews with 27 experts representing a cross section of industry voices. Classifications of case study's interviewees fall into two main categories, namely: (1) Government experts – including the General Directors and H&S officers from the MPWH's Government Tendering Department as well as the Ministry of Labor's Health and Safety Department; and (2) Industry experts - including the managing directors, owners, project and technical managers, and safety engineers of 23 construction contractor firms. Interviewee participants from the Government Tendering Department were targeted as this body is responsible for providing transparent government procurement processes, managing tender for public projects, and developing the legislation that governs those tenders. Jordan's Ministry of Labor is responsible for enforcing H&S regulations across all industries (MOL 2010). These two government departments were selected for interview due to their unique perspective, role, influence and overview of H&S and related construction issues. The 23 contractor firms were selected based on their grading by the MPWH and JCCA, so that the sample provided a good representation of all grades (i.e., Grade 1 through to 5 contractors). The sample comprised of: four Grade 1 firms, seven Grade 2 firms, seven Grade 3 firms, three Grade 4 firms, and two Grade 5 firms. The contractors selected therefore varied considerably in relation to their registered capital, financial status, qualifications and experience. Additional characteristics about each firm are shown in Table 1. Due to differences between the government and industry roles, two semi-structured interview questionnaires were designed to elicit responses from government and industry representatives. Interviews with government representatives comprised questions concerning current procurement methods used in selecting contractors and executing projects, methods for collecting H&S data, and the availability of H&S performance measures. The contractors' interview questions consisted of two main parts, including: general information about the firm, and current H&S administration and practices. Specific questions were asked based on each contractor's safety policy, H&S management processes and incident and injury reporting procedures, safety measures and personal protection equipment (PPE), employee H&S training, and insurance systems.

In Stage three, data collected from the literature review, local research and interviews were used to identify a list of barriers to the implementation of good H&S practices in Jordan. This list was forwarded to the same group of participants interviewed in Stage two via office visits and email correspondence, after which each participant was asked to rank each barrier identified using 5-point Likert scale, where 1 is the least relevant and 5 is the most relevant. The results and analysis of the data from all three stages of research are presented in the following sections.

Literature Review: Currents of Change in Construction Health and Safety

As a reflection of the range of H&S problems in construction a large body of research exists which reveal key factors driving better H&S performance. Most significantly, safety regulations are often cited as crucial to reducing construction work accidents (Kartam *et al.* 2000). Organizational safety policies have also been identified as a major driver (Hinze and Wilson 2000). The Construction Industry Institute (CII) ranked safety training and safety incentives and penalties as two of five high-impact zero accident techniques targeted at improving safety performance at the project level (Hinze & Wilson 2000). Jaselskis *et al.* (1996) recommended increasing the number of formal safety meetings with supervisors, and Chi *et al.* (2005) argued that safety equipment is crucial to minimizing and preventing site accidents. Hinze & Wilson (2000) state that safety inspections and jobsite visits are required to improve safety performance. As worker attitude towards safety is considered as a root cause of accidents, Abdelhamid & Everett (2000) also reported that educating and training workers is an important factor that drives better H&S performance.

Many countries have established and released their own safety programs and systems of performance measurement and management. Key performance indicators (KPIs) and safety factors that pertain to H&S have been identified and enforced throughout construction industries of the developed world (Scotney 2000, John and Anthony 2001). From a procurement perspective, safety is also reported as an essential part of recommended regulatory frameworks (Farooqui *et al.* 2008).

Safety Performance Measures and Management

Several safety performance measures have been developed to describe on site H&S performance (Idoro 2011), and to evaluate a firm's safety performance (Jaselskis *et al.* 1996). For example, the experience modification rate and OSHA's reportable incidence rate, as well as the lost workday rate and lost-time rate, have been implemented to evaluate company safety performance over a number of years (Jaselskis *et al.* 1996, Levitt 1976, Idoro 2011). Jaselskis *et al.* (1996) propose quantitative strategies for achieving better safety performance differentiating between company and project levels. Among the more significant company-related factors are upper management support, time devoted to safety issues by the company safety coordinator, meetings with field safety representatives, informal safety inspections, safety training, and company safety expenditures. At the project level, several factors are identified by Jaselskis *et al.* (1996) namely: increasing project manager experience levels, reducing project team turnover, reducing worker penalties, and increasing budget allocation to safety awards.

Examples of safety performance measurement initiatives in developed countries are numerous. In the US, the CII released the Zero Injury Technique (Hinze and Wilson 2000), where five high-impact zero accident techniques have been identified. In the UK, the audit program has been implemented as a safety competition, whereby the results are then used for benchmarking a company's H&S management system (Colin 1999). In the UK the Health and Safety Executive have developed a tool to measure H&S performance and improvement over time within any sector which is based on H&S KPIs (Scotney 2000, Charles *et al.* 2007). In Australia, the Health and Safety Continuous Improvement Matrix, developed by the Construction Industry Development Agency in 1995, is used as a benchmarking system for the comparison of OHS performance across the Australian construction industry. The system allows the grading of a company's OHS between zero and five against 17 OHS system elements that are set out on the matrix (John and Anthony 2001).

Procurement

Traditionally, the criteria for contractor selection in the construction industries of developing countries are generally focused on tender cost (Bowen *et al.* 2002, Jill and John 2010). It is common practice for contractors to reduce their cost to win tenders, where cuts in H&S management processes are one of the first items to face cost cutting exercises, and as a result OHS can suffer (Jill and John 2010). In a joint publication by the UK's Institute of Civil Engineers (ICE) and Engineers Against Poverty (EAP) it was reported that the current procurement routes hinder the improvement of H&S practices (Jill and John 2010). Hinze (1988) (cited in John and Anthony 2001) found that injury rates tend to be higher on those projects that were competitively bid. In their briefing note, the ICE and EAP argue that there are two main reasons why H&S should be considered a serious issue during the procurement of construction services and works. Firstly, to ensure that the responsibilities to maintain a safer site are set out in the procurement terms, enabling H&S issues related to the project and services to be taken seriously by all participating parties. Secondly, that the procurement process, contract terms and conditions be seen as complementary mechanisms for ensuring compliance with existing legislation and/or terms and conditions of project finance (Jill and John 2010).

A number of countries are reforming their procurement systems with the aim to raise the standard of OHS on construction projects. For example, as a part of China's National Economic Reform, the construction industry is moving towards competitive procurement systems (Charles *et al.* 2007). In the UK, the traditional lowest price contractor selection process is shifting toward the use of multiple selection criteria and a best-value contractor selection framework (Charles *et al.* 2007). New Zealand is also following these trends, where the Department of Labor recently published "A Principals Guide to Contracting to Meet the

Health and Safety in Employment Act 1992". The guidance sets out a broad process for building H&S into construction contracts (DOL 2010).

Agencies

To ensure that on site worker H&S standards are maintained, a number of countries have established agencies with varying objectives and geopolitical structures. Three types of agency can be distinguished: government, independent and joint; with each potentially having three levels, namely international, regional and national. Example agencies range from those that aim to shape polices and produce standards covering the entire spectrum of work-related H&S issues, such as the International Labour Organization (ILO 1992), to agencies that support government H&S strategy by providing education and certification. For example the UK's Health and Safety Executive (Scotney 2000) and National Examination Board in Occupational Safety and Health (NEBOSH 2013), the US's National Institute for Occupational Safety and Health (NIOSH 2012), and National Occupational Health and Safety Commission. Joint agencies bring together government and industry initiatives, e.g. the European Agency for Safety and Health at Work (EASHW 2013). Agencies responsible for enforcing H&S in construction are predominantly joint initiatives.

In developing countries by contrast H&S agencies are often public bodies linked to government ministries and social security departments (Agyem-Bediako 2007). Thus, their construction sectors are predominantly characterized by top-down strategies rather than joint or bottom-up industry driven initiatives and thereby lack independence. Poor inter-agency cooperation and communication also persists in construction sectors of developing countries. This study confirms that this is the case in Jordan, where the main bodies responsible for H&S are the MPWH and the Ministry of Labour.

Findings – Part 1: Status of H&S Performance in Jordan

Research on construction H&S in developing countries provide ample evidence of lapses in H&S regulations and standards, management practice and awareness (Kheni *et al.* 2006). Many of these findings are confirmed in this case study. This section reports the findings from a two-part semi-structured interview and questionnaire process. Significantly, findings from interviews with key experts from the two government bodies responsible for H&S revealed a lack of regulations and standards specific to the construction industry. Further, the findings from the 23 contractor interviews reflect problems and shortcomings resulting from the dominance of SME contractors in Jordan. The practice-based issues identified echo existing research on H&S performance in other developing countries, confirming that SME dominance is related to more significant numbers of accidents and injuries (Gibb and Bust 2006). In Jordan the economy depends almost entirely on SME contractors to help drive productivity, where about 98% of all businesses are classified as SMEs (JEDCO 2011). Furthermore contactors, and in particular SME contactors - classified by the JCCA as Grade 3, 4 and 5 - failed to provide minimal requirements to achieve better safety performance as shown in Table 1. The case study highlights that although a few contactors have adopted effective H&S practices the majority rely solely on the specification of H&S procedures and standards in contracts but had no monitoring practices onsite.

Detailed findings from this study are presented in the following sections, where three issues are identified as the main factors that influence H&S practice and characterize the status of construction H&S in Jordan, namely; (1) Government institutions responsible for health and safety, (2) Procurement methods, and (3) Contractor attitudes. The constraints surrounding the implementation of better H&S practices in Jordan's construction industry as identified by the respondents are described.

Government Institutions

As a part of the Ministry of Labor, the Department of H&S ensures all workplaces maintain minimum standards of H&S. Apart from undertaking workplace inspections; the Department is responsible for administrating the labor code, Law Act No. 8 of 1996, as well as Sections 78 to 85 of the H&S standards describing workplace H&S regulations. Jordan's current H&S standards were adopted from the US' OSHA standards. The Social Security Department is responsible for publishing statistics about work injuries, accidents, and deaths. These statistics are published annually by the Ministry of Labor. Whilst the reported statistics cover work-related accidents, injuries and fatalities for all industries, there is no specific statistics for individual industries. Consequently there is a lack of detailed information on construction industry accidents, injuries and fatalities. The interview participant from the Department of Health and Safety reported that:

> *"There is no specific data for the number of accidents in the construction industry; this is one of the major weaknesses in our database and statistics which we are currently working on. However, we need government support and in particular the MPWH helping with this, as they should put pressure on the industry i.e. the contractors to regularly report accidents".* (H&S Officer, Department of Health and Safety, Ministry of Labor)

The H&S official also stated: *"These statistics represent only the number of accidents that are reported by the owner or contractor, for the purposes of social security cover and compensation, and does not represent the actual number of accidents in the construction workplace; it even doesn't comprise 10% of the real accident number".* (H&S Officer, Department of Health and Safety, Ministry of Labor)

The Ministry of Labor's Department of H&S issue labor certificates to contractors bidding for jobs to certify compliance with all labor laws. Another interviewee from the Government Tendering Department reported: *"It is however difficult to assess whether a contractor issued with a labor certificate continues to comply with working conditions outlined in Labor Law 1996"*. (Officer, Government Tendering Department, MPWH)

Furthermore, the Department of H&S deals directly with onsite construction accidents, after being notified of accidents via the Police Department (where accidents must be reported). The Department of H&S is then responsible for contacting the Department of Social Security where additional investigations are conducted for insurance purposes. In undertaking accident reports and other regular H&S procedures, the Ministry of Labor hires H&S engineers and inspectors who are responsible for conducting site inspections. In this regard, the H&S Officer stated that:

> *"In the process to improve H&S performance in the construction industry we increased the number of inspectors, however it is still not enough to conduct a daily site inspection, we have 50 safety engineers and inspectors who are working throughout Jordan at the moment, I believe we need at least triple this number if not more, if daily site inspections are going to be done, which we have a lack of"* (H&S Officer, Department of Health and Safety, Ministry of Labor).

Procurement Methods

The Officer from the Government Tendering Department reported that the main procurement method used when executing public projects is the traditional lowest-price tender method. Contractors are invited to bid via tender advertisements in local newspapers and via the Government Tendering Department website. Bids are secured in a 'closed envelop' and lodged with the MPWH in a secured box. The bids are read in the presence of all bidding contractors and a committee of consultants, MPWH decision makers, and any project-related government institutions and sponsors. In most public projects the winning contractor is typically the one who presented the lowest price tender.

The 'general' and 'special' conditions for contracting in Jordan were also highlighted. General conditions are valid for government buildings and housing. In such cases, the MPWH requires, as part of the contract's general requirements, the provision of minimum H&S requirements (adopted from the Labor Law 1996 for H&S at work), which includes for example, onsite safety personnel and PPE for all workers. However despite these measures there is no specific safety cost disclosure required as part of the tender contract documents, and the winning contractor is not subject to monitoring processes to ensure the project complies with H&S requirements. The Government Tendering Department interviewee highlighted these difficulties:

> "After *winning the contract it's the contractor's responsibility to monitor H&S onsite, however we are responsible for investigating accidents if they happen and if reported, we penalize the contractor for violation of H&S requirements".* (Officer, Government Tendering Department, MPWH)

In addition, the regulations were criticized by one interviewee due to their lack of detail and brevity. One H&S manager stated:

> *"Several major safety issues are not described in enough detail. For example, the specification of safety equipment, what, when and where to wear them, are not stated. This leaves such requirements to be solely judged by employers".* (H&S Officer, Department of Health and Safety, Ministry of Labor)

Special conditions are used when establishing contracts for public infrastructure projects, such as wastewater and sewage treatment plants, and highways. These projects are usually sponsored and funded by foreign international agencies such as the UN and the World Bank. Under the special conditions of public infrastructure projects, H&S is a major requirement when selecting local contractors and subcontractors. Sponsoring organizations typically require detailed safety plans, evidence of a H&S track record, and that all bidding contractors meet H&S prequalification requirements. In addition, project specific international H&S requirements may also be specified by the foreign sponsor and any participating foreign contractor.

Attitudes of Local Construction Contractors

To understand the attitudes of construction contractors toward H&S, semi-structured interviews were conducted with managing directors, owners, project and technical managers, and safety engineers working in all classes of companies. Interview questions were divided into two parts, the first sought information about the firm's profile and size; the second sought information about safety policy, safety reports and records, safety manuals, safety measures, accidents numbers and accidents reporting systems, the use of PPEs, insurance, the role of safety management officers. The interview also included questions related to the number of accidents and their bearing on the cost of claims. However, some respondents refused to answer these questions, believing that this information was confidential and may negatively affect their company's reputation. These questions were therefore excluded from analysis. Table 1 summarizes the results.

Overall the results highlight that attitudes of Jordan's contractors towards construction site H&S is poor due to lack of H&S awareness and knowledge. Poor attitudes to H&S were also reflected in the preference given to commercial profits over most other project objectives, with H&S objectives being given a low priority by SMEs. One Ministry of Labor interviewee from the Department of H&S described SME contractor attitudes as being compromised by financial considerations:

> *"Contractors, and in particular small enterprises, want to make the maximum profits and cannot provide the required personal protective equipment for their workers. They do not have safety policies in place... and they wouldn't want to spend their time, money and resources to train their workers to a certain standard of H&S."* (H&S Officer, Department of Health and Safety, Ministry of Labor)

Contactor Grade	No. of firms	Safety policy	PPE	H&S Training for workers	On site safety signs/measures	Safety Tool box	Insurance	Availability of safety director	Recording accidents	Reporting of accidents and incidents to the government
Grade 1	4	4	4	3	4	Weekly 2 Monthly 2 Occasionally 0	Social security 4 Other 0	Full-time 4 Part-time 0	Regularly 2 Sometimes 2 Never 0	Regularly 0 Accident driven 4
Grade 2	7	5	4	2	6	Weekly 1 Monthly 3 Occasionally 4	Social security 7 Other 0	Full-time 4 Part-time 3	Regularly 2 Sometimes 5 Never 0	Regularly 0 Accident driven 7
Grade 3	7	0	1	1	1	Weekly 1 Monthly 2 Occasionally 4	Social security 7 Other 0	Full-time 2 Part-time 3	Regularly 0 Sometimes 2 Never 0	Regularly 0 Accident driven 7
Grade 4	3	0	0	0	0	Weekly 0 Monthly 1 Occasionally 2	Social security 3 Other 0	Full-time 0 Part-time 1	Regularly 0 Sometimes 2 Never 1	Regularly 0 Accident driven 3
Grade 5	2	0	0	0	0	Weekly 0 Monthly 1 Occasionally 1	Social security 2 Other 0	Full-time 0 Part-time 0	Regularly 0 sometimes 0 Never 2	Regularly 0 Accident driven 2

Table 1 Contractors' interview results

While previous research has identified organizational safety policy, safety training, safety meetings, safety equipment's, safety inspections, safety incentives and penalties as amongst the key factors driving safety performance (Jaselskis *et al.* 1996, Abdelhamid and Everett 2000, Hinze *et al.* 2000, Charles *et al.* 2007, El-Mashaleh *et al.* 2010, Jill and John 2010), the results in Table 1 reveals additional shortcomings in relation to the lack of international conventions surrounding onsite H&S being followed by contractors in Jordan. Only nine out of the 23 companies investigated have safety policies in place, and all nine were either Grade 1 or 2 companies, i.e., large and experienced firms with high levels of registered capital, finance, and qualifications. On average 75% of the sites visited when conducting the case study provided no evidence of H&S practices being in place, such as the posting of safety signs and safety measures, including the absence of PPEs, worker training and "tool box" safety talks.The project managers and H&S directors interviewed reported that due to pressures from senior management and owners to complete projects on time and in budget, they did not feel able to interrupt works for H&S talks or safety meetings. One interviewee said:

> *"..Panic plus rush plus safety, does not work..., [...] The owner who is the manager of the company I'm working at tells the project manager: 'as much as you save money and time...I will pay you a bonus...'. The project manager then does not care about how safe his project is... he even does not allow me to stop the work for 10 minutes each week or even each month to do training or "tool box" safety talks with the workers".* (Contractor H&S Site Officer)

Most interviewees from industry reflected these sentiments and stated that implementing H&S procedures would substantially increase the cost of construction. Consequently, the cost implementation of safety procedures was not disclosed in the contractor's bid and generally only the insurance cost was considered for those items.

Findings – Part 2: Constraints on Construction H&S in Jordan

The study found that Jordan's construction industry is hindered by many challenges that adversely affect H&S performance. These findings form a number of significant constraints or barriers to implementing good H&S practice on construction sites and are grouped thematically into three classes as shown in Table , namely: 1) Government regulations, legislation, and policies, 2) Management commitment and involvement, and 3) Worker awareness.

Main Groups of Constraints	RII	Sub-constraints Factors	RII
Government Regulations, legislations and Laws	0.85		
		Current procurement methods and contractor selection criteria	0.82
		Lack of H&S regulations for construction	0.8
		Lack of relevant accident data, statistics and information	0.69
		Current labor laws	0.62
		Insufficient insurance schemes	0.54
		Insufficient penalties schemes	0.52
		Inadequate support for innovation, research and technology	0.4
Management Commitment and involvement	0.76		
		Inadequate supply of PPE for workers	0.8
		Insufficient accident recording and reporting systems	0.72
		Inadequate training and education programs	0.66
		Insufficient safety meeting time with workers	0.64
		Insufficient communication channels	0.61
Workers awareness	0.66		
		Unwilling to wear PPE	0.86
		Extensive use of foreign labor	0.86
		High percentage of uneducated/unskilled labor	0.77

Table 2 Constraints for good health and safety practice

For each group a list of underlying issues were identified from the literature review and case study interviews. Stage 3 of the study then asked participants to firstly rank the importance of the three main classes of constraints, and then rank each underlying factor using a 5-Point Likert Scale, where 1 is the least important and 5 is the most. The Relative Importance Index (RII) method (Kometa *et al.* 1994) was used to calculate the significance of each category's rank, revealing the significance of three main classes and 15 sub-constraints shown in Table . The RII is computed as:

$$\text{RII}=\frac{\sum W}{A\times N}$$

Where *W* is the weight given to each factor by the respondents and ranges from 1 to 5; *A* is the highest weight = 5; and *N* – the total number of respondents.

Regulations, Legislation and Policies

As the most significant constraint, government regulations, legislation and policies were seen to be the main barrier to implementing good H&S practice. At the procurement level, poor safety performance can be explained by contractor selection criteria based on principles of lowest cost, which was ranked as the highest sub-constraint to achieving better

H&S practices. Results show that both the tendering body and contractor fail to properly consider H&S costs. In addition, the current procurement methods primarily used in Jordan, and in particular in the public sector, fail to clearly outline each party's responsibilities. The absence of a unified set of regulations adversely affects the enforcement of H&S on the job site and was ranked as the second most significant sub-constraint. There are no specific regulations, standards, and laws pertaining to the site conditions of construction projects. There are also no established construction safety standards or manuals provided by the MPWH or the Ministry of Labour. In cases where construction projects are sponsored by international agencies or undertaken by foreign contractors they are governed by their own country or International Contractors' H&S Standards. However, international standards are not necessarily applicable to the work conditions and environment of Jordan since methods of practice in advanced and industrial countries can differ widely from those used in Jordan.

The interview results also show that there is a lack of national H&S indicators and measures, as well as a lack of official safety data and records of construction accidents on site, making H&S one of the last issues to be considered by contractors and clients. Another significant challenge identified was the lack of compliance and certification standards suitable for Jordan's construction environment. Participants noted the absence of benchmarking processes that encompass apposite KPIs.

While innovation is regarded as an engine of economic growth, and a driver for poverty alleviation in developing countries (UNESCO 2010), a significant finding from this study shows that a lack of innovation, and research and development (R&D) is ranked as the least significant barrier to better H&S practices in Jordan. This confirms the fact that in developing countries little attention can be paid to R&D by SMEs. The SME contractor participants claimed that due to low profit margins and lack of government funding and tax incentives, their firms could not invest in R&D. Whilst this factor was ranked lowest of all seven barriers, it is one of the most crucial for improving construction H&S performance. Further, in a previous study conducted by (Alkilani *et al.* 2012b) it was found that there is a low capacity amongst Jordanian SME contractors to adopt new innovations, such as new information technologies, was perceived by contractors as a key contributing factor in the shortfalls in productivity.

Management Commitment and Involvement

Overall, the respondents ranked 'Management Commitment and Involvement' as the second most significant barrier to implementing good H&S practice, with an RII of 0.76. Respondents revealed that the senior management levels of all the construction firms interviewed were ultimately responsible for increases in accident rates due to their lack of willingness to allocate resources to the implementation of the necessary H&S procedures onsite. The results show that clients and consultants stressed the importance of H&S issues before work commences, but as the work progresses each stakeholder's concern to meet deadlines takes priority and less attention is paid to safety.

Compounding this issue is the fact that the construction industry in Jordan is dominated by SMEs, which are characterized by combined owner-manager leadership arrangements and are typically family-run enterprises. In such firms, owner-managers were found to act as gatekeepers in the allocation of resources, where more often they were not willing to prioritize the application and monitoring of H&S procedures onsite because of a focus on businesses growth - where profits where reinvested into areas that increased construction capacity and competitive advantage. The attitudes of the SME contractors interviewed also differed in relation to those of larger firms. SMEs considered H&S as a waste of resources and tended to have a lack of awareness of the effectiveness of safety prevention programs in reducing costs and increasing productivity. The results also reflect the fact that a considerable number of workers were not trained to recognize and avoid the risks associated with their tasks. This leaves those workers more susceptible to potential accidents.

Worker Awareness

'Worker awareness' was ranked as the third most significant barrier to implementing good H&S practice in Jordan. This is in contrast to findings of previous research relating to H&S in developing countries, which highlights the adverse effects of using unskilled foreign labor on construction industry H&S performance. In a study by Kartam *et al.* (2000), it was shown that the extensive use of unskilled foreign labor in developing countries was one of the greatest threats to maintaining a safer construction sites. Despite this lower ranking overall, the RII values shown in for the three sub-constraints (used to define worker awareness) reveal that participants ranked them as being of similar significance resulting in a lower variance and range between these rankings (σ^2=0.0018 and R=0.09) when compared to the other two groups of sub-constraints (σ^2=0.0202 and R=0.42, and σ^2= 0.0045 and R=0.19 respectively). The majority of respondents agreed that workers were reluctant, or neglected, to wear PPE. With the absence of onsite safety equipment and its lax enforcement an obvious lack of commitment across all levels of the construction company was identified as a major constraint to the improvement of construction site safety.

The high level of employment of foreign labor has long been a defining characteristic of the Jordanian construction industry. Different cultures, traditions and attitudes to work impact on human relations, in particular in relation to differences in work habits and communication. The majority of construction workers in Jordan are unskilled, untrained, and inexperienced. Foreign workers often come from poor communities and are willing to work in inferior working conditions. The workers are emotionally vulnerable and sometimes preoccupied with personal economic problems since most of them are working in unsecured conditions and their employment is not tied to an employment sponsorship agreement. All of these conditions can affect the concentration and attention of the worker and may contribute to mistakes.

Discussion: Summary of Problems and Possible Solutions

To improve H&S performance in the Jordanian construction industry a number of programmes can be developed, adapted from existing initiatives undertaken in other countries and implemented in a way which is sensitive to the contextual variables of Jordan's construction sector. Research on other developing countries has placed similar deficiencies in H&S performance under the spotlight (Farooqui *et al.* 2008, Koehn *et al.* 1995, Kartam *et al.* 2000, Jill and John 2010). Academia and industry research initiatives (Gibb and Bust 2006, Farooqui *et al.* 2008, Jill and John 2010, Alkilani *et al.* 2012a) propose solutions and recommendations on a range of H&S challenges facing the construction industries of developed and developing countries. These have mainly focused on programs that are either led by government *or* industry. With the objective of highlighting the importance of joint initiatives between government and industry the authors propose that the existing proposed solutions be extended and adapted.

Joint Government-Industry led Initiatives

In order to realise tangible results in terms of improvement in H&S performance in construction, the formation of a hybrid agency is essential. By bringing Government and industry bodies together a number of important initiatives can be achieved including: (1) revision of H&S codes and standards, (2) reforms to current procurement processes, and (3) promotion of H&S best-practice and culture, as well as the development and implementation of training and education programs for workers.

- Revising H&S Codes and Standards: The revision and development of new regulations that are relevant and up-to-date is essential. In undertaking such major reform, the MPWH together with the Ministry of Labour must work to draft new H&S regulations for the administration of safety. . A more comprehensive set of requirements that target the specific conditions of the construction industry is required so as to take into consideration the environmental, human, technological and social aspects,

- Procurement Reform: Current tender applications and contractor selection processes should be improved such that H&S is more clearly integrated. It is essential that safety responsibilities be specified in the contract as well as ways in which H&S measures are implemented and monitored onsite. Contractual specifications should also establish guidelines in order to prevent risk and identify hazards,
- Promotion of best-practice and development of H&S culture: Lack of best practice and culture surrounding H&S is related directly to the uncertainty, ambiguity and lack of awareness of H&S in the construction industry. H&S education is therefore critical to increasing the industry's performance and to fostering more positive attitudes towards H&S issues.

To support a joint initiative a number of government-led and industry led programs could also be implemented.

Government-led Initiatives

Government-led initiatives and programs include (1) the introduction of incentives, (2) the adoption of H&S indicators, and (3) provision of leadership in procurement. These initiatives should however take into consideration the conditions of Jordan's construction industry in the following ways.

- Introduction of incentives: Incentives that seek to improve H&S, e.g., via the use of appropriate tax incentives and penalties that consider the financial situation of all five grades of construction firm.
- Adoption of H&S Indicators: Government can assist the construction industry reconsider the importance of H&S in order to improve the industry's image (Ofori 2001) . The staggered introduction of a set of indicators to track the Jordanian construction industry's safety record would facilitate this objective. These could include:
  a. Construction accidents frequency rate,
  b. Severity rate of accidents,
  c. Injuries (# of persons suffering minor injury, permanent disability, fatality), and
  d. Enforcement action for non-compliance such as fines, suspensions or de-registration.
- Leadership in Procurement: As a minimum requirement, the MPWH Government Tendering Department contract specifications for public projects should contain specific guidelines to control all anticipated hazards by:
  a. Identifying persons responsible for overseeing contractor's performance,
  b. Requiring contractors to prepare and submit an acceptable project hazard prevention plan that defines supervisory and employee safety training,
  c. Identifying specific published safety standards and hazard prevention requirements, and
  d. Listing qualifying requirements for eligible contractors to ensure bidders are restricted to those whose past performance demonstrates key H&S competencies.

Industry-led Initiatives

Industry-led initiatives and programs include (1) risk and role sharing, (2) increasing available knowledge and developing H&S resources, (3) organisation-based KPIs for H&S, (4) increasing the involvement and accountability of senior management in H&S, and (5) increasing efforts to adopt new information technologies. Like government-led programs each initiative must take the conditions surrounding Jordan's construction industry into consideration in their development and implementation.

- Risk and role sharing: Developing partnerships between those involved in concept, design, construction planning, construction work, and maintenance. Jordanian developers and clients in particular should be encouraged to play a greater role in driving H&S by defining safety objectives in project scoping documents, selecting safe contractor and participating in onsite safety management,
- Increasing available knowledge and developing H&S resources: Launching H&S information banks and databases so as to make publicly accessible appropriate statistical information, accident records, and accident prevention methods,
- Organisation-based KPIs for H&S: Performance indicators that focusing on H&S practices that are sensitive to the context of Jordan should be established and agreed to by project stakeholders. National KPIs concerning H&S are critical to an industry-led solution, and should be be utilised by all five grades of construction firms,
- Increasing involvement of senior management: this requires giving safety a higher priority in company meeting, involvement by senior management in safety activities, open communication channels between workers and management on safety issues, and an emphasis on workers training and education programs,
- Increasing efforts to adopt new information technologies: Increasing the use of computer-based H&S applications and technologies.

Conclusion

The findings from the investigation provide a broad perspective of the legal, institutional, business and cultural frameworks shaping H&S management of indigenous construction across both the public and private sectors in Jordan. The study reveals that H&S management and measurement in Jordan is still in its infancy. There is a lack of government commitment exemplified by the inability to enforce regulations, policies and legislation which limits the operational efficiency of government departments responsible for OHS management and hinders the development of good H&S practice. The study highlights several key constraints that challenge the development of good H&S practice among construction contractors. Whilst the findings stemming from this study and the possible solutions recommended here focus on the Jordanian construction industry, the literature and anecdotal evidence obtained by interviewees suggests that the situation is not significantly different in other developing countries.

Acknowledgments

Special thanks to safety personnel at Ministry of Labour, Ministry of public works and Housing, and construction contractors who contributed to this research.

References

Abdelhamid, T. S. and Everett, J. G. 2000. Identifying root causes of construction accidents. *Journal of Construction Engineering & Management,* **126** (1), 52.

Agyem-Bediako, S. F., 2007. *Management commitment to safety among multinational construction companies in developing countries: A case of Ghana.* 3260116 (Sc.D.). University of Massachusetts Lowell.

Alkilani, S., Jupp, J. and Sawhney, A., 2012a. Promoting Health and Safety in Construction in Developing Countries: A Case of Jordan. *CIB W099 International Conference on Modeling & Building Health & Safety.* Singapore.

Alkilani, S., Jupp, J. and Sawhney, A., 2012b. Readying a Developing Economy for National Performance Measurement and Benchmarking: A Case Study of the Jordanian Construction

Industry. *Visions for the future of housing: Mega cities.* Turkey: IIAHS World Congress on Housing.

Bowen, P., Tookey, J. and Hardcastle, C., 2002. Construction Procurement and Management in the UK & South Africa. *10th Symposium Construction Innovation and Global Competitiveness.* CRC Press, 476-497.

Charles, C. 2007. Construction Safety anf Health Factors at the Industry Level: The Case of Singapore. *Journal of Construction in Developing Counreies,* **12** (2).

Charles, M., Pillay, J. and Ryan, R., 2007. *Guide to Best Practice for Safer Construction: Literature Review from Concept to Completion.* Australia: Cooperative Research Centre for Construction Innovation.

Chi, C.-F., Chang, T.-C. and Ting, H.-I. 2005. Accident Patterns and Prevention Measures for Fatal Occupational Falls in the Construction Industry. *Applied Ergonomics,* **36** (4), 391-400.

Chileshe, N. and Dzisi, E. 2012. Beneﬁts and Barriers of Construction Health and Safety Management (HSM): Perception of Practioners within Design Organisations. *HJournal of Engineering, Design and Technology,* **10** (2).

Chiocha, C., Smallwood, J. and Emuze, F. 2011. Health and Safety in the Malawian Construction Industry. *Acta Structilia,* **18** (1).

Colin, F. 1999. Benchmarking Health and Safety Performance Through Company Safety Competitions. *Benchmarking: An International Journal,* **6** (4), 325-337.

DOL, 2010. *A principal's guide to contracting to meet the Health and Safety in Employment Act 1992.* NZ.: Department of Labour.

EASHW, 2013. *The European Agency for Safety and Health at Work (EU-OSHA)* [online]. EU. Member States. Available from: https://osha.europa.eu/en/about.

El-Mashaleh, M., *et al.* 2010. Safety Management in the Jordanian Construction Industry. *Journal of Civil Engineering,* **4** (1).

Fang, D. P., Huang, X. Y. and Hinze, J. 2004. Benchmarking Studies on Construction Safety Management in China. *Journal of Construction Engineering & Management,* **130** (3), 424-432.

Farooqui, R., Arif, F. and Rafeeqi, S. F., Safety Performance in Construction Industry of Pakistan. ed. *First International Conference on Construction in Developing Countries*, 2008 Karachi, Pakistan.

Gibb, A. and Bust, P., 2006. *Construction healtha and safety in developing countries.* UK: European Construction Institute.

Hamalainen, P. i., Takala, J. and Leena Saarela, K. 2007. Global Estimates of Occupational Accidents. *Safety Science,* **44** (16), 137-156.

Hinze, J., Coble, R. J. and Haupt, T. C., 2000. *The Management of Construction Safety and Health.* Rotterdam ; Brookfield, VT: A.A. Balkema.

Hinze, J. and Wilson, G. 2000. Moving Toward A Zero Injury Objective. *Journal of Construction Engineering & Management,* **126** (5), 399.

Idoro, G. 2011. Comparing Occupatioal Health and Safety (OHS) Management Efforts and Ferformance of Nigerian Construction Contractors *Journal of Construction in Developing Counreies,* **16** (2), 151-173.

ILO, 1992. *Health and safety in construction, an ILO code of practice.* Geneva, International Labor Office, report no, 08.10.1.

ILO., 2010. *The Role of Worker Representation and Consultation in Managing Health and Safety in the Construction Industry.* Geneva.

Jannadi, O. A. and Bu-Khamsin, M. S. 2002. Safety factors considered by industrial contractors in Saudi Arabia. *Building and Environment,* **37** (5), 539-547.

Jaselskis, E. J., Anderson, S. D. and Russell, J. S. 1996. Strategies for Achieving Excellence in Construction Safety Performance. *Journal of Construction Engineering & Management,* **122** (1), 61.

JCCA, 2012. *The Current Status of Construction Sector in Jordan.* Amman: Jordanian Construction Contractors Association

JDOS, 2010. *Jordanian Construction Sector Statistical Survey 2008-2009.* Amman: Jordanian Department of Statistics.

JEDCO, 2011. *SME Development in Jordan* [online]. Jordan: Jordan Enterprise Development Corporation. Available from: http://worldsmeexpo.hktdc.com/pdf/2011/SeminarRoom_C/Dec_3/1630_1800_24/2_AAl-Ahmad.pdf.

Jill, W. and John, H., 2010. *Promoting Construction Health and Safety through Procurement: A briefing note for developing countries.* Institution of Civil Engineers

John, L. and Anthony, M. 2001. Measuring the occupational health and safety performance of construction companies in Australia. *Facilities,* **19** (3), 131-139.

Kartam, N. A., Flood, I. and Koushki, P. 2000. Construction Safety in Kuwait: Issues, Procedures, Problems, and Recommendations. *Safety Science,* **36** (3), 163-184.

Kheni, N., Gibb, A. and Dainty, A., The Management of Construction Site Health and Safety by Small and Medium-Sized Construction Businesses in Developing Countries: A Ghana Case Study. ed. *22nd Annual ARCOM Conference*, 2006 Birmingham. UK, 295-304.

Koehn, E. E., Kothari, R. K. and Pan, C.-S. 1995. Safety in Developing Countries: Professional and Bureaucratic Problems. *Journal of Construction Engineering & Management,* **121** (3), 261.

Kometa, S. T., Olomolaiye, P. O. and Harris, F. C. 1994. Attributes of UK construction clients influencing project consultants' performance. *Construction Management & Economics,* **12** (5), 433.

Konkolewsky, H.-H., 2004. Actions to Improve Safety and health in Construction. *Magazine of the European agency fro safety and Health at work.* Belgium.

Larcher, P. and Sohail, M., 1999. *Review of safety in construction and operation for the WS&S secto: part 1.* London. UK.: London school of hygiene and tropicla medicine, UK.

Levitt, R. E. 1976. Reducing Construction Accidents-Top management's Role. *Journal of the constrcution Division,* **102** (3), 465-478.

MOL, 2010. *Anual report.* [online]. Amman. Jordan: Ministry of Labour. Available from: http://www.mol.gov.jo/default.aspx?tabid=175.

Murie, F. 2007. Building Safety-an International Perspective. *International Journal of Occupational and Environmental Health,* **13** (1), 5-11.

NEBOSH, 2013. *The National Examination Board in Occupational Safety and Health* [online]. UK. Available from: http://www.nebosh.org.uk/About_Nebosh/.

NIOSH, 2012. *The National Institute for Occupational Safety and Health* [online]. Washington, D.C. US. Available from: http://www.cdc.gov/niosh/about.html.

Ofori, G. 2001. Indicators for measuring construction industry development in developing countries. *Building Research & Information,* **29** (1), 40 - 50.

Ofori, G., 2011. *New perspectives on construction in developing countries.* Abingdon, Oxon ; New York: Spon Press.

Paivi, H., Jukka, T. and leena, S. K. 2006. Global etimates of occupational accidents. *Safety Science,* **44**, 37-156.

Petrovoc-Lazarevic, S. and Perry, M., 2004. *Occupational Health and Safety Shortcomings in the Australian Construction Industry.*

Scotney, V., 2000. *Development of Health and Safety Perfromance Measurement Tool.* UK: Health and Safety Excutive, report no, 309/2000.

Teo, E. A. L., Haupt, T. and Feng, Y., 2008. Construction Health and Safet Performance in Developing and Developed Countries: A Parallel Study in South Africa and Singapore. *CIB W99 International Conference.* Gainesville, Florida, 485-499.

Toole, T. M. 2002. Construction Site Safety Roles. *Journal of Construction Engineering & Management,* **128** (3), 203.

UNESCO, 2010. *Measuring R&D: Challenges Faced by Developing Countries.*: UNESCO Institute for Statistics.

Weil, D. 2001. Assessing OSHA Performance: New Evidence from the Construction Industry. *Journal of Policy Analysis & Management,* **20** (4), 651-674.

Sustainable Innovation for Queensland's Housing Design: a Case Study

5

Michael Johnston & Mirko Guaralda (School of Design, Queensland University of Technology, Australia)

Sukanlaya Sawang (QUT Business School, Queensland University of Technology, Australia)

Abstract

This research provides an assessment tool that assists the selection process of sustainability in detached suburban housing. It investigates the implications of using different design and construction methods including architecturally designed houses, developer housing and prefabricated houses. The study simulates one example of the three types of houses that have been chosen to fulfil a real client brief on a real site on the Sunshine Coast, Queensland Australia. Criteria for sustainability assessment are formulated based on literature reviews, exemplar designs and similar research projects for which the houses can be adequately evaluated. This criterion covers aspects including energy use, materials and thermal performance. The data is collected using computer models and sustainability assessment software to compare and draw conclusions on the success of each house.

Our study indicates that architecturally designed housing with prefabricated building techniques are a better alternative to generic developer style housing. Our research provides an insight into the implications of three key elements of sustainability including energy use, materials and thermal performance. Designers, builders, developers and home-buyers are given an insight into some options currently available on the housing market and how the choices made during early design stages can provide a more positive environmental impact.

Keyword: Sustainable housing, prefabricated, perception, attitude, innovation.

Introduction

Buildings consume enormous quantities of resources and can cause any number of negative impacts to their occupants (Kibert 2012, Arif & Syal 2013). As the importance of sustainability has become increasingly evident in the built environment industry, designers and construction companies aim to improve building techniques, while maintaining their appeal to customers and within reasonable cost (Oluwole Akadiri and Olaniran Fadiya 2013). Our study aim is to evaluate and compare the design and construction methods of three different houses: architecturally designed, commercial/developer design and prefabricated housing. The purpose is to develop a framework that demonstrates the differences in design and construction methods and the varying environmental implications of each one.

Residential buildings are estimated to contribute 13% of Australia's greenhouse gas emissions, a figure set to rise 1.3% per year (Wood 2009). There are currently three main ways in which detached housing is designed and constructed in Australia. The first way is to engage an architect who will design the house based on clients' needs, site conditions, budget, aesthetics and their own expertise (Discoe, Qunii & Banish 2008). The house then goes to a builder who follows specific directions in the architect's construction details and specification, relying on the architect to manage the project.

The second and more common method is to buy a predesigned house; the clients select a dwelling from a catalogue or in a display village, which is then built at a fixed cost (Zhang et al. 2014). Minor changes can be made to the new house including extra rooms, facade changes, colour changes, kitchen layout and so on, but the basis of the house remains the same (Johnson 2014).

The final method is to use prefabricated modular housing, which is predesigned and arranged to the buyers' requirements and then built in a factory, before being transported and assembled on site. This method is cost effective, quick and claims to be more environmentally friendly than the previous two methods due to controlled construction environments, modular design and eco-friendly materials (Linner & Bock 2012; Davies 2005).

Prefabricated Housing Background

Prefabricated housing began in the 1830s when a London carpenter designed and built one for his son to ship to Australia to live in (Blismas & Wakefield 2009). By 1960 15% of American homes were mobile prefabricated houses, but interest in architecturally designed prefabricated housing did not flourish until the mid-90s to 2000s. Leading architects such as Le Corbusier, Walter Gropius, Frank Lloyd Wright, Richard Rogers, Archigram and Kisho Kurokawa have all had a hand in designing and experimenting with prefab's design potential (Arif & Syal 2013).

In 1945 a series of case study houses began to arise for Arts and Architecture magazine, from which Case Study House #8 by Charles and Ray Eames emerged (Shulman, Smith & Gössel 2007). Not only was it a stand out from the other case study houses, but it went on to become one of the most influential and significant houses of the 20th Century due to its innovation of design, construction and functionality (Koenig, Eames & Eames 2005). Despite their advantages and appeal, most prefabricated modular houses are constructed in rural areas, rather than the typical suburban.

Figure 1: Eames House (case study house No.8),

Implications of House Location

The choice of house is not the only factor in establishing sustainable living and lifestyles. Kibert (2008) states that 'In addition to the resources required to build and operate individual buildings, a wide range of additional impacts are the consequence of decisions made as to how to distribute buildings across the landscape.' Leaders in sustainable architecture and research Van der Ryn and Cowan (2007) explain in their book *Ecological Design* that the placement and location of a house is the key to its success as a sustainable living solution.

There is continuous debate over where the most sustainable place to live is. One suggestion comes from the book 'Sustainability and Cities', which outlines the advantages of cities and urban density due to increased public transport and lesser dependence on automobiles (Newman & Kenworthy 1999). It advocates that planning should discourage car travel, and that a higher density is necessary for limiting road infrastructure which they say is continuing to destroy natural landscapes (Van der Ryn & Cowan 2007).

Despite the strong argument for density, many Australians still choose to live in detached housing and in suburban areas. Table 1 demonstrates that the majority of people in Brisbane live in houses, except in Brisbane's CBD. This does not suggest that suburbs are better, but simply that people are still choosing to live there. This desire to live in suburbs depends on many factors, some possibilities could include:

- Supply and demand
- Suburbs being developed knowing that people will buy, more profitable to a developer than apartments
- Easy to market
- More appealing to families than an apartment
- Aligns with the 'American dream' that Australian culture is well in tune with
- Aligned with Australia's car culture and a lack of apartments in suburban areas

Table 1: 2011 Suburb profiles for Brisbane

	Brisbane CBD	Brisbane (All suburbs in Brisbane City Council)
Population	13,377	**1,627,570**
Detached house	41%	80%
Flat	41%	11%
Semi-detached/ Townhouse	7%	5%

(Source: www.domain.com.au, accessed on 12 November 2011)

In an interview with Brisbane Times (Moore 2010), the chairman of Brisbane City Council Neighbourhood Planning, Amanda Cooper, stated that 'When families have got children, when they need backyards, they look for a suburban type of experience. And then, as their children grow older, they are thinking that perhaps we need to move back into the inner suburbs.'

The book 'Positive Development' by Janis Birkeland (2002) discusses in depth some of the contradictions of high density living in cities. Reduced car travel and nearby amenities are often associated with urban sustainability, but too many variables and contextual factors mask the truth about the social and environmental impacts of such developments. Urban sustainability has been focused on density and form, using zoning and land uses from an aerial planning view to dictate where buildings should go favouring modernist style structures that unify design to their most simplistic and efficient form (Kendall 1999). Birkeland uses shopping centres to clarify the contradictions, stating that bringing 'more residents into the city, while allowing more shopping and recreation facilities in the suburbs, has created a vicious circle', leaving city residents with more fast food and less fresh food supermarkets. The question is then raised: 'what is the critical variable in urban sustainability, if not increased density?' to which Birkeland simply answers 'design'.

Part of this design process is assessing the needs of the client, in particular the size and room requirements of the house (Öst 2012; Schneider & Till 2005). The 2013 median land size reported by the Urban Development Institute of Australia (2014) for new houses across the five major capital cities in Australia, is $423m^2$. In Brisbane, the average land size for new houses in 2013 is $451m^2$, which is down from the 2012 average of $515m^2$ and considerably smaller than the 2006 average of $600m^2$. Professor and author of sustainable construction, Charles J. Kibert, believes the environmental implications of the building industry need rethinking, especially the needs and size requirement of buildings. Kibert raises two key questions: 'is the building actually needed, or is adequate space already available?' and 'Can the building be made smaller?' when discussing the initial steps to reducing a buildings carbon footprint (Kibert 2012).

Using the infrastructure that is currently in place, what difference would it make simply changing the initial design and construction technique of a house? Landscape architect and regional planning writer Ian McHarg (1981) suggests that existing urban areas should remain, as they are already established with infrastructure and need to ensure that open space is maintained and kept public. What kind of changes can be made with selecting different materials, changing mind sets and altering life choices even slightly? If each family were to make more sustainably conscientious decisions when building a house in suburbia, carbon emissions could be greatly reduced.

Exemplar Housing Projects

A pilot research on three exemplar projects has informed the research techniques and framework used in the main study. A comparative analysis of these case studies has exemplified current research limitations in this field. The exemplar projects include (a) architecturally designed house, (b) commercially designed house and (c) prefabricated house.

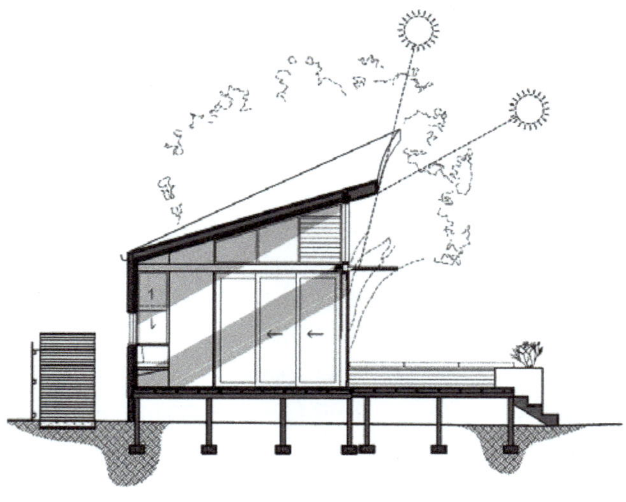

Figure 2: Daylight study for Brighton Compact (Callery & Bartak 2010)

Architect House Exemplar

Brighton Compact green house by Zen Architects is an example of an architecturally designed house that employs simple passive solar design principles and material choices to create a highly efficient building (Figure 2). The architects utilised energy rating systems early in the design process that allowed design alterations like room arrangements, ceiling heights, adding insulation, double glazing and thermal mass to be tested. Wary that the current rating systems are designed to bring houses that perform badly to a minimal standard, the architects aimed for a higher environmental performance beyond Greenstar or NABERS rating tools, with the help of computer software. The software used to rate the Brighton house favours less windows and more thermally efficient box designs, but other considerations such as views, day-lighting and architectural delight are not taken into consideration by the computer. This house includes passive solar design, and rooms are north orientated to maximize the solar gain.

Commercial Designed (Developer) House Exemplar

GreenMode, an organisation focused on economic and environmental improvements, performed a study that involved improving the sustainability of 35 'typical' South-East Queensland homes (Divecha, Muffet & Grant 2008). The study was carried out in alliance with RMIT University, Queensland Environmental Protection Agency and developers Delfin Lend Lease. The results found that drastic improvements could be made with proprietary changes, without having to redesign the house or spend a lot of money. Software called FirstRate 5 and SimaPro was used to simulate a variety of improvements made by the following list of changes (Figure 3):

1. Space heating and cooling - best of two orientation for home, double glazing and auto window shades, insulation and heat reflective paint, roof and room ventilation, 6 star air conditioning, ceiling fans and kitchen & bathroom extractor fans
2. Domestic Hot water - gas boost solar hot water
3. Appliances - high efficiency appliances, 3 star improvement in all appliances
4. Lighting - energy efficient lighting
5. Renewable onsite energy - 1kw solar PV system
6. Home construction - NRG Greenboard style wall construction, 60% concrete recycling rate at end of building life

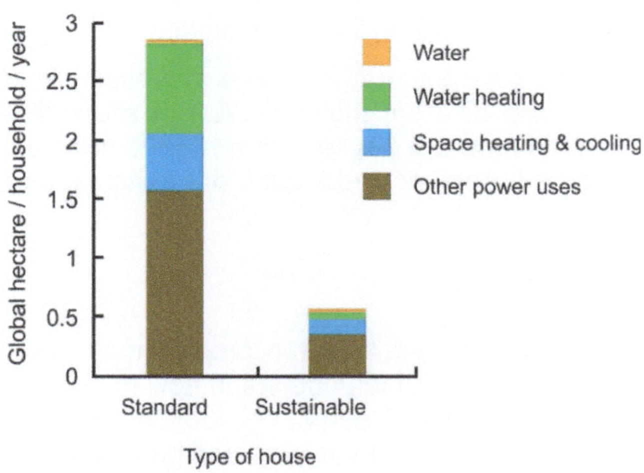

Figure 3: House footprint results - average across 35 typical project home designs (Divecha, Muffet & Grant 2008).

Prefabricated House Exemplar

Breezehouse by Michelle Kaufmann is an early example of 21st century prefab housing. The highly praised house features environmentally friendly materials, high ceiling and windows for natural light as well as unique glass sliding doors developed to bring the outdoors in (Kaufmann & Remick 2009). One of the major embodied principles is the use of 'smart design', which, just as Kibert suggests, involves minimising the size of the house to drastically reduce its carbon footprint (Figure 4). 'Design big, don't build big', as it is suggested that greater internal heights, logically placed rooms and a connection to the outside environment will make a house feel bigger naturally (Kaufmann & Remick 2009).

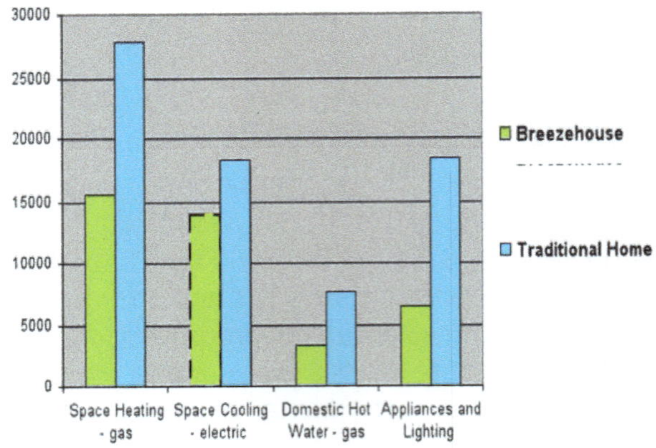

Figure 4: Breezehouse sustainability rating (Kaufmann and Remick 2009)

Existing Research Limitations

As the exemplar projects have demonstrated, current research is limited to studies between architecturally designed houses vs non-architecturally designed houses, traditional vs prefabricated housing and developer housing vs developer housing with sustainable improvements. For designers, builders and home-buyers this disjointed research can be confusing, in-depth and difficult to access, usually available from the companies attempting to promote their own product. The gap between those interested in architecturally designed housing and off-the-shelf housing is bridging further apart and many people are left unaware of the choices available to them and the impacts these decisions can make. There is currently a lack of independent research that relates to design and construction methods that reduce the carbon footprint and improve on living sustainably.

The research for this project is a comparative analysis between 3 different houses, each designed and constructed using different methods. A real site with a real client sets the constant variable for which the architect house was designed and constructed. The other 2 houses, a developer house and prefabricated house, are matched as closely as possible to fulfil the clients brief.

Framework Development

In formulating criteria for assessing this research project, many principles of sustainable design and construction were researched to understand how to assess the sustainability of a house. Along with the list outlined in the GreenMode study, aspects of the following two lists have ultimately informed the final criteria for this research project.

Kibert (2005) believes that the construction industry is currently too focused on making low cost, rapidly constructed buildings without any attention for quality or sustainability. He states that changing mind-sets is a huge challenge and that the norm of demolishing and landfilling buildings at the end of their useful life needs major rethinking.

Kibert's book *Sustainable Construction: Green Building Design and Delivery,* outlines the seven principles of sustainable construction:

1. Reduce - Reduce resource consumption
2. Reuse - resources
3. Recycle - use recyclable resources
4. Protect nature

5. Eliminate toxins
6. Economics - apply life-cycle costing
7. Quality - materials and finish

House designers from Eco-Logical Building (http://www.ecologicalbuilding.com.au) provide a list of factors that they incorporate into their designs for clients who want a house optimised for energy efficiency. According to the designers from Eco-Logical Building, these factors are specific to ecological residential design:

1. Orientation
2. Low impact construction techniques using sustainable building materials wherever possible
3. Passive solar design ensuring reduced greenhouse gas emissions and energy cost savings
4. Appropriate glazing to windows, providing thermal protection and higher acoustic ratings
5. Higher levels of insulation
6. Grey water recovery systems
7. Solar/wind energy options
8. Efficient site waste management

The criteria for this research will focus on aspects of each of the lists, including:

1. Energy consumption, as per Greenmode's first point concerning space heating and cooling.
2. Materials according to Kibert's list and Eco-logical's second point. This will consist of the major components of house construction.
3. Thermal performance, as outlined by points 1-5 in Eco-logical's list.

Methodology (Case study)

Site and Houses for this Study

The site is located in Noosa, Southeast Queensland, on a small suburban development 160km north of Brisbane. Noosa is located in a sub-tropical climate, with temperatures ranging between17 to 28°C in summer and 7 to 22°C in winter. The site is narrow with a frontage of less than 20m, tapering to 12m at the back and 25m long, with views of parklands to the north at the back. The total area of the site is 475m^2 including setbacks. The dimensions and details of the house and site are provided by the architect. The brief is provided courtesy of the architect who took on the client and designed the house, known as 'architect house' in this study.

1. Meticulously designed contemporary house where the design encompasses comfort, casual lifestyle, class and sophistication.
2. Modern with a minimal 'boxy' look.
3. Designed for a couple with 2 children.
4. Living areas should be open plan to maximise natural ventilation (it is likely that air-conditioning will be required but designed sensitive to architecture and neighbours.
5. Positioned to take advantage of parklands and best orientation.
6. Living areas should open onto outdoor entertainment areas and pools where stipulated.

7. First class kitchen is required and should be the centre of the living experience.
8. Main bedroom/ensuite/robe area interfaces to be 'special' i.e. separate and with views and unique materials.
9. Double garage and double storey.
10. Welcoming entry.
11. House site cover, setbacks and heights governed by Council approved conditions.

This house is chosen for its flat site and relatively simple design that could be closely matched with 2 other types of houses. The design is fairly 'typical', with no ground-breaking sustainable technologies and the only unique feature being the use of full height glazed curtain walls. The house instead relies on clever design in terms of orientation, window location, room arrangements, playing with heights and a clean minimal palette of materials that keep it special without over embellishing. House information is provided in Table 2.

Implementing Software

As with the Brighton Compact case study and GreenMode study, the method for comparing the three houses is based on using specific software. The evaluation software produces results in a common unit of measure which can be compared and evaluated. The three software packages include:

- *Archicad* is used to create a Building Information Model (BIM) of each house from existing 2D plans, including heights, materials, quantities and orientation. The BIM is exported to other software packages to calculate sustainability analysis.
- *Ecodesigner* is an add-on within Archicad, able to produce energy balance evaluations based on the BIM model. This produces energy consumption quantities and energy costs in the common unit of measure, CO2e.
- *eTool* evaluates each house based on the BIM, able to measure and calculate the quantity of materials and their carbon emissions. eTool takes into consideration a number of factors for each material that affects its performance, including assembly, operational, transport and recurring. The final output for each house is given in the common unit of measure, CO2e, as well as kWh/a and Australian dollars (AUD).

From the case studies and background research, a list of criteria has been devised to rate the performance of each house, including:

1. Energy Consumption (i.e. heating and cooling).
2. Materials (i.e. floor, wall, internal, roof, reuse, recycle, reduce).
3. Thermal (i.e. orientation, ventilation, solar gain/exclusion).

Under these three aspects, the results generated from the software are laid out graphically in diagrams and charts, followed by a discussion outlining reasons for the results, iscrepancies, and comparisons.

Table 2: Comparative information of three house exemplars

	Architect House	Developer House	Prefabricated House
Plan: Lower level	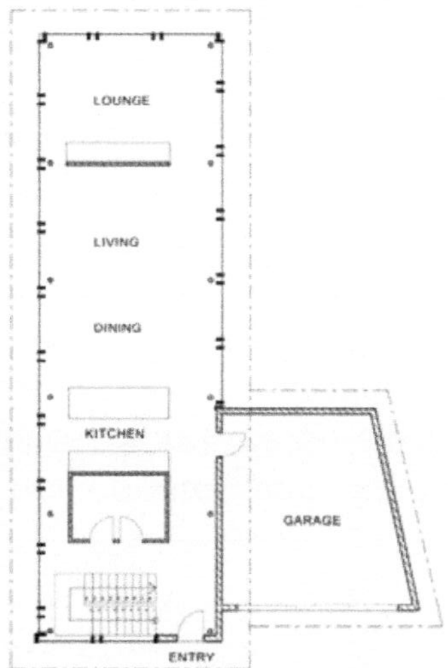 	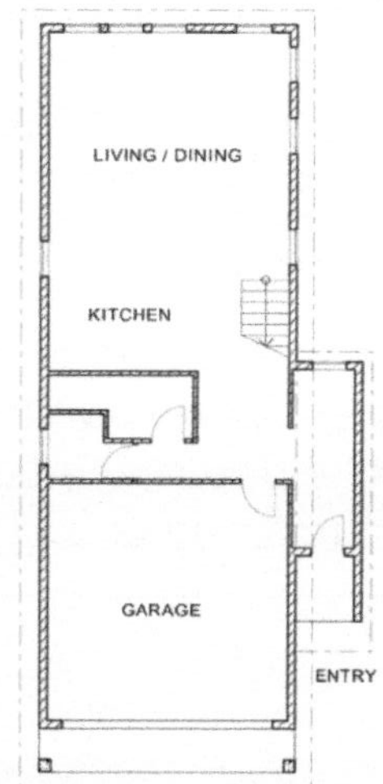 	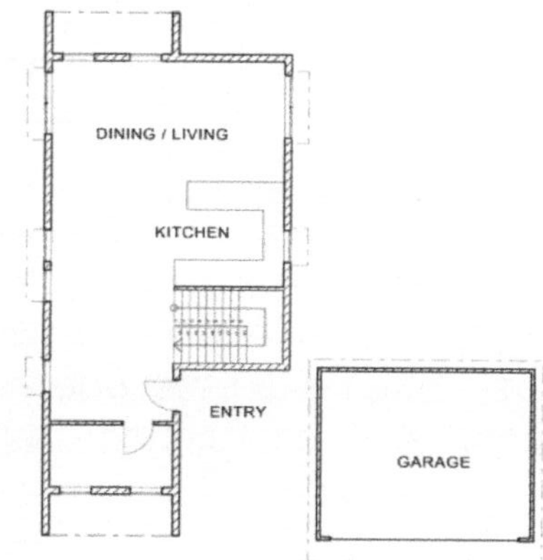
Upper Level	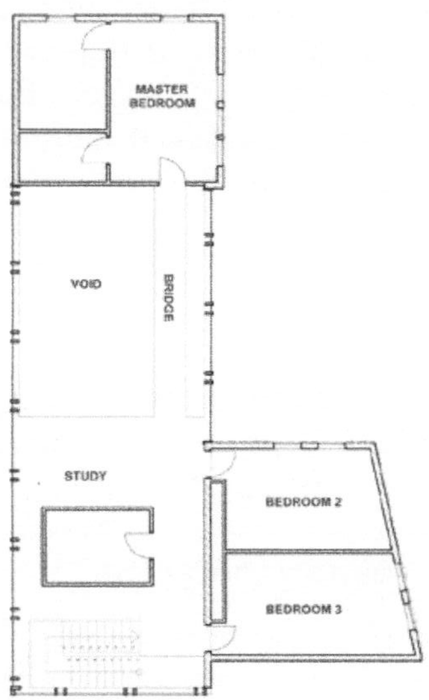 	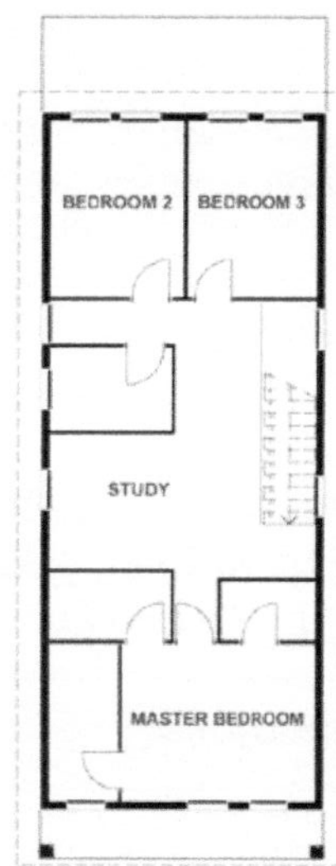 	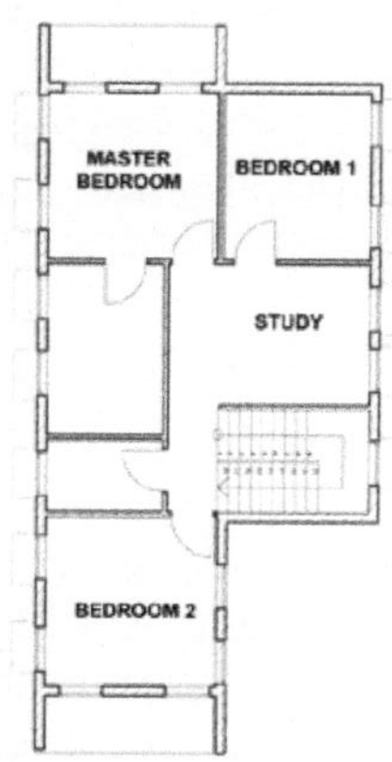

	Architect House	**Developer House**	**Prefabricated House**
Design:	The architect house sets the benchmark for this project; an architect was engaged to design and project-manage the construction of this house, working closely with the client to ensure every feature was to their liking. The documentation is traditional style, with a set of tender drawings being issued to find a builder acceptable to complete the project at a reasonable price.	A number of house developers' designs have been evaluated and the selected one is comparable to the solution provided by the architect case study. The developer has 4 house designs that are narrow enough to fit on the site, and has one with a reasonable layout. A few facade options are available, for consistency the 'boxy modern' style is chosen. The builder and designer are all within one company, so no tender process is necessary and the price is fixed upfront.	The prefab house is chosen due to its range of available modules, the ways in which the modules can be arranged and site suitability. The house modules have been designed by a renowned and highly awarded architecture firm located in Brisbane and supplied through a local company who make design alterations and provide a fixed price upfront.
Construction:	The construction is traditional style; some elements of the house are prefabricated, while others are cut to size on site.	The construction is traditional style; some elements of the house are prefabricated, while others are cut to size on site.	The construction is completely prefabricated in a controlled factory into modules and then delivered on site, where it is assembled by a small team.
Size:	325 m2	235 m2	230 m^2
Cost:	$640,000	$350,000	$370,000
Cost per m2:	$1970	$1490	$1608

	Architect House	**Developer House**	**Prefabricated House**
Materials:			
Structural Walls	Structural Steel	Structural pine frame	Structural pine frame
Cladding	Timber battening (garage and bedrooms) and zinc (master bedroom), glass curtain walls	Brick Veneer from Austral	Plywood cladding - stained
Roof	Custom Orb Colorbond	Custom Orb Colorbond	Custom Orb Colorbond
Floors	Tiles downstairs and wet areas, hardwood upstairs	Tiles throughout, carpet in bedrooms	Hardwood throughout, tiles in wet areas
Doors	Timber	Timber	Timber
Windows	Timber	Aluminium powder coated	Rosewood from New Guinea
External Treatment	Natural stained	Brick	Natural stained
Internal Finish	Plasterboard with matt paint finish and curtain walls	Plasterboard with acrylic paint finish	Plasterboard with matt paint finish
Insulation	Yes	Yes	Yes

Findings and Discussions

The study simulated one example of the three types of houses that were chosen to fulfil a real client brief on a real site on the Sunshine Coast, Queensland Australia. Criteria for sustainability assessment were formulated based on literature reviews, exemplar projects and similar research projects for which the houses could be adequately evaluated. This criterion covered aspects including energy use, materials and thermal performance. The data was collected using computer models and sustainability assessment software to compare and draw conclusions on the success of each house as below.

Energy Consumption

The following three graphs are results from Ecodesigner based on the BIM in the specified location of Noosa Queensland. Figure 5 shows that the developer house uses more energy each year than the other two, despite being almost the same size as the prefabricated house and 90 m2 smaller than the architect house. To make comparative assessments between the energy use and materials, the results will now be displayed in CO2e, the common unit of measure. The cooling and heating energy are combined into a total amount.

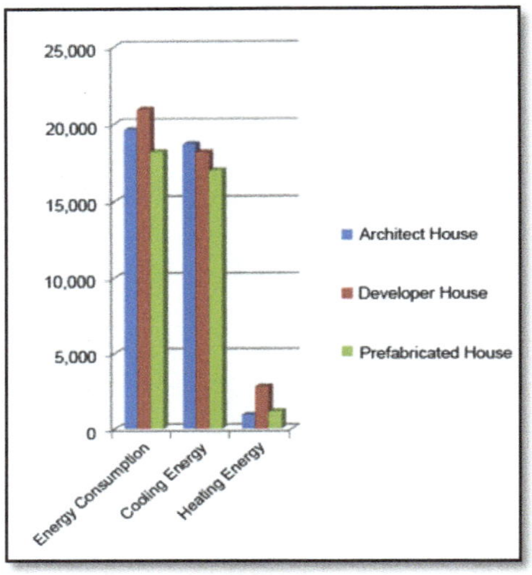

Figure 5: Energy consumption measured in kWha2

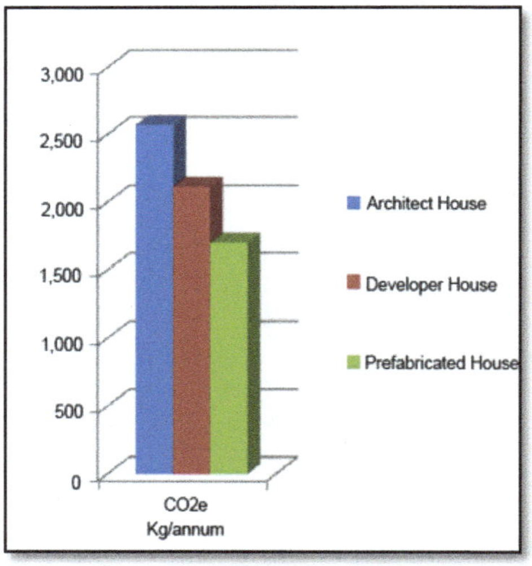

Figure 6: CO2e (kg/annum) of each house

Figure 6 shows expected results, as the architect house is 90m2 larger than the other two houses. The architect house uses about 20% more energy per year than the developer house, and the developer house uses about 20% more energy per year than the prefabricated house. By displaying the information in kg/m2 the difference in size is taken out of the equation and a fairer comparison can be made of the performance. If the architect house were the same size as the other two houses it would perform better than the developer house, but still not as well as the prefabricated house.

Materials

Figure 7 shows the results of the analysis from eTool in factors of carbon emissions (CO2e) over a 100 year life of the building. The total amount is broken into the four major contributors of the buildings carbon emissions, notably the floors and walls. The results show that the prefabricated house is the most sustainable house overall. The developer house and the architect house have similar total CO2e's, despite the architect house being 90m2 bigger. By breaking down the major components of the houses, the reasons for the total amount become clear.

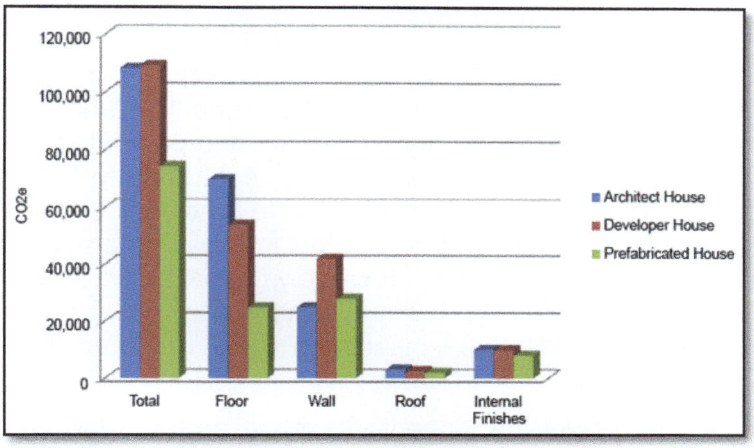

Figure 7: CO2e of materials in each house

The architect house and prefabricated house have been designed for south-east Queensland by local Brisbane architects. The materials are appropriate for the climate and have been chosen specifically to suit to the site. The developer house on the other hand is not specific to the site or the climate, as the same house can be bought in other states of Australia. Houses situated in sub-tropical climates benefit from light construction techniques with low thermal mass. The use of concrete slabs is less desirable than elevating the house with timber floors on stumps as seen with the prefabricated house.

a) Floor:

Concrete is used most extensively in the architect house due to its larger size, but could be reduced to the amount of the developer house by reducing its floor area. The floor in the prefabricated house is constructed from timber and elevated off the ground, as opposed to a concrete slab used by the other two houses. This results in a significant CO2e reduction due to timber's sustainability and recyclability. Timber could be improved by utilising recycled timber in the floor structure, walls and finishes, an opportunity missed by every house.

b) Wall:

What the architect house loses in floor construction is made up for in wall construction, as it uses timber cladding and thin sheet metal instead of the brick veneer that is used to clad the developer house. The majority of the architect house is glass curtain wall and louvres, a system which is not structural and can be seen in many modern skyscrapers. Instead, most of the structure is supported by steel columns spaced at 4m centres and drastically reduces the need for stud walls with external and internal cladding. The prefabricated house has a timber frame structure like the developer house, but is clad in timber boards instead of brick, reducing its CO2e significantly.

c) Internal Finishes:

Each house has a similar amount of internal finishes including plaster on the walls and ceiling, paint on the walls and ceiling, and tiles in the bathrooms. The architect house and prefabricated house use timber floors throughout the majority of the house including the bedrooms, while the developer house uses tiles throughout and carpet in the bedrooms resulting in a higher CO2e.

d) Reuse:

To reuse something is to utilise the product in its original form as a whole, for example using old speakers on a new amplifier (MacMillan 2006). Kibert (2005) explains reuse in sustainable architecture as 'Keeping materials in productive use, which also implies keeping buildings in productive use'.

Many components from the architect house can be dismantled and reused, including the windows, curtain walls, doors, cladding, timber upstairs and part of the steel posts. The only piece remaining would be the concrete slab, which would have to be dismantled in many pieces and recycled, not re-used. An example of how to reuse is the large expanses of glass becoming a shop front or partitions in an office. This is a level of creative thinking that goes beyond the analysis of software and is only possible with the house designer's intent, careful planning and recommendations of ways to recycle the house. The thought that goes into the initial design process will provide more inspiration when it comes time to dismantle the house.

The developer house has fewer components to reuse as it is slab on ground with brick veneer. It is very difficult to dismantle brick veneer, as it is fixed together and cannot be taken apart brick by brick. The house has an advantage of being made up of standardised components, such as windows and doors, which are used in many developer houses. The timber frame could essentially be re-used, but is more likely to be recycled into something non-structural.

The prefabricated house can be deconstructed and carted away to a different site for another use, as modular pieces are the easiest to disassemble and reuse. The prefabricated house would have to be dismantled in the same way it was assembled and taken away as complete modules. These can be easily relocated to prolong the life of the building and will leave behind nothing but footings and a small slab for the carport.

e) Recycle:

Recycling is different to reuse as to recycle something is to destroy the product and reuse it as parts, for example taking the parts of an old mobile phone and melting down the metals to make a new phone (MacMillan 2006).

Recycling would be less likely for the houses as most of the parts can be reused instead. The concrete slabs are very difficult to recycle as they are permanently fixed to the ground, but could be broken up and made into new recycled concrete. Other elements that would normally be reused could be recycled, such as the glass could be broken down and made into new glass. The zinc cladding from the architect house could be destroyed and made into a new cladding or used for something entirely different.

f) Reduce:

Ranging between 230m^2 - 325m^2, each house is well reduced in size and situated on a block of 475m^2, which is within the Brisbane average size of new blocks for 2012-2013. Although, it is important to note that the house was completed in 2006, when the average block for

Brisbane was 600m^2, making the block quite small for its time. As highlighted previously, the architect house uses glass curtain walls that reduce material use, avoiding cladding, structural timber, plaster and paint. Another example of reduction is the use of a flat roof, which does not require large trusses to achieve the pitch of the developer house, using less timber and less roofing material. Simple techniques like this assist in reducing the amount of materials that go into building a house.

Thermal Performance

Thermal performance will be calculated based on site and plan analysis. The architect house is the only house designed from the ground up to suit the site and climate, while the other two houses are limited on how they can adapt to take advantage of the orientation and surrounding areas. The first rule for sustainable designs in Van der Ryn and Cowan's Ecological Design is 'Solutions grow from place', a design principle that means each house must, in a way, be unique (Van der Ryn & Cowan 2007).

A sustainable house cannot succeed without some level of site specific customisation, as the more connected a house is to its site, the more it will thrive in capturing and using nature. This can be achieved from design by architects and building designers, partly from prefabricated modular design and not at all from generic predesigned houses. So much of houses' sustainable success derives from thermal performance which is directly related to orientation, sunlight and ventilation.

Architect House[1]

a) Orientation/Room arrangements:

The architect house has been designed to maximise thermal performance by carefully arranging each room based on orientation, use and relationships with other rooms. The main circulation space is placed near the entry on the south side, which includes entry from the street, entry from the garage, vertical circulation to the bedrooms and study via the stairs, concealed laundry and guest bathroom, and direct access to the large double height open plan living area. The open living area faces north-east and west, but has been positioned on site so that the outdoor area is on the east and features glass sliding doors. The west side is more sheltered from the sun because of the close proximity to the fence line and neighbouring house, and has a facade that allows less heat in. The lounge room at the northern most point of the house enjoys the best orientation and is flooded with light, and is cross ventilated east to west. Because of the form, construction techniques and triangulated site, the garage has been placed on the east side, when generally the best place for a garage is the west side as it is rarely occupied and can act as a thermal barrier.

The second and third bedrooms are above the garage which gives them a north - north-east aspect, a benefit achieved from the garage also being on this side. This wing of the building gives privacy from the street to the rest of the house including the glass clad eastern facade of the open plan living area and the outdoor area. A bridge spans the double height void over the open plan living and connects to the master bedroom. The bedroom overhangs the rest of the house and is isolated via the bridge, giving it the 'special' aspect that the client required. The room faces north east and has an ensuite and wardrobe on the west to block sun on this side.

b) Ventilation:

The window to wall ratio of the architect house is significantly higher than the other two, with majority of the open plan area having full height glazing and a large double height atrium

---

[1] Elysium house by Woods Bagot used with permission by Mark Damant

over the open plan living area. Much of this glazing includes full height louvres that allow breezes and cross ventilation to pass through the house very easily.

Due to large quantities of louvres and cross ventilation, mechanical ventilation is not needed for the open living areas. All the bedrooms are facing north-north-east and have ceiling fans for when there is little wind.

The double height void allows hot air to rise rapidly and vent through open louvres, rather than linger down at human height.

c) Solar:

Large expanses of glass allow a great amount of light into the architect house, which reduces the need for artificial lighting. This can have negative connotations, as too much sunlight can bring excessive heat into the house, increasing the need for mechanical ventilation. To overcome this, the house has 1m roof overhangs all around that block sunlight in summer, and allow desirable amounts in during winter, due to the sun being lower in the sky. The louvres on the upper level of the house are heavily tinted on the east and north sides, and metal on the west to completely block out the harsh western sun. The occupants have the freedom to change the angle of the louvres to allow or block certain levels of light and wind, which can be done depending on the direction on particular days.

Developer House[2]

a) Orientation/Room arrangements:

Choosing a developer house avoids the design stage, and many people will base their layout decisions to suit their lifestyle, without considering the parameters their site provides. If the site is not running in the direction that the pre-designed house is suited for, there will be complications. For example, the developer house chosen for this study had the living room leading out to an outdoor area on the north side, but due to the client's needs this had to be filled in to become a second living space. The outdoor alfresco area could no longer exist there and had to move to the east side, acceptable for this design but if there was a requirement for a covered outdoor area then it would have to be built outside the building envelope increasing costs and adding extra roof. A problem with only being able to mirror predesigned houses along one axis means outdoor and living areas can end up in undesirable orientations, potentially increasing the need for heating or cooling.

To suit the site, the developer house design has been mirrored along the north-south axis, putting bathrooms, laundries and kitchen on the western side and the bedrooms on the east. The whole house fits into a single building envelope apart from the entry foyer. The garage sits on the southern side, and the layout of the bottom floor is similar to that of the architects design. The house lacks the double height void and the large expanses of windows, opting for standard sliding windows and one sliding door leading to the outdoor area. This area remains exposed as the garage is not blocking it from the street.

Upstairs features the same number of rooms as the architect house, but are more compressed and have a different layout. The second and third bedroom take the desirable north side, with one of them having to be exposed to the western side. The master bedroom is placed on the southern side, blocked on the undesirable west side by the ensuite, and features a balcony facing onto the street. This is a missed opportunity for the master bedroom to be on the most desirable aspect with a balcony overlooking the parklands to the north, and has therefore not met the clients' exact wishes. This could be altered but would incur extra fees and require substantial changes to the layout of the bathrooms, stairs and roof.

2 Pine Rivers 235 by GJ Gardener http://www.gjgardner.com.au/data/designs/12617_14.pdf

b) Ventilation:

The open plan living area has opportunities for cross ventilation on three sides which help keep the house cool, but only through small windows and doors. The void over the stair may assist hot air escaping upwards through its small window, but not as effectively as the large void in the architect house. The bedrooms miss out on cross ventilation because windows are only placed on a single wall rather than both available sides. As bedroom 4 was removed to fulfil the clients need of an upstairs study, the upstairs area receives mild east-west cross ventilation through its two modest windows.

c) Solar:

The lack of windows helps to reduce heat transfer from too much glass, but also reduces the amount of light, increasing the need for artificial light. Brick veneer is a good insulator and helps to keep the house cool, but is not ideal for the sub-tropical climate as it can retain heat during summer months. The concrete slab contains a great deal of thermal mass, which would not heat up from the sun as much as in the architect house, but also does not have an opportunity to vent from the lack of windows. 600mm roof overhang helps reduce excess heat from the sun entering the house, but is insufficient for shading the downstairs area as well. This needs to be close to double in width similar to the architect house, or have awnings over the windows like the prefabricated house.

Prefabricated House[3]

a) Orientation/Room arrangements:

As the prefabricated house is made up of modules, the design team has more flexibility with the placement of rooms and can place them essentially anywhere to suit the house. The modules can only be connected in certain ways, and any elements that have to be overly customised will add cost. The layout of the house downstairs is similar to the other two houses, with the garage placed separately on the east side to better reflect the architect house. This does not have as many advantages as the architect house because the bedrooms cannot be built on top to take advantage of the north-east orientation, and is rather a decision made simply because of the site's shape. The stair would have been better placed on the west side with the laundry and bathroom, but the modular nature of the stairs makes it difficult to achieve as it is a prefabricated piece. Other than these few facts the layout downstairs is quite good, but could benefit from more windows on the north side.

Upstairs is similar in layout to the developer house, with a study, 3 bedrooms and 2 bathrooms. Due to the location of the stair, two bedrooms have been placed on the west side and one on the east. The bathrooms are on the west side, but the ensuite is not in a position to shield the master bedroom from the western sun. Protruding blade walls on the west side assist in protecting this half of the house. In many ways the clients brief has not been satisfied, and is overall a cramped house.

b) Ventilation:

Cross ventilation is achievable downstairs, but is completely neglected upstairs except for the southern bedroom. The stair is able to vent hot air out but not to the extent of a double height void as seen in the architect house. Each bedroom has at least 2 windows and due to the light construction, would be a breeze filled house. The raised timber frame floor allows the underneath of the house to ventilate, which is a technique seen in the 'Queenslander' design common to Brisbane and surrounding areas.

---

[3] By Michael J. based on Happy Haus' Dhan series by Donovan Hill http://www.happyhaus.com.au/uploads/PLANS/Donovan%20Hill/Base%20Plans/Donovan%20Hill%20Happy%20Haus%20Floor%20Plans.pdf

c) Solar:

A lack of concrete slab means the prefabricated house has a very low thermal mass, and heat is able to escape easily from the floor. Timber clad walls all around mean heat is not retained within the house, but can still be sufficiently insulated to provide maximum efficiency. The roof has no overhangs except over the blade walls on the north and south sides. Instead, window hoods cover each window and can be different widths depending on the orientation of the window.

Conclusion

This focused research provides an insight into the implications of three key elements of sustainability; energy use, materials and thermal performance. Designers, builders, developers and home-buyers are given an insight into some options currently available on the housing market and how the choices made during early design stages can provide a more positive environmental impact.

It may be an ambitious proposition, but could site specific design that utilised prefabricated technology become the norm for suburban houses in south-east Queensland? By better utilising the knowledge of architects, research and software, all home-buyers will benefit from the advantages of choosing to live in a sustainable home. Software is a powerful tool for evaluating houses from the early design stages through to post-construction, and is being continuously updated to reflect more accurate results (Roy, Brown & Gaze 2003). This software is becoming easier to use and more widely available so that the average person is able to make educated decisions with the help of their own computer.

Measuring the performance of a building is a complicated process due to the wide variety of impacting factors (Schneider & Till 2005). Finding a common unit to quantify the results is difficult as different methods and equations are needed to convert the information to the same unit.

Analysing buildings with software can be extremely intricate and detailed. There are several software applications that can perform detailed analysis and there are companies that will carry out a study for a fee. This level of analysis is generally used for large scale buildings so detached housing misses out due to insignificant size.

This study has a common unit of measure as CO_2e, which is not compatible for every aspect of building analysis, and can be very limiting when using software as a means of research. The study relies on the algorithms and equations built into the software to achieve the results in CO_2e, so the accuracy is ultimately up to the program designers and their prior research and ongoing development.

The construction component has been taken out; this study does not look at the methods of construction that is prefabricated vs traditional construction, instead only looking at the design aspect and ongoing affects that people may be unaware of. The eTool is able to add construction specific details including machinery, man power, time, glues, nails, hammers, drills, saws, waste, bins, rainy days, delays etc. This would take a long time and would require more research into what is involved in the construction process of a specific house, but could add depth in future research.

Another important aspect of evaluating a building performance is the thermal comfort levels, as these figures can accurately determine whether mechanical heating or cooling is needed in the building. The software is very complicated, beyond the reach of this study, and would be suitable for future research of a similar kind. Free software is available, but there are currently compatibility problems between the modelling software and the analysis software. This issue will hopefully be resolved in the future, as there are already signs of it with exporters and file converters currently available. Some software still benefits from users modelling the building in their own format for optimisation and accuracy, which requires a new modelling skill set and misses out on being able to import already documented buildings.

Appliances and lighting has been exempted from this study so as to focus on the 3 main criteria. This information can be added to the appliance and lighting section in eTool and can test the difference made by efficient appliances. Ecodesigner is also able to add average figures to the analysis and be customised to test different efficiency ratings. By using energy efficient appliances, water saving taps, solar hot water, solar panels and only using electricity when needed, the house carbon footprint will be drastically improved.

The location of where the materials are manufactured can be added to the eTool software to create a more accurate CO2e, as the transportation of materials increases its carbon footprint. This could alter the results if the prefabricated house was, for instance, built in Sydney and transported via multiple trucks to Noosa for assembly.

Transport and how it is used will affect the ongoing performance of a house (Snyder 2005; Staib, Dörrhöfer & Rosenthal 2008). This can be attributed to pre-design stages, when the site location of the house is being decided. Building a house in an area without public transport or which is great distances away from work, school, shops etc. will leave private transport as the only option. Further studies could include the transport required for the family living in the house to get to work, education, shopping and leisure, and will be greatly altered by distance and the types of transport available and utilised.

Costs are an important aspect of choosing a house, and as shown with the evaluated houses, the architect designed house is almost double the price of a developer house (Winstanley, Thorns & Perkins 2002). Future research could include a breakdown of costs that explore the reasons why the architect house is more expensive than the developer house even after design fees have been excluded. The initial cost of implementing sustainable features is outweighed by the savings made by those sustainable features. What is the merit of designing a house the client actually wants, that may cost slightly more, but has less impact on the environment and ultimately will save money because of its performance, efficiency and cost saving over a specified period?

Based on the data from Ecodesigner and used in this study, though its accuracy is questionable, it shows an example of how currently an environmentally friendly house that has higher initial costs could end up saving money over a life span of 50 years. The higher initial cost may not be true for other prefabricated house designs and may be reduced in the future. It would also depend in part on builders overcharging for the use of technologies they do not commonly apply, but when the use of materials and technology becomes more established, costs are also reduced (Manseau & Shields 2005).

A limitation of this study is its reliance on rating systems. Although there is no consensus on the value and efficiency of these tools (Gowri 2004), this research has used the selected software as a way to compare, in an objective way, three different design propositions. Future research might address the issue of the difference between planned and actual performances of buildings. This study has attempted to gather a better understanding of how different design strategies and methodology could ideally perform in a specific location.

The results of the study indicate that architecturally designed housing, especially when married with prefabricated building techniques, are a better alternative to generic developer style housing. Rethinking the types of materials houses are constructed with, and the ways in which they are constructed will ensure the future development and improvement of sustainable design, with technological advances becoming accessible for the average Australian.

References

Arif, M. & Syal, M. 2013, 'Addressing Global Changes for a Sustainable Built Environment: Implementing Sustainable Construction', Construction Innovation, 13.

Birkeland, J. 2002, Design for sustainability: a sourcebook of integrated, eco-logical solutions, Earthscan, London.

Blismas, N. & Wakefield, R. 2009, 'Drivers, constraints and the future of offsite manufacture in Australia', Construction Innovation: Information, Process, Management, 9, 72-83.

Callery, B. & Bartak, E. 2010, 'Brighton Compact Green House: Case Study', Houses Magazine.

Davies, C. 2005, The prefabricated home, Reaktion, London.

Discoe, P., Quinn, A. & Banish, R. 2008, Zen Architecture: The Building Process as Practice, Gibbs Smith Publishers.

Divecha, S., Muffet, L. & Grant, T. 2008, 'Mawson Lakes Footprint Report', Strategic Matters, GreedMode and RMIT.

Eco-Logical Building. n.d.. Available from http://www.ecologicalbuilding.com.au.

Gowri, K. 2004, 'Green building rating systems: An overview', ASHRAE Journal, 46, 56.

Johnson, L. 2014, 'Accommodating Australians. Commonwealth Government Involvement in Housing: Designer Suburbs. Architects and Affordable Homes in Australia', Journal of Australian Studies, 38, 120-3.

Kaufmann, M. & Remick, C. 2009, Prefab Green, Gibbs Smith, Layton, UT, USA.

Kendall, S. 1999, 'Open building: an approach to sustainable architecture', Journal of Urban Technology, 6, 1-16.

Kibert, C. J. 2008. Sustainable construction: green building design and delivery (2rd Ed), Hoboken, NJ, John Wiley & Sons, Inc.

Kibert, C.J. 2012, Sustainable construction: green building design and delivery, John Wiley & Sons, Inc. Hoboken, NJ.

Koenig, G., Eames, C. & Eames, R. 2005, Charles & Ray Eames 1907-1978, 1912-1988: Pioneers of Mid-century Modernism, Taschen, Cologne, Germany.

Linner, T. & Bock, T. 2012, 'Evolution of large-scale industrialisation and service innovation in Japanese prefabrication industry', Construction Innovation, 12, 156-78.

Manseau, A. & Shields, R. 2005, Building tomorrow: innovation in construction and engineering, Aldershot, Hants, Ashgate, England.

McHarg, I.L. 1981, 'Human ecological planning at Pennsylvania', Landscape Planning, 8, 109-20.

MacMillan, A. 2006. How to recycle anything. Prevention, April, 119-121

Moore, T. 2010, 'Brisbane's suburbs of growth and decline', Brisbane Times, http://www.brisbanetimes.com.au/business/property/brisbanes-suburbs-of-growth-and-decline-20101103-17e71.html.

Newman, P. & Kenworthy, J. 1999, Sustainability and cities: overcoming automobile dependence, Island Press, Washington, DC.

Oluwole Akadiri, P. & Olaniran Fadiya, O. 2013, 'Empirical analysis of the determinants of environmentally sustainable practices in the UK construction industry', Construction Innovation, 13, 352-73.

Öst, C.E. 2012, 'Housing and children: simultaneous decisions?—a cohort study of young adults' housing and family formation decision', Journal of Population Economics, 25, 349-66.

Roy, R., Brown, J. & Gaze, C. 2003, 'Re-engineering the construction process in the speculative house-building sector', Construction Management & Economics, 21, 137-46.

Schneider, T. & Till, J. 2005, 'Flexible housing: opportunities and limits', Architectural Research Quarterly, 9, 157-66.

Shulman, J., Smith, E.A.T. & Gössel, P 2007, Taschen, Case Study Houses, Cologne, Germany.

Snyder, V. 2005, 'Refabricating Architecture: How Manufacturing Methodologies are Poised to Transform Building Construction', Journal of Architectural Education, 59, 51-2.

Staib, G., Dörrhöfer, A. & Rosenthal, M. 2008, Components and Systems: Modular Construction Design, Structure, New Technologies, Walter de Gruyter, Berlin.

Urban Development Institute of Australia 2014. UDIA Submission to the Senate Inquiry into Affordable Housing, 31 March 2014.

Van Der Ryn, S. & Cowan, S. 2007, Ecological design, Island Press, London.

Winstanley, A., Thorns, D.C. & Perkins, H.C. 2002, 'Moving house, creating home: Exploring residential mobility', Housing Studies, 17, 813-32.

Wood, R. 2009, 'Structural decomposition analysis of Australia's greenhouse gas emissions', Energy Policy, 37, 4943-48.

Zhang, S., Pan, Y., Li, N. & Sun, Q. 2014, 'A Comparative Study of the Direct Costs Between Prefabricated Housing System and the Traditional Construction Technology—A Case Study of Precast Concrete Wall Panel', Proceedings of the 18th International Symposium on Advancement of Construction Management and Real Estate, Springer, 349-58.

A Conceptual Approach to Studying the Organisational Culture of Construction Projects

6

Thayaparan Gajendran, (The University of Newcastle, Australia)
Graham Brewer, (The University of Newcastle, Australia)
Andrew Dainty, (Loughborough University, UK)
Goran Runeson, (University Technology of Sydney, Australia)

Abstract

Studying the culture of project organisations requires a robust theoretical framework, which provides a platform for generating understanding. It is proposed that cultural analysis frameworks are most effectively conceptualised from multiple philosophical and multi-methodological positions. The goal of the present paper is to describe a cultural analysis framework for studying construction project organisations, based on a synthesis of the culture literature. Four key aspects underpinning organisational cultural frameworks are explored: the paradigms used to conceptualise organisational culture, the methods by which individuals represent and assess cultural dimensions, the cultural perspectives assumed by the observer when defining and describing culture, and the managers' orientation toward culture in their organisations. The proposed framework comprises three synthesised cultural philosophical positions: integration-technical, differentiation-practical and fragmentation-emancipation. These philosophical positions span the polar extremes defining the cultural paradigm continuum, which and when combined provide researchers and organisational managers with a sound foundation from which to study the culture of project organisations.

Keywords: Culture, Cultural Analysis, Project Organisation, Integration, Fragmentation

Introduction

Attention to cultural issues in the construction industry, both in the form of permanent and temporary project organisations, has gained significant momentum in recent times (Fellows 2010). Moreover, the increasing internationalisation of construction also warrants better understanding of cross-national cultural practices (Tijhuis and Fellows 2012). Industry reviews (e.g. Latham 1994; Egan 1998) have cited culture as an impediment to the industry's functioning and wellbeing. In the light of such reviews, researchers have focused on studying issues perceived as being connected to culture with the view to improving industry performance. However, the tendency to treat culture solely as a variable that can be manipulated for improvement is not an amenable enough conceptualisation of culture to shed meaningful insight into the ills of the construction industry (Green 2011).

The culture as a variable within an organisation or project, the '*Functional*' paradigm, is amenable to control and thereby could contribute to organisational performance. Evaluation of selected literature reveals that the key motivation of cultural studies in construction was to develop empirical understandings of the impact of culture upon different construction industry issues, to assist with some form of performance improvement. Although performance focus is critical for organisational development, studies focussing solely on manipulation of culture for performance improvement will not develop adequate and meaningful understanding of project organisations (Alvesson 2002).

One of the cultural research issues focused on since the nineties is managing the adversarial attitudes arising out of procurement (Cheng et al. 2004; Walker et al. 2001) and contractual (Phua & Rowlinson 2003) approaches to improve the landscape of the construction industry. Impact of the differences in the cross-national cultures on construction organisations (Pheng & Yuquan 2002; Ling, Ang & Lim 2007; Zou et al. 2007; Ankrah, Proverbs, & Debrah 2009) is another key research issue that is focused on project improvement. Cultural studies also centre on developing an understanding of professional cultures (Ankrah & Langford 2005), including behaviour of professions (Liu & Fellows 2008) and ethical disposition of professions (Liu, Fellows & Ng 2004).

Cultural issues associated to improved knowledge transfer in project organisations (Brochner, Rosander & Waara 2004), topologies of corporate culture of construction firms (Igo & Skitmore 2006) and motivational aspects in the construction workplace (Smithers & Walker 2000; Hartmann 2006) have also been studied. In the same period, influence of culture on ICT system implementation (Anumba et al. 2006) and construction safety (Root 2005; Nielsen 2007) was explored. Most of the above studies, although having mobilised various methodological approaches, essentially seek to understand, or change, cultures (Cheung et al 2011; Price & Chahal 2006) for improved performance of project organisations; a position closely associated to the functional paradigm.

Despite the empirical contributions of these recent studies to the understanding of the impact of culture in project organisations, it can be argued that use of a non-functional approach (Adriaanse and Voordijk 2005; Rooke, Seymore & Fellows 2003, 2004) to study culture, could provide an attentive lens.

Essentially, researchers assume a philosophical position(s), explicitly or implicitly, when conducting cultural studies. Their position(s) will influence the choice of methodologies employed to assess culture. As way of example, research that uses existing frameworks such as Hofstede's cross-cultural analysis framework (see Hofstede 2010) and Carmon and Quinn's competing values framework (2006), are grounded in functional paradigm. Although proven popular, these frameworks are set in a pre-determined singular philosophical position, offering limited flexibility for researchers wishing to combine multiple philosophical positions. Moreover, Nakata (2009) suggests that it is time to look beyond Hofstede for constructing alternative cultural frameworks for global marketing and management.

This paper offers a conceptual 'cultural analysis' framework that serves dual purposes in analysing organisational cultures. Firstly, it simplifies and guides cultural studies of organisations, providing a framework against which cultures can be defined, evaluated and understood. Secondly, it assists in designing a suitable cultural assessment approach, from multiple cultural philosophical positions, to study cultures in practice. It will be argued that a more in-depth and comprehensive understanding of organisational cultures within construction project organisations can be garnered from such a cultural analysis approach.

Cultural Assessment: a Background

Complicating this cultural framing is the fact that a single definition of culture does not exist, and the characteristics of culture have not yet been consistently agreed upon. Despite this lack of agreement, consistent efforts have been made in recent years to transcend the intellectual debate through a broad conceptualisation of culture (e.g. Martin, Frost & O'Neill 2004) focusing on the underlying assumptions, beliefs and values (referred to as 'cultural manifestations') of groups. These factors are transmitted as cultural meanings that guide the way a group thinks, feels and acts. The culture of an organisation can then be characterised, based on to what extent the 'underlying beliefs' are shared and other associated cultural manifestations within a group and among groups (Schein 2004; Alvesson 1993a, 1993b, 2002). The level of harmony or ambiguity will be reflected in the environments in which the group(s) operates.

Cultural analysis accommodates multiple philosophical positions, allowing researchers to engage with culture by drawing on differing theoretical standpoints (Martin 2004). This paper constructs such a synthesised cultural analysis framework by drawing from a number of philosophical positions, an approach which assists in the investigation of cultures as part of project organisations. Initially, two cultural paradigms, 'functional' and 'non-functional', are discussed to explain the fundamental difference in the way culture is conceptualised in organisational studies.

Various levels of cultural manifestation are identified, including their characteristics and how they represent organisational culture, resulting in a number of definitions of culture being presented. The discussion then leads to three cultural perspectives: integration, differentiation and fragmentation, that conceptualise culture according to the ways in which cultural manifestations are shared in organisations. The three cultural philosophical orientations (a) technical (b) practical and (c) emancipatory, are introduced in order to illustrate how each perspective tends to deal with culture. Finally, these three perspectives and orientations are synthesised to form a cultural analysis framework, philosophical positions that provides a unique and comprehensive approach to studying specific project organisations.

Culture in Organisations: a Review of the Cultural Paradigms

The literature on organisational culture shows an enormous variation in the conceptualisations, definitions and approaches of research, due to the influence from various disciplines and schools of thought (Alvesson 2002; Martin, Frost & O'Neill 2004; Rousseau 1990). Philosophical conceptualisations of culture are generally divided between; (a) the concept of real, objective and functionalist 'out there' phenomena (Davis 1985; Deal and Kennedy 1988; Denison & Mishra 1995; Ouchi 1981; Peters and Waterman 1982) and (b) a non-functional paradigm for thinking about the social world or social reality in organisations (Smircich 1983b, 1985; Smircich & Calas 1995; Morgan 1980; Alvesson 1987; Denison 1996).

The difference in cultural research methodologies is underpinned by the expected outcomes of cultural studies; that is, whether the outcomes are 'context specific' or 'generalisable to the population'. Generally, researchers who ascribe to quantitative research (e.g. statistical inference, large sample sizes, reliability and validity measures) seek generalisable cultural analysis outcomes. However, researchers who undertake qualitative research and context specific analysis (e.g. ethnography) argue that the goal of cultural analysis is to provide an interpretative frame within a cultural context (Martin 2004). These methodological divisions are also reflected in conceptualisations of organisational culture. As indicated previously, these concepts are divided into two polar categories—the functional and non-functional view, as Smircich (1983a: 354) describes below:

> Some researchers give high priority to the principles of prediction, generalisability, causality, and control; while others are concerned by what appear to them to be more fundamental issues of meaning and the process by which organisational life is possible.

Arguably, combining these perspectives through a blend of both approaches offers a more complete understanding. Although this raises issues around the commensurability of the knowledge that they respectively generate, such a suggestion is broadly in line with wider assertions as to the need to mobilise multiple perspectives in project management research in the pursuit of 'middle range' theories (Söderlund 2004; Green and Schweber 2008).

Culture as a Variable vs. Root Metaphor: the Difference between Functional Paradigm and non-Functional Paradigm

Viewing culture as a variable subject to conscious manipulation, provides a simplified and functional approach to cultural understanding. From this perspective, it is possible to manage culture and to link culture to organisational performance implying a causal relationship. Researchers who embrace the 'variable' view of culture rely on a traditional functionalist framing of social reality, approaching culture as one variable among many others (communication, organisational structure, strategy, etc.) in an organisation. This approach also asserts that organisational operations are embedded in cultural symbols (e.g. rituals, values or norms) and that through the modification of member behaviours, changes in organisational outcomes can be achieved. Therefore, studies of culture that adopt this view propose that culture is linked to organisational performance. Moreover, positive virtues, attitudes and behaviours are claimed to be useful in achieving corporate goals and these relevant dynamics can, and should, be manipulated for corporate performance (Deal and Kennedy 1988; Ouchi 1981).

In the functional view, instrumental reasons dominate the discussion of culture, with the emphasis on economic growth, advancement of technology and the exploitation of nature (Deal and Kennedy 1988; Ouchi 1981; Ouchi & Wilkins 1985; Wilkins 1983; Wilkins & Ouchi 1983; O'Reilly, Chatman & Cadwell 1991). According to Alvesson (2002), this has led to research on corporate culture, which focuses on values that directly relate to the organisation's effectiveness and state of success. He argues against equating the 'strong' cultures, which are characterised as advantageous to firms, to those 'valuable' cultures that foster 'good' performance in general (e.g. values supportive of excellence, team work, profitability, honesty). The consequence of the functionalist approach is that culture is reduced to those limited aspects that are perceived to influence organisational efficiency.

Researchers who avoid such notions of a causal link between culture and organisational performance argue that this view is simplistic and seriously underestimates the theoretical potential and value of culture (Alvesson 1993b, 2002; Meyerson 1991a, 1991b, 1994; Meyerson & Martin 1987; Geertz 1975). They favour a non-functional approach to culture whereby culture is 'thickly', or more comprehensively, described. Such 'thick descriptions' do not merely attempt to explain human behaviour, but its context as well, such that the behaviour becomes meaningful to an outsider. This notion of 'thick descriptions', originally coined by Clifford Geertz to describe his ethnographic methodology (Geertz 1975), has gained widespread recognition in the social sciences and beyond. In organisational studies, this paradigm supports the study of the informal aspects of organisations—aspects that may be obscured, paradoxical and contradictory. Thus, such descriptions offer radically differing perceptions of culture and its consequences within organisations, as has been revealed within a project context (de Bony 2010).

Contrary to the view that culture is something that an organisation can possess and command, proponents of the 'root metaphor' view stress that the organisation is itself a culture, or a collection of cultures. This philosophical position has its roots in anthropology and sociology (Geertz 1975). The proponents of a 'root metaphor' approach argue that cultures are organisations—whereby culture is a context in which social events, behaviours, and institutions materialise. Smircich (1983a, p 348) describes this approach below:

> Culture as a root metaphor promotes a view of organisations as expressive forms, manifestations of human consciousness. Organisations are understood and analysed not mainly in economic or material items, but in terms of their expressive, ideational, and symbolic aspects.

According to this view, the social world is constructed by people and reproduced by the networks of symbols and meanings that people share (Burrell & Morgan 1979). The root metaphor conceptualisation of culture emphasises a more general understanding of, and reflection upon, cultural settings. Advocates of the root metaphor approach oppose the view that organisational effectiveness can be attained through direct cultural manipulation, as this fails to address the negative features of peoples' behaviour such as resistance to change. Those who adopt this approach shun simplistic, objectivist, functionalist views of culture which link culture to performance, preferring a practical/blended conceptualisation of culture. Such a perspective has been mobilised in project management research with respect to managing cultures (see Marrewijk 2007; Marrewijk et. al. 2008).

Cultural Manifestations

The two key paradigms of culture described above are also associated with certain cultural manifestations and definitions. Manifestations are a critical component of the conceptualisation of culture as they are the aspects that enable culture to be felt, observed and deciphered. The literature suggests that culture can be assessed via different levels of cultural manifestation (e.g. shallow to deep). The level of manifestation also impacts on the ability to decipher and study culture. Marrewijik (2007) and Cheung, Wong and Wu (2011) have discussed the use of cultural manifestations in relation to project management research and emphasised the need for using appropriate manifestations to decipher culture.

Levels of Cultural Manifestations

Researchers hold varying views on the manifestations that represent culture. This has led to the classification of cultural manifestations into a number of levels, each based on how deeply and accurately they decipher cultures. Hofstede (1998a, p. 479) indicates that:

> Culture is a characteristic of the organisation, not individuals, but it is manifested in and measured from the verbal and/or non-verbal behaviour of individuals—aggregated to the level of the organisational unit.

The cultural model in Figure 1(a), developed by Hofstede et al. (1990, 2001), identifies symbols, rituals and heroes as manifestations. He believes that these manifestations are exposed or brought to the surface by practices adopted by organisations. However, he proposes 'value' as a deep manifestation of culture and uses it to decipher cultures (Hofstede 1998a; Hofstede et al.1990). Schein (2004) identifies a number of attributes that are used to describe culture. They include observed behavioural irregularities, group norms, espoused values, shared meanings, root metaphors, formal rituals and celebrations. However, Schein (1984, 2004) articulated all attributes into three levels of cultural manifestations, namely 'artefacts', 'espoused values and beliefs', and 'underlying assumptions'. A model developed by Rousseau (1990), Figure 1(b), is consistent with Schein's model, but identifies two more layers of cultural manifestations, namely 'behavioural norms' and 'patterns of behaviour'.

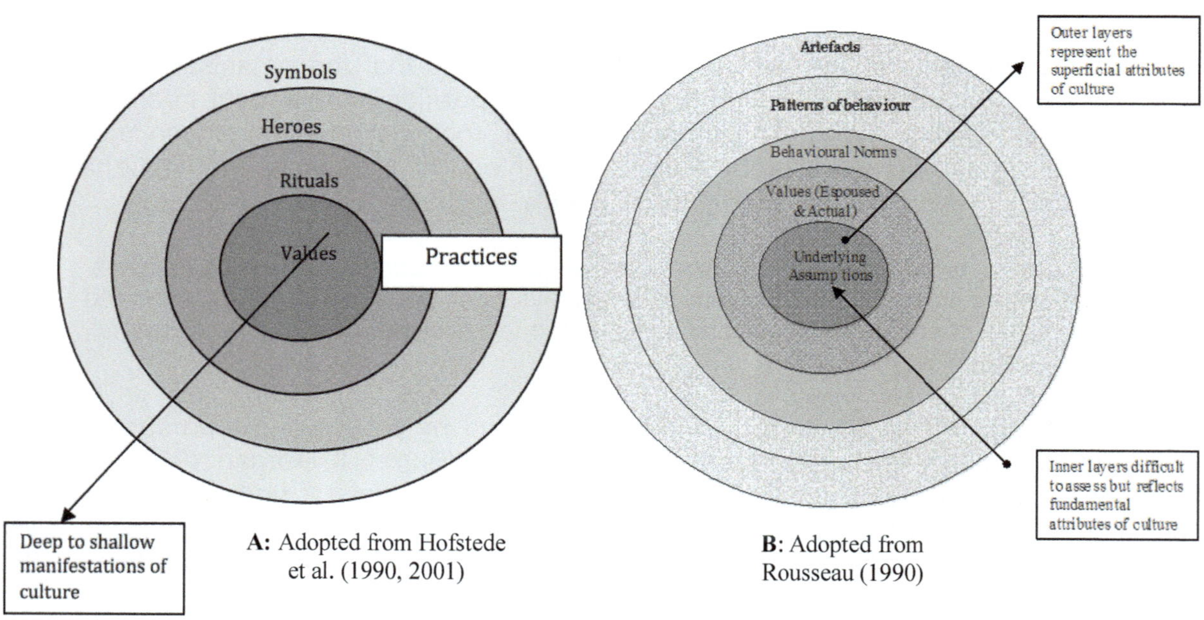

Figures 1(a) & 1(b) Different layers of cultural manifestations proposed by Hofstede et al. (1990) and Rousseau (1990)

It is believed that artefacts, such as patterns of behaviour and behavioural norms, are the more visible parts of culture. Values (that is, espoused values) are generally reflected in the goals and strategies of an organisation. The underlying assumptions/beliefs generally represent unconscious or taken-for-granted perceptions or beliefs. Taken-for-granted assumptions are powerful and deep-seated, and are thus more difficult to detect than espoused values. Based on their philosophical position, researchers employ different cultural manifestations to decipher culture.

Use of Cultural Manifestations in the Assessment of Culture

The most appropriate manifestations to use to decipher culture are highly contested within the literature. Early popular cultural studies (Deal & Kennedy 1988; Peters & Waterman 1982; Ouchi 1981) focused on the outer layers of manifestations, such as artefacts and patterns of behaviour, and linked culture to organisational performance. These studies favour the 'functional-variable' paradigm of culture. Schein (1993, 2004) and Rousseau (1990), however, have stressed the importance of in-depth understanding, labelling artefacts as 'superficial'. Indeed, Schein (1984, 1990, 2004) discourages the analysis of organisational culture through the 'visual artefacts' of the constructed environment (e.g. the architecture, technology, office layout, dress code, behaviour patterns, public documentation, employee orientation and stories). Alvesson (2002) also agrees that employing artefacts in cultural analysis can be misleading. For example, he suggests that trying to understand organisational culture by studying the coffee breaks, dress code, and meeting arrangements, may not give the best insight into culture. Both Schein (2004) and Alvesson (2002) argue that the surface level cultural data are easy to obtain but difficult to analyse and decipher. Although this level of analysis describes behavioural patterns, it does not explain why a group behaves in a particular way.

To explain why the artefacts and patterns of behaviour have certain characteristics, studies need to analyse the underlying assumptions/beliefs and values. Therefore, it is suggested that a study of culture should focus on the underlying, or 'taken-for-granted' assumptions and orientations, rather than the customary behaviour of a group of people. Understanding the underlying assumptions will enable identification of the reasons for conflict, ambiguity and self-contradiction in organisations and groups (Meyerson & Martin 1987; Alvesson 1993b). Extracting deep and inner layers of manifestations generally involves ethnographic methodologies and interviews with key members of an organisation (Morse & Richards 2002; Schein 1993).

Cultural Perspectives and Orientations

In addition to manifestations of culture, there is also a range of cultural perspectives that emerge from the paradigms of culture. The cultural definitions can be clustered into three cultural perspectives integration, differentiation, and fragmentation as proposed by Martin (1992, 2002, 2004) that span the two extreme intellectual cultural paradigms. The three perspectives act to distinguish cultures based on how cultural manifestations are shared among the unit/organisation (the 'shared nature of culture'). The cultural definitions, by associating cultural perspectives, also differ based on their view of how cultural manifestations should be, or are, shared within a cultural unit (Martin 2004).

In revealing organisational cultures, it is important to give attention to the role of the cultural perspectives as they provide an account of the mind-set or belief influencing different forms of cultural studies. The kind of cultural understanding developing in the project management domain is deeply rooted in the way the cultural studies are designed; essentially, which of the above perspectives is adopted by the researchers. The key aspects that distinguish the three cultural perspectives are; how culture is shared, how boundaries are conceived, and how ambiguity is associated in organisations.

In addition to the perspectives, culture literature also discusses three types of cultural orientation to demonstrate approaches; technical, practical and emancipation, that organisations employ to intervene with culture. These three cultural orientations distinguish intervention approaches e.g. 'control' to manipulate culture or 'emancipate' to transform cultures based on the assumptions of how cultural manifestations should be shared among the unit/organisation (Alvesson 2002). Shifting from technical and practical, to emancipation orientations also relates to a move from 'culture is controllable/pro-management of culture' to 'culture is not controllable/anti-management of culture' (Alvesson and Willmott, 1992).

A Philosophical Framework for Cultural Analysis

Based on the above discussion, it is clear that the most effective cultural analyses attempt to decipher culture from deeper manifestations and assess whether they are consistent (shared) or inconsistent (not shared) with other cultural manifestations in organisations. Project organisational culture can be comprehensively described as follows:

> Culture is a set of deeply held beliefs or underlying assumptions possessed by the individuals that make up groups in organisations, which provide a means to understand the values, behaviours and artefacts exhibited by the group. The underlying cultural beliefs that manifest themselves in groups in an organisation may be in harmony (shared) or in conflict (unshared), sometimes displaying apparent ambiguity, paradox or contradiction. The level of harmony or ambiguity among beliefs, values, behaviours and artefacts will impact upon the emergent culture and shapes organisational environments in which the group operates.

In support of the above cultural description, a cultural analysis framework, developed as part of a synthesized outcome of this review, identifies three philosophical positions in relation to conceptualising, interpreting and managing culture.

Overview of the Conceptual Framework for Cultural Analysis

The framework presented in Table 1 presents three philosophical positions from cultural studies: integration-technical, differentiation-practical and fragmentation-emancipatory. In essence, two philosophical paradigms proposed by Smircich (1983a) are aligned to Martin's three cultural perspectives to provide a platform on which to build a cultural analysis framework. Three cultural perspectives, namely Integration, Differentiation, and Fragmentation are used to explain how culture is conceptualised and how cultural environments are described. The alignment of Smircich's paradigms with Martin's concept of cultural perspectives will assist in interpreting cultural environments and the outcomes of cultural analysis.

Martin (1992, 2002)	Smircich (1983a)	Willmott (1997) Knights & Willmott (1987) Alvesson (2002)	Hofstede (1998a) Schein (2004, 1993) Rousseau (1990)
Cultural Perspective	**Cultural Paradigm**	**Cultural Orientation**	**Cultural Manifestations**
Integration Culture shared across entire unit **Differentiation** Culture shared by parts of the unit **Fragmentation** Culture may or may not be shared	Culture as variable (Functional ⇓ Culture as a root metaphor (Non-functional)	**Technical** Removal of formal irrationality **Practical** Removal of misunderstanding **Emancipatory** Removal of socially unnecessary suffering	Artefacts (Symbols, rituals, heroes) Patterns of Behaviour Behavioural Norms Espoused Values Underlying Assumptions/ Beliefs

Table 1 Cultural analysis of organisations: a synthesis and research framework

In Table 1, three types of cultural orientations outlined by Willmott (1997) and Alvesson (2002) are aligned with Martin's (2002, 2004) three perspectives of culture The three cultural orientations, namely Technical, Practical, and Emancipatory are used to explain how to manage or make meaning of cultural environments. The first orientation relates to the functional/variable paradigm while the last relates to the non-functional/metaphorical paradigm. Along with these cultural positions, differences also emerge in the manifestations used to decipher culture and the ontological/methodological approach taken to assess culture of project organisations (Hofstede 1998a; Schein 1993, 2004; Rousseau 1990).

Perspective/ Attributes	**Integration**	**Differentiation**	**Fragmentation**
Relationship to manifestation	Consistency exists among cultural manifestations. Various levels in the hierarchy display similar viewpoints. Culture is monolith, integrated and homogeneous	Consistency and inconsistency among cultural manifestations exists at different levels. This promotes differentiation and diversity at group and individual level.	Lack of clarity of consistency or inconsistency among cultural manifestations.
Degree of consensus	Organisation wide consensus on issues among all members of the organisation. **Culture as shared across the organisation**	Consensus exists within subcultures but not between them. **Culture as shared within groups but not across an organisation**	Issue-specific consensus and confusion among individuals. Overall lack of consensus. **Culture unshared in the organisation**
Orientation to ambiguity	Excludes ambiguity. Culture is defined in a way that denies ambiguity. It assumes that culture is about agreements because values are shared.	Channels ambiguity to outside subculture. Culture is often emphasised by disagreements.	Acknowledges ambiguity. Ambiguity is accepted as inevitable and continues. It is part of the usual way of doing business.
Perspectives/ Attributes	**Integration**	**Differentiation**	**Fragmentation**

Orientation/ Attributes	**Technical**	**Practical**	**Emancipatory**
Focus	The focus is to identify and manipulate cultural variables to generate the intended culture.	The focus is to generate and interpret symbolic communication to assist cultural understanding	The focus is to expose domination and exploitation aspects of a culture
Process	The emergence/ management of culture is influenced by a 'Calculation' process that enhances prediction and control	The emergence/ management of culture is influenced by an 'Appreciation' process that improve mutual understanding	The emergence/ management of culture is influenced by a 'Transformation' process that develops more rational social relations
Outcome	The outcome of this orientation is the removal of formal irrationality in organisations	The outcome of this orientation is the removal of misunderstanding	The outcome of this orientation is the removal of socially unnecessary suffering
Orientation/ Attributes	**Technical**	**Practical**	**Emancipatory**

Figure 2 Three philosophical positions on culture: A snapshot (Martin 1992, 2002; Willmott 1997; Alvesson 2002)

The cultural analysis framework presented in this study linking the three cultural perspectives (Martin 2002) to manifestations (Denison 1996; Hofstede et al. 1990; Schein 1990) and to paradigms (Alvesson 2002; Smircich 1983a) provides a comprehensive basis to analyse culture. This framework can be employed to study a cultural situation whereby the cultural environment is conceptualised through harmony (integration), conflict between groups (differentiation) and webs of ambiguity, paradox, and contradiction (fragmentation). Conceptual blind spots in each of these perspectives on culture (for example, the integration view is blind to ambiguities, and the fragmentation and differentiation views are blind to that which most cultural members share) are dealt with when a study is approached with a combination of all three perspectives (Martin 2002). Figure 2 (an elaboration of Table 1) summarises the essence of the cultural descriptions attributed to the three cultural perspectives, while linking them to cultural orientations. The following discussion compares and contrasts the alternative philosophical positions and synthesises a cultural analysis framework.

The above conceptualisation of culture can also assist in designing the empirical component of a study. This can assist in developing research instruments (e.g. survey questions, observations, and interview) by identifying the boundaries of theoretical constructs and the sample selection by identifying the members and physical scope of the units of analysis. Moreover, these conceptualisations can assist in identifying the appropriate blend of methods to study the relevant constructs.

This framework could enable researchers to contextualise their research direction and practicing managers to contextualise the proposed solutions to problems. It is argued that all three philosophical positions identified in the framework play a role in finding practicable solutions to construction industry problems (see Gajendran and Brewer 2007). The following section characteristics of project organisations as a base to discuss the application of the cultural analysis framework to study culture in construction.

The Integration-Technical Philosophical Position

Although studies in construction culture have not argued for an explicit unitary project culture, they do seek to develop and extend shared culture by the way of focusing on integration within organisations. Researchers have assumed this position in an attempt to remove irrationalities in the construction process and seek improved project outcomes. Both in construction (e.g. Baiden, Price & Dainty 2006; Dainty, Briscoe & Millett 2001a; London 2008; Winch 1989, 2001; Winch, Ushani & Edkins 1998) and mainstream literature (e.g. Berggren, Soderlund & Anderson 2001), there is a focus on the need for greater integration across project organisations to deal with fragmentation. These studies generally focus on coordination and monitoring of the diverse groups of participants and their operations in a construction project, proposing systems to monitor the progress made towards the achievement of the stakeholders' goals. Moreover, they suggest that shared expertise is a key characteristic of an integrated team, promoting free sharing of information among different units and developing the ability to predict project time and costs accurately. Therefore, integration-technical position is assumed in studies that focus on fragmentation within teams, firms and across project supply chains, in order to promote aligned and improved project outcomes.

Integration Perspective

The 'integration' perspective belongs to the functional paradigm. It seeks to understand the shared cultural manifestations in an organisation that hold it together, i.e. the 'glue' that holds different and diverse groups together in an organisation (Alvesson 2002). Therefore, 'Integrated' perspectives favour definitions of culture that view cultural manifestations as shared across an organisation/unit in its entirety, where ambiguity is non-existent or ignored. From this perspective, culture is considered as something that is shared by, or is unique to, a given group or organisation. The opposing 'fragmentation' perspective, belonging to the non-functional paradigm, argues that organisations, by nature, are paradoxical, hindering development of a unitary shared culture. It assumes that organisations cannot exist without ambiguity and that any ambiguity should not be ignored.

In many studies that support an integration view, culture is associated with peoples' customary behaviour and their habitual ways of seeing the world (Wilkins 1983; Schein 2004). In this way, Schein (2004, p. 17) characterises culture as

> a pattern of shared basic assumptions that was learned by a group as it solved its problems of external adaptation and internal integration, that has worked well enough to be considered valid and, therefore, to be taught to new members as the correct way to perceive, think, and feel in relating to those problems.

Schein (1984, 1993, 2004) further suggests that groups learn culture by forming 'shared assumptions'; they solve problems through interpreting and acting upon the environment while integrating internal operations. In a similar vein, Wilkins (1983, p. 26) proposes

> culture primarily as the 'givens' or the taken-for-granted and shared assumptions that people make about how work is to be done and evaluated and how people relate to one another (...) the way we see the world is applied unthinkingly as the underlying assumptions develop, and most others in the organization or the unit also apply it in the same manner - so it does not get challenged.

Davis (1984, p.1) further argues culture is 'the pattern of shared beliefs and values that give members of an institution meaning, and provide them with the rules for behaviour in their organization.'

In essence, any idealistic organisational integration studies will be rooted in the belief that a project organisation should develop a shared understanding aiming for a unitary culture, implying that the lack of such shared understanding is a problem. It is argued that supporting integration in organisations, through modifying rules of behaviour will deliver strong cultures that enhance organisational profitability (Kotter & Heskett 1992; Peters & Waterman 1982; Denison & Mishra 1995; Latham 1994; Egan 1998). Although the integration perspective, in terms of a unitary culture, is not widely endorsed by cultural researchers, using tools to manage cultures for enhanced performance is popular.

Culture–Performance Link

Most of the proponents of the integration perspective propose that culture can be used as a 'tool' to improve the performance of an organisation. Some argue that many of these studies use outer layers of cultural manifestations (refer Figure 1) to interpret culture (Simircich 1983, Alvesson 2002). For example, Deal and Kennedy (1988) focus on interpreting the shared espoused values of top management (e.g. mission statements) and Ouchi (1981) focuses on studying the shared formal or informal practices such as communication or decision-making. Popular literature on organisational culture in the 1980s generally subscribes to this integration perspective, relating ostensible artefacts, symbols and stories to the generation of organisation-wide consensus of cultures and values (Barley 1983; Schein 1984; Martin 2004; Schein 2004). Significant proportions of cultural studies relating to project organisations also make culture-performance links (see Cheung, Wong and Wu 2011; Ankrah, Proverbs and Debrah 2009; Andersen 2003; Zhang and Liu 2006; Baiden and Price 2011).

The unconscious attraction to adopt the 'integration' perspective in project cultural studies resonates from the argument for more integrated projects, which is closely associated to achievement of performance gains. Moreover, the position that culture is controllable and can be manipulated for performance gains, conforms to the popular view and offers the appeal of tangible outcomes from cultural studies.

Some Critiques of Integration Perspective

In synthesis, the integration perspective may employ varying 'cultural manifestations' to decipher culture. However, its essence is in the consistency of cultural manifestation and the consensus about basic beliefs among the members of the group. There is also a tendency to focus on the leaders of the group as the creators and interpreters of the group's culture (Schein 1983; Meyerson & Martin, 1987). Alvesson (2002) argues that the integration perspective has a bias towards the positive functions of culture, that culture is controllable and has causal links to organisational performance.

The image of organisations presented in an integration perspective (an image of organisation-wide harmony and homogeneity) is difficult to sustain, given the prevalence of real-life paradoxes, inconsistencies, disruptions, conflicts and ambiguities in contemporary organisations (Martin 2004; Meyerson 1991 a & b). Indeed, one of the key criticisms of this perspective is the linear approach it takes to framing culture; including an over-reliance on symbols as functional elements of the organisation and the use of moral judgement to identify 'good' and 'bad' cultures (Alvesson 2002). The integration perspective views any deviation from the ideal culture as a problem, which needs to be fixed. It is argued that inconsistencies (or subcultures) exist due to poor communication and, therefore, all that is required to remedy the lack of integration is a clear, shared understanding of the organisation's vision/mission, careful supervision and a better performance measurement regime.

The critiques of integration are overshadowed by the 'fragmentation perceptive' that evades moral judgements on culture and acknowledges paradoxes and ambiguity as an integral part of organisations. Of course, not all integration studies deny deviation from the unitary culture (see Wilkins 1983; Schein 2004) and they take an intermediate position between integration and fragmentation- the 'differentiation' perspective. Some cultural studies in project management (e.g. Ankarah and Langford 2005) favour the 'differentiation' perspective by focusing on sub cultures. Although they shy away from arguing for a unitary culture within an entire organisation, they may assume unitary culture within each sub culture.

Technical Orientation to Culture

The intention of a technical orientation is to fulfil the role of rational instrument to enhance prediction and control of functions in an organisation. This orientation focuses on the identification and manipulation of variables that are perceived to impact on an organisation's performance (Martin 2002; Alvesson 2002; Willmott 1997). Therefore, most of the 'technically-oriented' literature on culture is optimistic, believing that managers can control the values of their subordinates to achieve effective behaviour—viewing culture as a resource for effective managerial action. For example, Ouchi and Wilkins (1985, p. 462) observed that organisations use the integrated notion of culture as a rational instrument to solve problems. Therefore, culture is designed by senior management to profile the employees to fit the organisation's espoused values. Willmott (1997) states that the 'technical' cognitive orientation is based on 'empirical-analytic science', one that removes all aspects of formal irrationality from organisations, denies ambiguity and seeks to develop organisation-wide shared understanding. The expected outcome of 'technically' orientated intervention is to create an integrated culture. Therefore, the technical orientation aligns with the integration perspective. Although cultural studies in organisations do not explicitly assume technical orientation, their approach to culture is influenced by removal of irrationality (see Young and Pheng 2008; Wong, Wong and Heng 2007; Höök and Stehn 2008; Price and Chahal 2006).

Argument for the Integration-Technical Philosophical Position

Although studies in construction culture have not argued for an explicit unitary project culture, they do seek to develop and extend shared culture by the way of focusing on integration within organisations. Researchers have assumed this position in an attempt to remove irrationalities in the construction process and seek improved project outcomes. There is a focus on the need for greater integration across project organisations to deal with fragmentation in both construction (e.g. Baiden, Price & Dainty 2006; Dainty, Briscoe & Millett 2001a; London 2008; Winch 1989, 2001; Winch, Ushani & Edkins 1998) and mainstream literature (e.g. Berggren, Soderlund & Anderson 2001). These studies generally focus on coordination and monitoring of the diverse groups of participants and their operations in a construction project, proposing systems to monitor the progress made towards the achievement of the stakeholders' goals. Moreover, they suggest that shared expertise is a key characteristic of an integrated team, promoting free sharing of information among different units and developing the ability to predict project time and costs accurately.

The Differentiation-Practical Philosophical Position

Differentiation Perspective

The differentiation perspective, unlike the integration perspective, accepts some level of diversity (in shared assumptions, or consensus) in the member groups of an organisation. That is, cultural manifestations are shared within sections of an organisation. Therefore, the emerging cultural environment is 'differentiated'- an intermediate philosophical position between 'integration' and 'fragmentation' perspectives.

Consequently, 'differentiated' perspectives favour definitions in which cultural manifestations are shared within parts of an organisation (e.g. departments, groups etc.) and are unitary within those units/groups. In this perspective, ambiguity is acknowledged but confined to the subgroups (e.g. Smircich 1983b; Denison 1996; Mills 1988; Schein 1996; Martin 2002). For example, Denison (1996) suggests that values, beliefs and assumptions are the roots for the deep structure of organisations. Meanings are established through socialisation that produces a symbolic world. This leads to the stable or fragile nature of culture based on the cognition and action of individuals. Mills (1988) supports this view and suggests culture in an organisation is based on defined conditions, which could foster contradictions and conflict among segments. Each segment, however, develops a shared understanding of the cultural manifestations. Smircich, (1983b p.56) suggests culture is shared within a group based on how they develop a worldview:

> In a particular situation the set of meanings that evolves gives a group its own ethos, or distinctive character, which is expressed in patterns of belief (ideology), activity (norms and rituals), language and other symbolic forms through which organisation members both create and sustain their view of the world and image of themselves in the world. The development of a worldview, with its shared understanding of group identity, purpose, and direction is a product of the unique history, personal interactions, and environmental circumstances of the group.

The differentiation perspective also acknowledges cultural variations within an organisation through growth or new ventures, and engagement with groups who are geographically dispersed and technically/functionally varied. Therefore, as the age and size of the organisation grows, or makes connections with outer entities, differentiation occurs (Gregory 1983). This view argues for the recognition of subcultures; however, it holds that subcultures have unitary or consistent shared understanding. Schein (1990, p. 117) describes this perspective as follows:

> (...) once the group has many subcultures, its total culture increasingly becomes a negotiated outcome of the interaction of its subgroups. Organisations then evolve either by special efforts to impose their overall culture or by allowing dominant subcultures that may be better adapted to changing environmental circumstances to become more influential.

Clearly, it is difficult to establish whether this perspective falls directly under the functional or non-functional paradigm, as research from both the integration (functional) and fragmentation (non-functional) paradigms acknowledge aspects of the differentiation paradigm, with some restrictions. The differentiation paradigm offers pragmatism pragmatic position in enabling researches to position their study between the extreme ideological views (that is 'integration' and 'fragmentation').

Boundaries and Subcultures in the Differentiation Perspective

The differentiation perspective promotes the view that organisations are generally composed of a diverse set of subcultures that share some integrating elements of a dominant culture. The mix of sub cultures will depend on the nature of the environment in which the organisation operates and its internal composition (Martin, Sitkin, and Boehm, 1985). In this view, an organisation is described as a collection of subcultures based on arbitrary boundaries. These boundaries, from a cultural point of view, are not defined in legal or formal terms but are based on identification, interaction and the development of shared meaning and ideas (Alvesson 2002).

The commonly used boundaries in cultural studies are physical location (e.g. head office, site, precast yard, etc.) and bodies (people are embodied with culture and they carry it) (Santos & Eisenhardt 2005; Moore & Dainty 2001). Parker (2000) sees the origin of subcultures in different sources: spatial/functional (geographical location of work function), generational, (age and length of time in the organisation) and occupational/professional training.

Hofstede (1998b), Schein (1993) and Jones (1983) identify a number of sub cultural classifications. Schein (1996) characterises three types of subcultures, namely: operator, engineer, and executive and discusses the tensions between them. Hofstede (1998) identify three sub cultures, namely professional, administrative, customer interface, along the six dimensions (Process vs. results, employee vs. job, parochial vs. professional, open vs. closed, loose vs. tight, normative vs. pragmatic). The three subcultures discussed by Jones (1983) include product, bureaucratic and professional. Kumaraswamy et al. (2002) identified organisational, operational, professional and individualistic sub-cultures in the context of construction project organisations (c.f. Ankrah, Proverbs and Debrah 2009).

According to Meyerson (1994) different professions generate distinct ideologies due to the embodiment of different cognitive and symbolic systems producing different rules, conventions and meanings. The ideologies and approaches of the different professional subsystems come into contact when they work interdependently within organisations and as the actions and interpretations of their members negotiate and create order. As behaviours, interpretations, and justifications become adopted, they create an order that constitutes the particulars of that subculture, which reconstitutes the institutional system in a particular time and context. This explains the dynamic and complex sub-cultural taxonomies that emerge in construction organisations.

Martin (2002) argues that some literature assumes that the boundaries of a collective coincide with the boundaries of a culture. For example, an organisation's culture is coincident with the boundaries of that organisation. However, this view is considered to be problematic as most collectivities may be holding boundary roles, temporary or virtual roles, etc. In the context of a project, boundaries of architectural, engineering, quantity surveying and contracting firms and their cultures may not provide a clear separation on how their cultures engender the culture of a project organisation. The virtual communication does, to some extent, express a culture without bodily contact between group members. Moreover, temporary workers also violate traditional assumptions concerning subcultures in organisations, displaying an ambivalent relationship to the project and the associated organisation(s) (Usdiken, Soxen & Enbiyaoglu 1988). Evidently, the boundaries of cultures and project organisations cannot be simply equated or clearly defined; boundaries need to be viewed as dynamic and changing.

In effect, boundaries have significant impact on emergent sub cultures in project organisation, at least from professional (see Ankarah 2005; Liu, Fellows and Ng 2004; Akiner and Tijhuis 2007, operational (see Lansley and Riddick 1991), and geographical boundaries (see Bony 2010; Pheng and Yuquan 2002; Toor and Ogunlana 2008). In addition, the temporary nature of project organisations elevates the tendency for more dynamic and transient boundaries (Turner and Muller 2003; Turner 2006). The emergence of new and hybrid procurement arrangements generates different natures of boundaries in each project, specifically from an operational point of view. Ultimately the inherent nature of the project environment poses significant challenges in developing a coherent and static understanding of sub cultures, pointing cultural studies towards the fragmentation perspective.

Ambiguity and Conflict in Differentiation

Earlier discussion on boundaries highlighted the acknowledgement of the presence of inconsistencies as well as consistencies in cultural manifestations among sub cultures. Therefore the differentiation perspective does not totally deny ambiguity in project organisations. Instead, it channels ambiguity and conflict to subcultures, so that each subculture responds to a small part of the organisational complexities and uncertainties. The differentiation view acknowledges that subcultures may conflict with the dominant organisational culture or may be orthogonal to a dominant culture, reflecting differences in functional, national, and occupational or project affiliations.

A 'conflict and claims culture', partly arising from ambiguity, is an area that has concerned the construction sector. Most studies channel conflicts arising from ambiguities to sub cultures: such as client, contracting, quantitative surveying, estimator sub cultures (see Rooke, Seymour and Fellows 2003; Loosemore 1999a; Ankrah and Langford 2005). The ambiguity focus of these studies remains on those sub-cultural aspects that are ostensible. Research assuming differentiation perspective is not too far removed from the oversimplifications of the integration perspective. Consistency and clarity still predominate within a culture, and ambiguity is only prevalent among subcultures.

In essence, the differentiation perspective in cultural studies arises from the view that cultural manifestations are shared within subgroups, but shared among partly among the groups. Therefore, cultures are not considered monolithic, but are treated as a collection of subcultures (Martin 2004; Schein 2004; Ankrah, Proverbs and Debrah 2009). In the differentiation view, multiple sources influence culture generation rather than an individual leader, as in the integration view. This creates multiple subcultures that share common histories or assumptions within an organisation (Martin 2004). Studies on the differentiation perspective have revealed that in an organisation, tensions could exist between subsystems of subcultures; for example: managers versus site workers, young versus older workers, consultants versus contractors, contractors vs. sub-contractors (Awakul and Ogunlana, 2002).

The attraction to adopt this perspective in cultural studies resonates from the argument for managing boundaries and national/professional cultures for enhanced project understanding. More specifically, to deal with conflicts (that may arise from ambiguity) between different groups of members in project organisations- that is to deal with different sub-cultures. The belief that sub-cultures are controllable and can be manipulated for aligning the interests of the project organisation, offers the appeal of tangible outcomes from cultural studies, as in the integration studies. However, the nature of the project organisational environment, particularly the complex transient boundaries and uncertainties, advocates conceptualising culture of organisations from the fragmentation perceptive. Although, extreme non-functional fragmentation studies do not believe in the existence of any form of shared culture within organisations, most studies shy away from this extreme stance.

The Practical Orientation to Culture

The 'practical orientation' to culture does not take an optimistic view of culture as technical orientation. Moreover, it does not advocate that managers can control the values of their subordinates to achieve effective behaviour (Alvesson 2002). By acknowledging misunderstanding, this orientation accepts ambiguity and intends to remove it through symbolic communication. This non-functional orientation to culture contests the idea that management effectiveness can be attained through cultural manipulation and emphasises the importance of generating a more holistic understanding and reflection of an organisation's life. By assuming this emphasis, a 'practical orientation' to studying organisations involves observing and interpreting organisations through 'appreciation' rather than 'calculation' (Willmott 1997). This orientation, therefore, aligns with the 'differentiation' perspective on culture (Martin 2002).

In the differentiation perspective ambiguity is acknowledged, but confined to the subgroups and sub cultures (Martin 2004). Here, meanings are established through socialisation that produces a symbolic world and leads to the stable or fragile nature of culture that fosters contradiction and conflict among subcultures. The aim of the practical orientation is, 'to achieve common interpretations of situations [for] coordinated action' (Smircich 1983a, p. 351) and associates closely with removing misunderstanding among sub cultures in an organisation to achieve coordinated action. The 'practical' orientation recognises and appreciates 'questions of interpretation and description' and explores the inter-subjective experience of organisation thus removing unnecessary misunderstandings in organisations. Practical orientation, addressing communication issues to improve information flows (and thereby attempting to reducing misunderstanding), is widely adopted in construction project management (see Phua and Rowlinson 2003; Loosemore and Muslmani 1999; Ochieng and Price 2010).

Argument for the Differentiation/Practical Cultural Position

Research assuming this position focuses on improving communication in the construction process and to minimise misunderstandings. Studies acknowledging the elusiveness of total integration and aiming for partial integration with selected firms, tiers, or groups that fall into this cultural position. This position can be argued for studies dealing with (e.g. professional, trade, informal groups) sub-culture of groups, lack of collaboration and trust among groups and those seeking understanding of communications between different sub-cultures (Ankrah, and Langford, 2005; Loosemore and Muslmani, 1999). These studies can assist in improving understanding between sub-cultures that lead to conflict mitigation and enhanced project collaboration.

The Fragmentation-Emancipation Philosophical Position

Fragmentation Perspective

The third key perspective of organisational culture studies is 'fragmentation' (Martin et al. 1983; Meyerson and Martin 1987; Martin, 2004). This approach views culture as a disorder that contributes to dysfunctional aspects in organisational life—thereby, this approach aligns with the 'non-functional' cultural paradigm (Alvesson 1993b, 2002; Meyerson 1991a, 1991b; Feldman 1991). The fragmentation perspective accepts ambiguity, rather than denying or channelling it, and views organisational culture as inherently ambiguous leading to 'fuzziness' in organisational life (Meyerson, 1991a, 1991b; Alvesson 1993b; Sveningasson & Alvesson 2003).

The fragmentation approach to culture is described aptly by Meyerson (1991a p. 256) who notes that 'culture was the code word for the subjective side of organisational life (...) its study represented an ontological rebellion against the dominant functionalist or "scientific' paradigm".' Ignoring this perspective in organisational studies according to Meyerson (1991b) means 'to dismiss the [informalities and] ambiguities in favour of strictly what is clear and shared. [This] is to exclude some of the most central aspects of the members' cultural experiences and to ignore the essence of their cultural community' (p. 131-132). Recent literature in construction management has raised attention to the subjective informal, aspects of project organisations (see Chan & Raisanen 2009; Marrewijk 2008), highlighting the potential of such a perspective in developing a better understating of project organisations. Meyerson (1991b, p.131-132) argues that, in reality:

> (...) members do not agree upon clear boundaries, cannot identify shared solutions, and do not reconcile contradictory beliefs and multiple identities. Here culture is characterised as a web, where individuals are nodes in the web–temporarily connected by shared concerns to some nodes and not to others. When a particular issue comes to the surface, certain nodes and patterns of connections become relevant. The emerging patterns of connection in an organisation will consist of agreements, disagreements, and ignorance. Most importantly, from this perspective the inherently active and dynamic nature of culture is made apparent.

The characteristics of project organisations resemble some of the above features. Particularly, construction projects are characterised as loosely coupled systems with transient boundaries. In each project a web of temporarily connected organisations forms a project organisation to deliver a product or service (see Turner and Muller 2003; Dubois, and Gadde, 2002). Although members of organisations are connected via a transient web, they believe they belong to a culture. Meyerson (1991b, p.131-132) also argues that organisations:

> (...) share a common orientation and overarching purpose, face similar problems [e.g. project goals- delivering a project within the established cost and timeframe with desired quality, while making profits], and have comparable experiences. However, these shared orientations and purposes accommodate different beliefs and incommensurable technologies, these problems imply different solutions, and these experiences have multiple meanings (...) Thus, for at least some cultures, to dismiss the ambiguities in favour of strictly what is clear and shared is to exclude some of the most central aspects of the members' cultural experienced and to ignore the essence of their cultural community.

Primarily, Meyerson's view is that organisational members do not necessarily display a high level of shared understanding, but can still share cultures through common orientations and purposes which accommodate different beliefs and ambiguities. Unlike the 'integration' perspective, awareness of ambiguity is not experienced as a temporary stage during an organisational change process; rather, ambiguity is seen as a permanent state—'the way things are'. It is argue that this perspective adds a sound basis to cultural analysis of organisations by acknowledging ambiguities could be inherent and perpetual in organisations (Meyerson & Martin, 1987; Martin 2002; Alvesson 2002). Therefore, cultural manifestations are not necessarily shared across or within sections of the organisation, creating a cultural environment that is 'fragmented'. The approach fundamentally challenges the dominant idea that a culture (subculture) is a clear and known entity that creates unity and harmony within an organisation (or a group within it). In this perspective, irreconcilable interpretations are simultaneously entertained, paradoxes are embraced and the lack of a shared/integrated set of values is accepted as normal.

Here, relationships between manifestations are characterised by a lack of clarity caused by ignorance or complexity. Any attempt to design organisational processes to resolve irreconcilable conflicts in organisations is regarded as a temporary and superficial 'smoke screen'. The confusion emanating from inconsistent viewpoints and disagreements makes it difficult to draw cultural and subcultural boundaries. The boundary around the organisation then becomes amorphous and permeable, as various 'feeder' cultures from the surrounding environments fade in and out of attention (Martin 2002).

The difficulty in articulating the ever-changing patterns and ambiguities of culture makes the fragmentation perspective the most difficult to conceptualise (Martin 2002). Fragmentation studies see ambiguity as the defining feature of cultures in organisations. As such, Martin (2004) argues that understanding ambiguities should be the central component of any study of culture. However, the very nature of the fragmentation approach has resulted in very few researchers adopting it as a framework for organisational studies, due to its emphasis on uncertainty, contradiction and confusion and for acknowledging that some ambiguities cannot be eliminated (Martin 2004). Those who embrace the fragmentation perspective could argue the interpretations of the recommendations proposed by industry reviews (e.g. Latham 1994 and Egan 1998), in terms of embracing integration by controlling, manipulating and predicting cultures to address construction industry issues, is fundamentally flawed. This does not mean cultures cannot be changed. Here, cultures are transformed via emancipation that liberates members from exploitation by making them aware of their traditionally held unconscious values. Therefore, this perspective discredits the practice of manipulating and controlling cultures.

Defining Ambiguity and Sources of Ambiguity

In organisations and projects, managers frequently emphasise an informal, family-like community spirit. However, everyday practices and relationships often deviate from this ideal (Alvesson 1990). In most instances, the term 'organisational culture' conjures the image of an organisation as being:

> typically unitary and unique, characterized by a stable set of meanings (...) arguably [however] cultural manifestations are far from always neatly organized, values not easily ranked and cultural ideas may be unsystematic and incoherent.(Alvesson 2002, p. 145)

Meyerson (1991a) argues that in some cultures, boundaries are not agreed, shared solutions are non-existent, and members do not reconcile contradictory beliefs and multiple identities. This leads to lack of clarity in organisations with ambiguity as a figure (or anti-figure) that emerges from this context. Two types of ambiguities are identified in the literature, arising from:

- a lack of clarity - whereby ambiguity is felt as an internal state and occurs due to the absence of information - once the information is available it vanishes; e.g. incomplete drawings/specifications could lead to ambiguity with respect to how a detail should be constructed. This kind of ambiguity is removed when the required information becomes available. However, not supplying the information within a reasonable period of time could lead to conflict and subsequent litigation (Meyerson, 1990; Meyerson & Martin, 1987).
- irreconcilable contradictions – in which ambiguity is the result of conflicts between individuals (or groups) which embrace two or more meanings that are irresolvable. E.g. irresolvable contradictions could emerge when a client and contractor attach different meaning to a product, process or situation (Meyerson, 1990; Meyerson & Martin, 1987).

In organisations, there is both unity and division in work relations depending upon the complexity of the surrounding environment, group membership and the degree of ambiguity in organisational rules and policies (see Ankrah, and Langford, 2005). Although the fragmentation perspective is considered to be important, some researchers question the extensive use of this perspective, as fragmentation could be interpreted as no culture in an organisation (Alvesson 2002).

One could argue that fragmentation could be one of the perspectives for exploring cultures in organisations in the construction industry. However, construction organisations are underexplored from this perspective. Adoption of this perspective could provide a meaningful approach to understand culture in construction.

The Emancipatory Orientation

The 'emancipatory' cognitive interest is based on 'critical science' and intends to develop more rational social relations through realisation (Alvesson 2002; Knights & Willmott 1992; Willmott 1997). The focus of this mode is the exposure of domination and exploitation, therefore, it takes a 'transformation' orientation. That is, it endeavours to make members of a group aware of the emancipatory approach to investigate the negative features of organisational life and to develop an understanding of, and to counteract, the taken-for-granted beliefs and values that exploitation and domination that characterise relationships in their culture, while encouraging them to transform the way they operate.

One outcome of this mode of analysis concerns the removal of 'unnecessary suffering' in organisations (Willmott 1997). The main focus imposes limits on actions and interactions. This orientation provides insight into organisational life by grasping the traditional patterns and the repressive aspects of culture and may contribute to liberating the organisation from its traditional cultural environment. It includes understanding irreconcilable tensions between opposites, sometimes described as ironies, paradoxes, or contradictions (Martin 2002). Ambiguity studies, generally, view such occurrences as normal and salient, and part of the organisation's function.

This orientation, therefore, aligns with the 'fragmentation' perspective on culture in that ambiguity is acknowledged as an inherent part of an organisation, and views organisational culture as inherently ambiguous leading to 'fuzziness' in organisational life. Therefore, the organisations face contradictions and paradoxes (e.g. through information conflicts, un reasonable risk exposure) contributing to unnecessary suffering of members. Such suffering, from a construction point of view, includes the culture of claims, litigation, unsafe practices and other forms of exploitation etc. By liberating the members of the organisation from taken for granted assumptions, unnecessary suffering can be minimised.

Although the emancipatory approach does not relate to business results directly, it nevertheless, influences people in relation to their ways of constrained thinking and acting. Therefore, the purpose of emancipatory cultural studies is about liberating human potential or, more defensively, illuminating the obstacles to emancipation. Encouraging critical reflection on beliefs, values and understanding of social conditions becomes the prime task of this cultural orientation. This orientation is underexplored in construction project management.

Argument for the Fragmentation-Emancipatory Cultural Position

Clearly, when culture is discussed in the project literature, fragmentation is most often associated with transient boundaries, ambiguity and conflicts. However, studies of project organisations cultures based on a Fragmentation - emancipation position are not common in construction. The complex characterisation of culture along with no explicit link between project performance enhancements makes this perspective unattractive. Nevertheless, burgeoning interest can be sensed in certain elements of these approaches in studying issues such as conflicts (e.g. Awakul & Ogunlana 2002; Loosemore, Nguyen & Denis 2000; Rooke, Seymour & Fellows 2003), disputes (e.g. Loosemore 1999a, 1999b), claims (e.g. Rooke, Seymour & Fellows 2004), and general fragmentation emanating from industry structure, information and attitude (Marrewijk et al 2008).

This position does not believe that culture can be fully manipulated or controlled to achieve desired outcomes; rather cultures are transformed by emancipation. Change management literature suggests that managers are neither in total control of cultures nor without any influence for transforming cultures. Moreover, acknowledging that some of the initiatives to remove irrationality are smoke screens that are not sustained with time- but rather seek solutions acknowledging paradox and chaos can provide a new range of initiatives in project management. This means, empowering members of a culture to transform traditionally held beliefs resulting in exploitative outcomes into new egalitarian beliefs producing co-constructed outcomes. There is a need for future cultural studies to explore this perceptive and to understand the process of cultural transformation in inherently fragmented organisational contexts.

Contextualising Cultural Analysis in Construction Organisations

This paper argues that cultural studies can adopt a single philosophical position or mix of positions. The characterisation of organisational issues to be studied is fundamental to identify a suitable cultural position(s) to conduct meaningful cultural studies. It is arguable that the worldview (e.g. objectivist-subjectivist) of the researchers/managers/authors characterise the way in which the organisational issues are approached. However, adopting the pragmatism view using multiple or mixed research methods (Creswell and Clark 2007) enable to maintain a balance among different ideologically extreme worldviews.

Construction organisations are inherently unique and fragmented- subjected to boundaries and multiple stakeholders (Lam et al. 2007; Rahman & Kumaraswamy 2005; Gajendran 2010). The stakeholders play a critical role in characterising the culture: by having organisational goals that are not shared across the project organisation, stakeholders can create unnecessary boundaries between members and, thus, bring fragmentation (Usdiken, Soxen & Enbiyaoglu 1988; Cox 1996; Cox & Thompson, 1997; Walker & Wing 1999; Dainty, Briscoe & Millett 2001a, 2001b). Indeed, Hansfield and Nichols (1999) point to poor alignment of organisational cultures as the factor adversely affecting integration, a point broadly supported in the literature (see Baiden, Price & Dainty 2006; Dainty, Briscoe & Millett 2001a; Voordijk, Meijboon & Haan 2006).

Studies into project organisations have proposed numerous initiatives for integration, to address the ills caused by fragmentation (see Latham 1994 and Egan 1998). Extensive focuses on removal of irrationality or manipulation of culture to improve performance may not yield the expected outcomes. By nature, construction project organisations are paradoxical and contradictory, generating ambiguity and leading to conflicts (Chan and Raisanen 2009; Marrewijk et. al. 2008). Such inherent characteristics of project organisations, e.g. paradoxes, boundaries, and ambiguity cannot be totally eliminated; rather, project organisations need to be studied and managed within those constraints. It is important to note that implementation of any initiatives or solutions to deal with fragmentation are also subject to the fragmentation of the construction industry- to some extent beyond the control of organisations. Not contextualising the cultural studies without appropriate acknowledgment to inherent fragmentation could lead to failed or ineffective outcomes. Therefore, with any integration initiative, due consideration needs to be given to the project characteristics that generate fragmentation, in order to facilitate more effective integration (or reduction of fragmentation) (Charoenngam et al. 2004; Chatman & Jehn 1994).

Conclusions

Theoretical perspectives should drive project management practice as much as empirical enquiry. Although cultural studies have mobilised various methodological approaches in the project management domain, the point of departure for this research is that they have predominantly been conceptualised within the 'functional' paradigm, and driven by empirical inquiry. These approaches invariably treat culture as a feature of an organisation or project, which is amenable to conscious control and manipulation towards performative outcomes. Moreover, these offer limited flexibility by prescribing a singular philosophical position from which to observe culture. In contrast, by reviewing cultural studies literature as applied to organisations this paper has synthesised a robust theoretical platform for understanding the culture of projects and organisations. It concludes that when investigating cultural environments adopting the 'non-functional' paradigm: facilitates the better understanding of individuals' behaviour within group settings, therefore allowing project participant behaviour to be contextualised within a temporary project organisation, and; provides a richer, more compelling explanation of project performance/outcomes as an emergent feature of the complex network of relationships that is inherent in project organisations.

This paper presents a conceptual framework that serves dual purposes. Firstly, it simplifies and guides the cultural analysis of organisations, providing a framework against which cultures can be defined, evaluated and understood. Secondly, it reveals the three philosophical perspectives (Integration, Differentiation, and Fragmentation) and three orientations (Technical, Practical and Emancipation), which necessarily underpin all cultural analyses. This paper suggests that a hybrid functional/non-functional cultural analysis framework will be more effective, allowing researchers and practicing managers to identify and operate from one or more possible positions. In this way they can embrace a pragmatic approach to the study of organisational cultures, using both positivist and subjectivist methods to obtain authentic outcomes.

References

Adriaanse, A. and Voordijk, H. (2005) 'Inter-organisational communication and ICT in construction project: a review using meta-triangulation', *Construction Innovation*, **5**, 159-177

Alvesson, M. (1987) 'Organisations, culture and Ideology', *International Studies of Management & Organisations*, **VXII** (3), 14-18

Alvesson, M. (1990) 'On the popularity of organisational culture', *Acta Sociologica*, **33** (1), 31-49

Alvesson, M. (1993a) *Cultural Perspectives on Organisations*, Cambridge University Press, Cambridge

Alvesson, M. (1993b) 'Organisations as rhetoric: Knowledge-intensive firms and the struggle with ambiguity', *Journal of Management Studies*, **30** (6), 997-1015

Alvesson, M. (2002) *Understanding Organisational Culture*, SAGE Publications, London

Alvesson, M. and Willmott, H. (1992) 'On the idea of emancipation in management and organisational studies', *The Academy of Management Review*, **17** (3), 432-464

Andersen, E S (2003) 'Understanding your project organisation's character', *Project Management Journal*, **34** (4), 4-11

Ankrah, N. A. and Langford, D. A. (2005) 'Architects and contractors: a comparative study of organisational cultures', *Construction Management and Economics*, **23** (6), 595-607

Ankrrah, N. A. Proverbs, D. and Debrah, Y. (2009) 'Factors influencing the culture of a construction project organisation: An empirical investigation' *Engineering, Construction and Architectural Management*, **16** (1), 26-47

Anumba, C. E. H. Dainty, A. Ison, S. and Sergeant, A. (2006) 'Understanding structural and cultural impediments to ICT systems integration: A GIS based case study', *Engineering, Construction and Architectural Management*, **13** (6), 2006

Awakul, P. and Ogunlana, S. O. (2002) 'The effect of attitudinal difference on interface conflicts in large scale construction projects: a case study', *Construction Management and Economics*, **20**, 365-377

Baiden, B. K. Price, A. D. F. and Dainty, A. R. J. (2006) 'The extent of team integration within construction projects', *International Journal of Project Management*, **24**, 13-26

Baiden, B. K. and Price, A. D. F. (2011) 'The effect of integration on project delivery team effectiveness', *International Journal of Project Management*, 29 (2), 129-136

Barley, S. R. (1983) 'Semiotics and the study of occupational and organisational cultures', *Administrative Science Quarterly*, **28**, 393-413

Berggren, C. Soderlund, J. and Anderson, C. (2001) 'Clients, contractors, and consultants:

the consequences of organisational fragmentation in contemporary project environments' *Project Management Journal*, **32** (3), 39-48

Brochner, J. Rosander, S. and Waara, F. (2004) 'Cross-border post-acquistion knowledge transfer among construction consultants', *Construction Management and Economics*, **22** (4), 421-427

Burrell, G. and Morgan, G. (1979) *Sociological paradigms and organisational analysis: Elements of the sociology of corporate life,* Heinemann, London

Cameron, K. S. Quinn, R. E. and DeGraff, J. (2006) *Competing Values Leadership: Creating Value in Organisations,* Edward Elgar, Lancaster

Cameron, K. S. and Quinn, R. E. (2006) *Diagnosing and changing organisational culture,* CJossey-Bass, San Francisco

Chan, P. W. and Raisanen, C. (2009) 'Editorial: informality and emergence in construction', *Construction Management and Economics*, **27** (10), 907-912

Charoenngam, C. Ogunlana, S. O. Ning-Fu, K. and Dey, P. K. (2004) 'Re-engineering construction communication in distance management framework', *Business Process Management Journal*, **10** (6), 645-672

Chatman, J. A. and Jehn, K. A. (1994) 'Assessing the relationship between industry characteristics and organisational culture: How different can you be?', *The Academy of Management Journal*, **37** (3), 522-553

Cheng, E. W. L. Li, H. Love, P. and Irani, Z. (2004) 'A learning culture for strategic partnering in construction', *Construction Innoation*, **4** (1), 53-65

Cheung, S.O., Wong, P.S.P. and Wu, A.W.Y. (2011) 'Towards an organisational culture framework in construction', *International Journal of Project Management*, **29**, 33-44

Coffey, V. (2010) *Understanding organisational culture in the construction industry,* Spon, London

Cooper, C. L. and Rousseau, D. M. (1999) *The Virtual Organisation,* Wiley, West Sussex

Cox, A. (1996) 'Relational competence and strategic procurement management', *European Journal of Purchasing & Supply Management*, **2** (1), 57-70

Cox, A. and Thompson, L. (1997) ''Fit for purpose' contractual relations: determining a theoretical framework for construction projects', *European Journal of Purchasing & Supply Management*, **3** (3), 127-135

Creswell, J. W. and Clark, V. L. P. (2007) *Designing and conducting mixed method research*, Thousand Oaks: Sage

Dainty, A. Green, S. and Bagilhole, B., Eds. (2007), *People and culture in construction: A reader*, Taylor & Francis, Oxen

Dainty, D. R. J. Briscoe, G. H. and Millett, S. J. (2001a) 'New perspectives on construction supply chain integration', *Supply Chain Management*, **6** (3/4), 163-173

Dainty, D. R. J. Briscoe, G. H. and Millett, S. J. (2001b) 'Subcontractor perspectives on supply chain alliances', *Construction Management and Economics*, **19**, 841-848

Daly, J. P. Pouder, R. W. and Kabanoff, B. (2004) 'The Effects of Initial Differences in Firms' Espoused Values on Their Post-merger Performance', *The Journal of Applied Behavioral Science*, **40** (3), 323-343

Davis, S. (1984) *Managing Corporate Culture,* Ballinger, Cambridge MA

de Bony, J. (2010) 'Project management and national culture: A Dutch-French case study', *International Journal of Project Management*, **28**, 173-182

Deal, T. E. and Kennedy, A. A. (1988) *Corporate cultures: the rites and rituals of corporate life*, Penguin Books, London

Denison, D. R. (1996) 'What is the difference between organisational culture and organisational climate? A native's point of view on a decade of paradigm wars', *Academy of Management Review*, **21** (3), 619-654

Denison, D. R. and Mishra, A. K. (1995) 'Towards a theory of organisational culture and effectiveness', *Organisation Science* 6 (2), 204-223

Dubois, A. and Gadde, L.-E. (2002) 'The construction industry as a loosely coupled system: implications for productivity and innovation' *Construction Management and Economics*, **20**, 621-631

Egan, J. (1998) *Rethinking Construction*, Department of the Environment, London.

Feldman, M. S. (1991) 'The meaning of ambiguity: Learning from stories and metaphors', Peter J Frost, L. F. M., Meryl Reis Louis, Craig C Lundberg, Joanne Martin (Ed.), *Reframing Organisational Culture*, Sage. California, 145-164

Fellows, R. (2006) 'Understanding approaches to culture', *Construction Information Quarterly*, **8** (4), 159-166

Fellows, R. (2010) 'Book Review: Understanding Organisational Culture in the Construction Industry', *Construction Management and Economics*, **28** (8), 898-900

Gajendran, T. (2010) *The role of Culture in the Integration of Information and Communication Technology (ICT) in construction projects*, The University of Newcastle, Doctor of Philosophy, Newcastle, 469

Gajendran, T. and Brewer, G. (2007) 'Integration of information and communication technology: Influence of the cultural environment', *Engineering, Construction and Architectural Management*, **14** (6), 532-549

Geertz, C. (1975) *The Interpretation of Cultures*, Hutchinson, London.

Gregory, K. L. (1983) 'Native View Paradigms: Multiple cultures and culture conflicts in organisations', *Administrative Science Quarterly*, **28**, 359-376

Hansfield, R. B. and Nichols, E. L. (1999) *Introduction to supply chain management*. Prentice-Hall, Englewood Cliffs NJ

Hartmann, A. (2006) 'The role of organisational culture in motivating innovative behaviour in construction firms', *Construction Innovation*, **6**, 159-172

Hofstede, G. (1998a) 'Attitudes, values and organisational culture: Disentangling the concepts', *Organisational Studies*, **19** (3), 477-492

Hofstede, G. (1998b) 'Identifying organisational sub cultures: an empirical approach', *Journal of Management Studies*, **35** (1), 1-12

Hofstede, G. Neuijen, B. Ohayv, D. D. and Sanders, G. (1990) 'Measuring organisational cultures: A qualitative and quantitative study across twenty cases', *Administrative Science Quarterly*, **35** (2), 286-316

Hofstede, G. Hofstede, G. J. and Minkov, M. (2010) *Cultures and organisations: software of the mind*, McGraw Hill, London

Hofstede, G. (2001) *Culture's consequences: Comparing values, behaviors, institutions and organisations across nations*, 2nd ed, Sage, Thousand Oaks

Höök, M. and Stehn, L. (2008) 'Applicability of lean principles and practices in industrialized housing production', *Construction Management and Economics*, **26** (10), 1091-1100

Jones, G. R. (1983) 'Transaction cost, property rights, and organisational culture: An exchange perspective', *Administrative Science Quarterly*, **28**, 454-467

Kabanoff, B. Waldersee, R. and Cohen, M. (1995) 'Espoused values and organisational change themes', *Academy of Management Journal*, **38** (4), 1075-1104

Knights, D. and Willmott, H. C. (1987) 'Organisational culture as management strategy: A critique and illustration from the financial services industry', *International Studies of Management & Organisations*, **17** (3), 40-63

Kornelius, L., Wamelink, J.W.F (1998) 'Case Study: The virtual corporation learning from construction', *Supply Chain Management*, **3** (4), 193-202

Kotter, J. P. and Heskett, J. L. (1992) *Corporate culture and performance*, Maxwell Macmillan, New York.

Kumaraswamy, M., Rowlinson, S., Rahman, M. and Phua, F. (2002) 'Strategies for triggering the required 'cultural revolution' in the construction industry', in Fellows, R. and Seymour, D. (Eds), *Perspectives on Culture in Construction*, CIB Report, CIB, Shizuoka, 275

Lam, K. C. Want, D. Lee, P. T. K. and Tsang, Y. T. (2007) 'Modelling risk allocation decision in construction contracts', *International Journal of Project Management*, **25**, 485-493

Latham, M. (1994) *Constructing the Team*, HMSO, London

Ling, F. Y. Y. Ang, A. M. H. and Lim, S. S. Y. (2007) 'Encounters between foreigners and Chinese: Perception and management of cultural differences', *Engineering, Construction and Architectural Management*, **14** (6), 501-518

Liu, A. and Fellows, R. (2008) 'Behaviour of quantity surveyors as organisational citizens', *Construction Management and Economics*, **26** (12), 1271-1282

Liu, A. Fellows, R. and Ng, J. (2004) 'Surveyors' perspective on ethics in organisational culture', *Engineering, Construction and Architectural Management*, **11** (9), 438-449

Loosemore, M. (1999a) 'Responsibility, power and construction conflict', *Construction Management and Economics*, **17**, 699-709

Loosemore, M. (1999b) 'A grounded theory of construction crisis management', *Construction Management and Economics*, **17**, 9-19

Loosemore, M. and Muslmani, H. S. A. (1999) 'Construction project management in the perian culf: inter-cultural communication', *International Journal of Project Management*, **17** (2), 95-100

Loosemore, M. Nguyen, B. T. and Denis, N. (2000) 'An investigation into the merits of encouraging conflict in the construction industry', *Construction Management and Economics*, **18,** 447-456

Marrewijk, Av. (2007) 'Managing project culture: The case of Environ Megaproject', *International Journal of Project Management*, **25**, 290–299

Marrewijk, Av. Clegg, S. R. Pitsis, T. S. and Veenswijk, M. (2008) 'Managing public–private megaprojects: Paradoxes, complexity, and project design', *International Journal of Project Management*, **26** (5), 591-600

Martin, J. (1992) *Cultures in organisations: Three perspectives,* Oxford University Press, New York

Martin, J. (2002) *Organisational Culture: Mapping the terrain*, Sage, Newbury Park, CA

Martin, J. (2004) Organisational culture, *Research Paper Series, Stanford*, Research Paper No 1847

Martin, J. Frost, P. J. and O'Neill, O. A. (2004) Organisational culture: Beyond struggles for intellectual dominance, *Stanford Research Paper Series*, No 1864.

Martin, J. Sitkin, S. B. and Boehm, M. (1985) 'Founders and the elusiveness of a cultural legacy', Frost, P. J. More, L. F. Louis, M. R. Lundberg, C. C. and Martin, J. (eds.), *Organisational Culture*, Sage. California, 99-124

Meyerson, D. (1994) 'Interpretations of stress in institutions: The cultural production of ambiguity and burnout', *Administrative Science quarterly*, **39** (4), 628-653

Meyerson, D. and Martin, J. (1987) 'Cultural Change: An integration of three different views', *Journal of Management Studies*, **24** (6), 623-647

Meyerson, D. E. (1991a) Acknowledging and uncovering ambiguities in cultures, Frost, P. Moore, L. Louis, M. Lundberg, C. and Martin, J. (eds.), *Reframing Organisational Culture*, Sage. California, 254-270

Meyerson, D. E. (1991b) ''Normal' ambiguity? A glimpse of an occupational culture', Frost, P. Moore, L. Louis, M. Lundberg, C. and Martin, J. (eds.), *Reframing Organisational Culture*, Sage, California, 131-143

Mills, A. (1988) 'Organisation, gender and culture', *Organisation studies*, **9** (3), 351-370

Moore, D. R. and Dainty, A. R. J. (2001) 'Intra-team boundaries as inhibitors of performance improvement in the UK design and build projects: a call for change', *Construction Management and Economics*, **19**, 559-562

Morgan, G. (1980) 'Paradigms, Metaphors and Puzzle solving in organisation theory', *Administrative Science Quarterly*, 605-622

Morse, J. M. and Richards, L. (2002) *Readme First for a User's guide to Qualitative Methods*, Sage Publications, California

Nielsen, J. (2007) 'Struggles for health and safety in the Danish construction Industry', *International Journal of Occupational and Environmental Health*, **13** (1), 21-27

O'Reilly, C. A. Chatman, J. and Cadwell, D. F. (1991) 'People and organisational culture: A profile comparison approach to assessing person-organisation fit', *The Academy of Management Journal*, **34** (3), 487-516

Ouchi, W. G. (1981) *Theory Z: how American business can meet the Japanese challenge*, Addison-Wesley, Reading

Ouchi, W. G. and Wilkins, A. L. (1985) 'Organisational Culture', *Annual Review of Sociology*, **11**, 457-483

Parker, M. (2000) *Organisational culture and identity*, SAGE, London

Peters, T. J. and Waterman, R. H. (1982) *In search of excellence: Lessons from America's best run companies*, Hamper & Row, New York

Pheng, L. S. and Yuquan, S. (2002) 'An exploratory study of Hofstede's corss-cultural dimensions in construction projects', *Management Decision*, **40** (1), 7-15

Price, A. D. F. and Chahal, K. (2006) 'A strategic framework for change management', *Construction Management and Economics*, **24**, 237-251

Ochieng, E. G. and Price, A. D. F. (2010) 'Managing cross-cultural communication in multicultural construction project teams: The case of Kenya and UK', *International Journal of Project Management*, **28** (5), 449-460

Phua, F. T. T. and Rowlinson, S. (2003) 'Cultural difference as explanatory variable for adversarial attitudes in the construction industry: the case of Hong Kong', *Construction Management and Economics*, **21** (7), 777-785

Rahman, M. M. and Kumaraswamy, M. M. (2005) 'Assembling integrated project reams for joint risk management', *Construction Management and Economics*, **23**, 365-375

Rooke, J. Seymour, D. and Fellows, R. (2003) 'The claims culture: a taxonomy of attitudes in the construction industry', *Construction Management and Economics*, **21** (2), 167-174

Rooke, J. Seymour, D. and Fellows, R. (2004) 'Planning for claims: an ethnography of industry culture', *Construction Management and Economics*, **22** (6), 655-662

Root, D. (2005) 'Creating a culture of safety on construction sites', *Risk management*, **52** (11), 56-60

Rousseau, D. M. (1990) 'Assessing Organisational Culture: The Case for Multiple Methods', Goldstein, I. L. (Ed.), *Frontiers of Industrial and Organisational Psychology*, Jossey-Bass Inc, California

Ruikar, K. Anumba, C. J. and Carrillo, P. M. (2005) 'End-user perspective on use of project extranets in construction organisations', *Engineering, Construction and Architectural Management*, **12** (3), 325-343

Santos, F. M. and Eisenhardt, K. M. (2005) 'Organisational boundaries and theories of organization', *Organisation Science*, **16** (5), 491-508

Sathe, V. (1985) *Culture and related corporate realities: texts, cases and readings on organisational entry, establishment, and change,* Irwin, Homewood IL

Schein, E. H. (1984) 'Coming to a new awareness of organisational culture', *Sloan Management Review*, **25** (2), 3-16

Schein, E. H. (1990) 'Organisational Culture', *American Psychologist*, **45** (2), 109-119

Schein, E. H. (1993) 'Legitimating Clinical Research in the study of organisational culture', *Journal of counseling and Development*, **71** (6), 703-708

Schein, E. H. (1996) 'Three Cultures of management: The key to organisational learning', *Sloan Management Review*, **38** (1), 9-20

Schein, E. H. (2004) *Organisational culture and leadership*, Jossey-Bass, San Francisco

Schneider, B., Ed. (1990) *Organisational climate and culture*, Jossey-Bass, San Grancisco

Schultz, M. (1994) *On Studying Organisational Cultures: Diagnosis and understanding*, W De G, Berlin

Smircich, L. (1983a) 'Concepts of Cultures and Organisational Analysis', *Administrative Science Quarterly*, **28** (3), 339-358

Smircich, L. (1983b) 'Studying organisations as cultures', Morgan, G. (Ed.), *Beyond Method*, Sage, Beverly Hills

Smircich, L. (1985) 'Is the concept of culture a paradigm for understanding organisations and ourselves', Frost, P. J. More, L. F. Louis, M. R. Lundberg, C. C. and Martin, J. (eds.), *Organisational Culture*, Sage, California, 55-71

Smircich, L. and Calás, M. B., Eds. (1995) *Critical perspectives on organisation and management theory*, Dartmouth, Brookfield

Smithers, G. L. and Walker, D. H. T. (2000) 'The effect of the workplace on motivation and demotivation of construction professionals', *Construction Management and Economics*, 18, 833-841

Sveningasson, S. and Alvesson, M. (2003) 'Managing managerial identities: Organisational fragmentation, discourse and identity Struggle', *Human Relations*, **56** (10), 1163-1193

Toor, S-u-R and Ogunlana, S O (2008) 'Leadership skills and competencies for cross-cultural construction projects', *International Journal of Human Resources Development and Management,* **8** (3), 192-215

Turner, J R (2006) 'Towards a theory of project management: The nature of the project', *International Journal of Project Management*, **24** (1), 1-3

Turner, J R and Muller, R (2003) 'On the nature of the project as a temporary organisation', *International Journal of Project Management*, **21** (1), 1-8

Usdiken, B. Soxen, Z. and Enbiyaoglu, H. (1988) 'Strategies and boundaries: subcontracting in construction', *Strategic Management Journal*, **9** (6), 633-637

Vakola, M. and Wilson, I. E. (2004) 'The challenge of virtual organisation: critical success factors in dealing with constant change', *Team Performance Management*, **10** (5/6), 112-120

Van-Maanen, J. and Barley, S. R. (1985) 'Cultural Organisation. Fragments of a theory', Frost P. J. (Ed.), *Organisational Culture*, Sage, California, 31-53

Voordijk, H. Meijboon, B. and Haan, J. d. (2006) 'Modularity in supply chains: a multiple case study in the construction industry', *International Journal of Operations & Production Management*, **26** (6), 600-618

Wainwright, D. and Waring, T. (2004) 'Three domains for implementing integrated information systems: redressing the balance between technology, strategic and organisational analysis', *International Journal of Information Management*, **24** (4), 329-346

Wakefield, R. L. Leidner, D. E. and Garrison, G. (2008) 'Model of conflict, leadership, and performance in virtual teams', *Information Systems Research*, **19** (4), 434-59

Walker, A. and Wing, C. K. (1999) 'The relationship between constriction project management theory and transaction cost economics', *Engineering, Construction and Architectural Management*, **6** (2), 166-176

Walker, D. H. T. Peters, R. J. Hampson, K. D. and Thompson, M. J. (2001) 'Achieving a responsive industrial relations environment for construction industry workers: a project aliancing case study', *Construction Innovation*, **1**, 211-225

Wilkins, A. L. (1983) 'The Culture Audit: A tool for Understanding Organisations', *Organisational Dynamics*, **12** (2), 24-38

Wilkins, A. L. and Ouchi, W. G. (1983) 'Efficient Cultures: Exploring the relationship between culture and organisational performance', *Administrative Science Quarterly*, **28**, 468-481

Willmott, H. (1997) 'Management and organisation studies as science', *Organisation*, **4** (3), 309-344

Winch, G. M. (1989) 'The construction firm and the construction project: a transaction cost approach', *Construction Management and Economics*, **7**, 331-345

Winch, G. M. (2001) 'Governing the project process: a conceptual framework', *Construction Management and Economics*, **19**, 799-808

Winch, G. Ushani, A. and Edkins, A. (1998) 'Towards total project quality: a gap analysis approach', *Construction Management and Economics*, **16**, 193-207

Wong, J. Wong, P. N. K. and Heng, L. (2007) 'An investigation of leadership styles and relationship cultures of Chinese and expatriate managers in multinational construction companies in Hong Kong', *Construction Management and Economics*, **25** (1), 95-106

Yong, K. T. and Pheng, L. S. (2008) 'Organisational culture and TQM implementation in construction firms in Singapore', *Construction Management and Economics*, **26** (3), 237-248

Zaghloul, R. and Hartman, F. (2003) 'Construction contracts: the cost of mistrust', *International Journal of Project Management*, **21**, 419-424

Zhang, S. B. and Liu, A. M. M. (2006) 'Organisational culture profiles and construction enterprise in China', *Construction Management and Economics*, **24** (8), 817-828

Zou, P. X. W. Fang, D. Wang, S. Q. and Loosemore, M. (2007) 'An overview of the Chinese construction market and construction management practice', *Journal of Technology Management in China*, **2** (2), 163-176

Maturity and Maturity Models in Lean Construction

7

Claus Nesensohn, David Bryde, Edward Ochieng and Damian Fearon,
(Liverpool John Moores University, UK)

Abstract

In recent years there has been an increasing interest in maturity models in management-related disciplines; which reflects a growing recognition that becoming more mature and having a model to guide the route to maturity can help organisations in managing major transformational change. Lean Construction (LC) is an increasingly important improvement approach that organisations seek to embed. This study explores how to apply the maturity models to LC. Hence the attitudes, opinions and experiences of key industry informants with high levels of knowledge of LC were investigated. To achieve this, a review of maturity models was conducted, and data for the analysis was collected through a sequential process involving three methods. First a group interview with seven key informants. Second a follow up discussion with the same individuals to investigate some of the issues raised in more depth. Third an online discussion held via LinkedIn in which members shared their views on some of the results. Overall, we found that there is a lack of common understanding as to what maturity means in LC, though there is general agreement that the concept of maturity is a suitable one to reflect the path of evolution for LC within organisations.

Keywords: Lean Construction maturity, Maturity models, Lean enterprise self-assessment tool, Construction, Project management.

Introduction

Recent developments in the field of maturity models (MMs) have indicated the need and opportunity to develop MMs for the Lean approach. In 2010, Perkins *et al.* published a paper in which they described the development of a MM for Lean in the aerospace industry at the Massachusetts Institute of Technology (MIT). This specific assessment tool, the *Lean Enterprise Self-Assessment Tool* (LESAT), claims to provide an understanding of the current organisational maturity (Leanness) and guide the transformation towards a more lean enterprise. Research has consistently shown that several industries applied MMs to effectively implement a change strategy and to assess the state of maturity in terms of a discipline or strategy (Huat and Low 1994, Perkins et al. 2010). In another major study, Amaratunga et al. (2002) found recognition in the construction sector that MMs are potentially useful. Work in relation to MMs in construction includes the development of the Standardised Process Improvement for Construction Enterprises (SPICE) model (Sarshar et al. 1999). It is further claimed that such models support organisations through setting directions, prioritising improvement actions, and initiating cultural change (Pennypacker 2005). Hence MMs are a well-established concept within many industries and management disciplines.

One type of potentially radical organisational change being undertaken in parts of the construction sector is the implementation of Lean Construction (LC) principles (McGrath-Champ and Rosewarne 2009), – in whatever way LC is conceptualised and operationalised by individual companies (Sage et al. 2012). In order to effectively deliver projects using LC, organisations have to develop new ways of thinking and integrate elements of production management and project management into a holistic system for construction (Koskela and Ballard 2012). In addition it is argued that LC often needs changes to the organisational culture and its people (Green et al. 2008). According to Nesensohn et al. (2013), the transformation towards LC requires long-term and deep-rooted cultural change; to embed a Lean philosophy as an integral part of how the organisation operates. Despite the current challenges in embedding LC within the industry, it is fairly well understood and has many advocates within the industry and the academic community.

The current debate in LC is linking to performance measurements and non-financial KPIs to support the transformation of LC (Sarhan and Fox 2013). However, very few, if any, authors have been able to draw on any research into the opinions of key informants about how to apply the concept of 'maturity' to LC. Specifically, there is a need to understand whether the concept of MMs, widely adopted in many industries but less so in construction, with the notable exception of SPICE, has a useful role in relation to LC. Furthermore there have been calls in the literature for further work in terms of exploring synergies between MMs and LC, see Nesensohn et al. (2013b), and this paper responds to this call. In doing so it seeks to enhance understanding of MMs amongst key industry informants in a LC context. Moreover it adds to the body of knowledge by reporting on empirical study of practitioner reflections which can help inform and develop new theory.

Theoretical Framework

The past two decades has seen a rapid development of publication of articles related to maturity (Wendler 2012). To test this hypothesis a systematic keyword search process was carried out focused on the occurrence of the word 'maturity' in abstracts of business management and engineering journals from 1990 to 2013. Four databases were used, namely: Business Source Complete, Emerald, Scopus, and Discover (a research tool of the higher education institute (HEI) of the authors). Data of the actual number of articles per database for each year were collated and used to calculate the growth rate per database and year; this is illustrated in *Figure 1*.

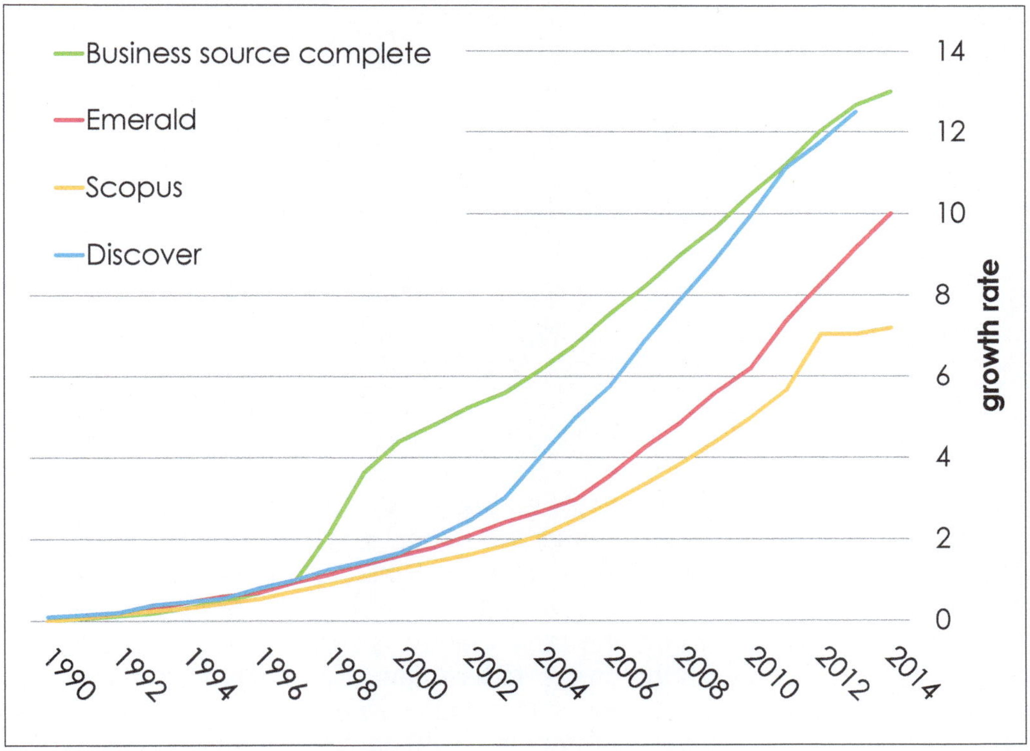

Figure1 Growth rate of maturity-related publications in a management context

The growth rate is calculated from a 1990 baseline. Hence Growth rate [x] = total number of publications [year]/ total number of publications [1990]. As shown in Figure 1 it is becoming increasingly difficult to ignore the increased attention given to the term *'maturity'* within business management and engineering journals since the 1990s. Growth rates ranging from 7 up to 12.7 emphasise the spurt in growth of papers with maturity-related topics.

A second search was carried out with a focus on four journals which are either recognised or encouraged by the International Council for Research and Innovation in Building and Construction (CIB). Again a search for the word maturity within the abstract and keywords of academic articles in each journal was undertaken and the growth rate for each year and journal was calculated. The results are shown in Figure 2. The pattern is similar to that in Figure 1 but there are some distinctions. The first journal that published articles with the occurrence of the word maturity within the abstract or keyword was the CME in 1994. This journal in particular has experienced a steady growth in pertinent articles over the period, with a growth rate between 1994 and 2013 of 16. The ECAM journal had a later focus on maturity-related topics, starting in 1998, but it saw a spurt of growth in 2008.

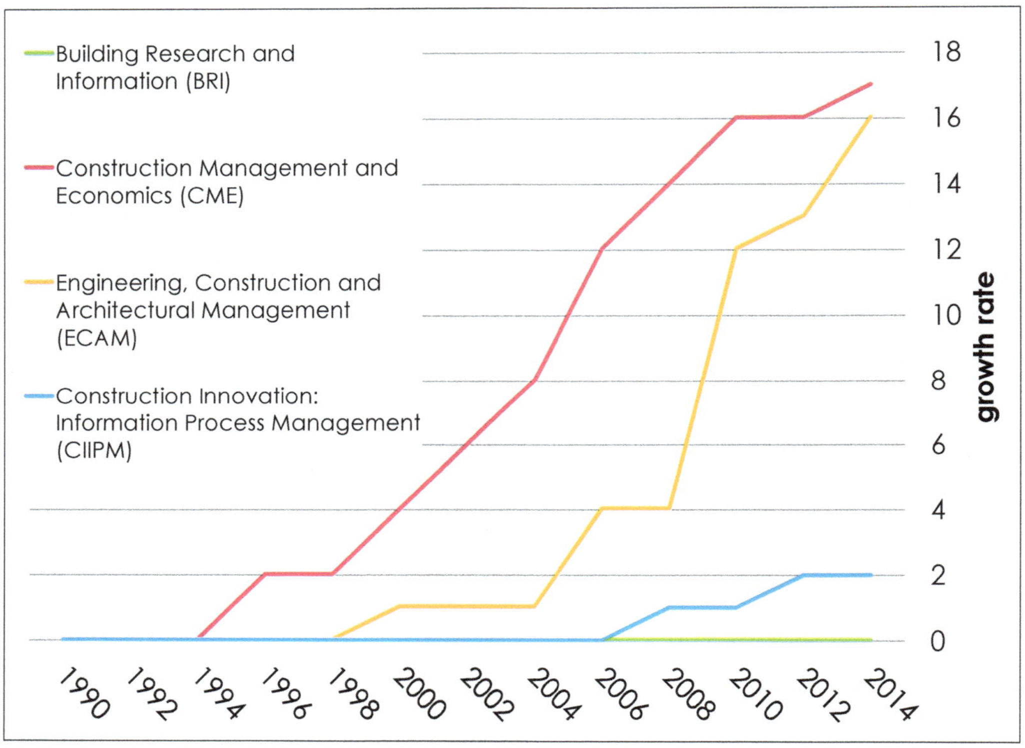

Figure 2 Growth rate of maturity-related publications in selected journals

Within a wider perspective we also calculated the growth for a number of relevant keywords in the construction-specific literature and the wider management literature. These two literature domains were distinguished by selecting the subject terms "construction" and "management" respectively in the Discover databases shown in Figure 1. The results are shown in Table 1. It is apparent from this table that the growth within the construction-related literature has not kept pace with that in the wider management literature. This corroborates findings from a recent study, which has shown that between 2009 and 2010 34 new MMs have been published predominately in general management journals (Wendler 2012). Yet many of the MMs now in the public domain are potentially relevant to construction contexts, with Eadie et al. (2012) identifying 53 particular MMs which are applicable.

Keyword	Construction-specific Literature			Wider Management Literature		
	No. articles in 1990	No. articles 2014	growth	No. articles in 1990	No. articles 2014	growth
Process maturity	1	94	93	5	1691	1686
Project maturity	0	145	145	3	785	782
Project management maturity	0	62	62	2	604	602
Cultural maturity	0	14	14	0	90	90
Organization maturity	0	60	60	3	1268	1265

Table1 Growth in the context of construction-specific literature and the wider management literature

'Process', 'Project Management' and 'Project' Maturity

One lens for viewing maturity through is in relation to the delineation of 'process', 'project management' and 'project' maturity. Process maturity describes the ability of delivering high performance in organisations (Hammer 2007). The concept of process maturity has its roots in the *Total Quality Management* (TQM) movement that identifies a mature process as an increasing level of performance, with high consistency in the operation of particular processes (Cooke-Davies et al. 2001). Process maturity is also pertinent to project contexts through the notion of continuously performed processes by the project organisation (Amaratunga et al. 2002, Siriwardena et al. 2005). Lockamy III and McCormack (2004), further showed that process maturity is measurable and can be illustrated through a defined lifecycle. Process maturity is therefore the extent to which an organisation is able to explicitly define, manage, measure and control a specific process (Lockamy III and McCormack 2004, Paulk et al. 1993a). The concept of a measurable lifecycle as process maturity, has been also developed and tested relative to the project management processes (Ibbs and Kwak 2000).

Project management maturity derives its understanding of '*maturity*' through the re-contextualising of the concept of organisational maturity (in software development processes) to project management (Cooke-Davies and Arzymanow 2003, Ibbs and Kwak 2000, Maier et al. 2012). The logic behind this is that a project needs a mature organisation and its support to be successful and effective (Richardson 2010, pp. 427). In this sense, a linked concept: organisational maturity, can be described as the application of maturity to an organisation, which is then, in relation to the degree of maturity, perfectly conditioned to achieve its objectives (Andersen and Jessen 2003). This also includes the extent to which the organisation has explicitly and consistently deployed their processes (Van Looy et al. 2011). Project maturity reflects the ability of the organisation to manage different types of projects (Phungula 2008). So maturity in a project organisation relates to how effective and efficient they are in achieving their project objectives (Andersen and Jessen 2003). Ibbs and Kwak (2000), illustrated in their study; project maturity through actions which are conceptualised by what organisations and people do. This notion can be articulated in terms of the receptiveness of an organisation (Skulmoski 2001); and furthermore, an organisation which combines operational action with competence, skills and attitudes (Andersen and Jessen 2003).

'Immature' v 'Mature' Organisations

Another useful lens through which to view the concept of maturity in management terms is to consider what distinguishes an immature organisation from a mature one. Immature organisations have no objective foundation to assess quality or to solve product or process problems (Paulk et al. 1993a, Sarshar et al. 2000). Further they are characterised by achieving less goals than planned; having poor quality assurance and higher actual costs compared to the estimated costs (Harmon 2004). However, it is important to note that immature organisations may occasionally deliver excellent results (Humphrey 1989) through the heroic efforts of individuals and by using approaches in a spontaneous and unplanned manner (Harmon 2004, Sarshar et al. 2000). Therefore project managers of immature organisations mainly focus on solving immediate issues and by constantly engaging in fire-fighting (OGC 2010).

By way of contrast, mature organisations are characterised by creating schedules for projects and keeping projects regularly on track (Harmon 2004). They have automatic behaviours, systematic processes, clear defined roles and responsibilities, clear ways of doing things and constantly achieve their planned goals with minor deviation (Harmon 2004, OGC 2010, Paulk et al. 1993a). In addition, they have a higher ability to attract and develop individual talents (resources) which perform consistently in their way of working as expected from their position (Curtis et al. 2001, Curtis et al. 2009). Hence project managers of mature organisations continuously monitor their capability, quality and customer satisfaction (OGC 2010, Paulk et al. 1993a). As an organisation matures it institutionalises its behaviour in such areas as corporate culture, practises, procedures, standards and structures (Paulk et al. 1993a).

Maturity Models (MMs)

It has been claimed that MMs are useful for organisations when they want to measure the current organisational capability or to implement a change or improvement strategy in an organised way, such as to transform their enterprise (OGC 2010). Hence MMs describe the evolution, over a defined period of time, of an organisation of people, technology, products and processes (Tapia et al. 2008, Wang et al. 2011). The most significant published MMs for the purpose of this study are shown in Table 2 below.

Research Method

The literature review above established the need to understand conceptions of MMs. Hence in order to make an empirical contribution to the body of knowledge of both MMs and LC the study explored and analysed perceptions of maturity in a LC context. The data for the analysis was collected through a sequential process involving three methods. First a group interview with seven key informants on LC, i.e. those working in the academic community and as practitioners was undertaken. The group were brought together under the auspices of the International Group of Lean Construction (IGLC). The IGLC is a well-known international network of professionals and researchers in architecture, engineering, and construction and the seven interviewees are all prominent within the IGLC. Hence it was a conscious decision to interview this group of people from this forum. The key informants were from three countries: US, UK and Brazil; with the individual breakdown being P#1 (USA), P#2 (UK), P#3 (USA), P#4 (USA), P#5 (Brazil), P#6 (USA), and P#7 (USA). They were all Panel Members at the IGLC. As such they are recognised as experts in the field with extensive experience of contributing to the theory and practice of LC.

A follow up discussion with the same key informants via email was undertaken to investigate some of the issues raised in the group interview in more depth. An online discussion held via LinkedIn was made available for all members of the Lean Construction Network group and the Lean Construction Institute group. In total 19 senior participants based in UK, USA, Australia, Russia, Chile and Norway participated in this study and shared their perception on some results of the group interview. The purpose of this online discussion was to validate the findings of stage 2 and 3 to the wider LC community.

The rational to use a group interview in the first stage, was the lack of construction key informants with experience in LC and MM. Hence the participants had to be introduced to

MM at the beginning of the group interview. The presentation of the concept of MM in a LC context stimulated a group dynamic on a basic shared understanding of MM. The initial group interview undertaken lasted approximately one hour. Contemporaneous notes were taken of the comments of the key informants as a basis for the construction of detailed notes afterwards. The construction of more detailed notes took place during the follow-up discussion. To clarify the veracity of the taken notes, each of the seven key informants were sent a copy after the discussion. All interviewees confirmed that they were an accurate reflection of the group interview.

Model	Developer	Description	Relevance
Capability Maturity Model Integrated (CMMI)	Software Engineering Institute (SEI) at Carnegie Mellon University (Paulk et al. 1993b)	Consists of 5 maturity levels, key process areas, best practices and goals which are characterised by activities and a way of behaving within the organisation.	This model provides the most common and widely adopted structure amongst MMs.
Organizational Project Management Maturity Model (OPM3)	Project Management Institute (PMI, US) (Project Management Institute 2003)	This model integrates the PMI Body of Knowledge with organizational maturity and addresses portfolio, program and project management. It consists of a list of over 600 best practices and 4 stages (maturity levels).	Shows a MM on the basis of an existing Body of Knowledge which uses extensive quantitative data to determine the current maturity.
Berkley Project Management Process Maturity Model	Dr. Williams Ibbs, University of California, Berkley (US) (Kwak and Ibbs 2000, Kwak and Ibbs 2002)	This model builds on the CMMI structure and industry best practices in PM. It consists of 5 maturity levels, key PM processes, organisation characteristics and focus areas.	This MM illustrates a practical adoption with several modifications of the CMMI structure to PM.
Portfolio, Program and Project Management Maturity Model (P3M3)	Office of Government Commerce (OGC, UK) (OGC 2010)	The P3M3 combines portfolio, program and project management, but it allows the independent use of each or more than one model. It consists of 5 maturity levels, process perspectives and specific attributes and generic attributes.	Again this MM illustrates a different practical adoption of the CMMI structure to PM.
Standardized Process Improvement for Construction Enterprises (SPICE)	Centre for Information Technology in Construction (CITC), University of Salford (UK) (Sarshar et al. 1999)	This MM is a CMMI adaption to the construction industry. It comprises 5 maturity levels, key processes, and process enablers.	The SPICE is a re-contextualization of the CMMI to construction processes which shows a unique and detailed adoption of the CMMI to the construction industry.

Table 2 Significant MMs for the purposes of this study

The data from this group interview and the follow up discussion were then analysed using the framework method©; a widely used method for qualitative data analysis developed from the National Centre for Social Research (NatCen) in the 1980s (NatCen Social Research 1999). This approach has been used in previous construction management research (Baiden et al. 2006, Whittock 2002), and is derived from the thematic framework, which provides the central element of the method (Ritchie et al. 2003). The method provides a systematic and comprehensive thematic approach to ordering and synthesising the data against key issues and themes which are developed from the data (Ritchie and Spencer 2002). The framework consists of the generation of charts or matrices (often as a DIN A3 page) which organise the data in the form of distilled key topics in cells for each participant which form the subthemes of the thematic framework (Whittock 2002). The framework method© comprises of six key steps (Ritchie and Spencer 2002), and its deployment is described as follows.

1. *Familiarisation* - by the immersion into the textual data through reading and reviewing the notes a number of times.
2. *Identify descriptive categories* - during the familiarisation process the thematic framework (descriptive category) is identified through recognising and note taking of important themes and sub-themes. Hence it is intended to aid an understanding of the meaning of the data through the development of the thematic framework. The output of these activities is the development of a summary theme and sub-themes from the data that capture key themes which are judged to be important to understand the conceptions of maturity and towards MMs in a LC context amongst the participants.
3. *Indexing* - not applicable as pilot charting (step 4) was carried out instead.
4. *Pilot charting* - to test whether the developed thematic framework was capable of holding all the collected data a pilot charting of examples of the textual data mapped against the framework was carried out. After a successful pilot charting a thematic framework which is appropriate to organise all the data can be derived. In this study it consisted of two sub-themes:

   Theme 1 Perception of maturity in Lean construction

   Sub-theme 1.1 Conceptions of maturity in a LC context

   Sub-theme 1.2 Conceptions towards maturity models in a LC context.
5. *Charting* - this process summarises the key points of the data and provides a distilled summary of meaning organised in the framework matrix (Ritchie et al. 2003). A DIN A3 matrix with both subthemes on the top of the matrix and the seven participants on the left side were created. Judgement is important in this process to find the right balance between summarising and condensing the data in a way that retains enough contexts and essence of the comments being made i.e. not relying on verbatim parts of textual data (ibid.).
6. *Investigation and interpretation* - The last step of the framework method© involves the interpretation of the data as a whole in order to define concepts, map the range and nature of the phenomena, create typologies, find associations and provide explanations of the findings (Ritchie and Spencer 2002). To facilitate this process the tool Mind Maps® was utilised, which is widely used within operational research and other management-related disciplines (Vidal 2006).

The philosophical consideration for this research can be viewed from two perspectives. The first was linked to essential requirement of investigating maturity models and LC. The second was linked to the need to select a research strategy that enabled the dissemination of MMs data from the group interview participants. The findings are presented below under the two sub-themes drawn from the analysis.

Findings and Discussion

The findings of the group interview revealed some scepticism towards MMs in a LC context. For instance questioned participant #4 (P#4) the existence of MM-driven improvement; stating that he '*never saw a successful MM*'. Another perception was that MMs could lead to reducing LC to a toolbox consisting of certain tools and practices. P #2 added to this that '*existing MMs have been criticised*', *with* the SPICE model gaining limited traction in the industry. Another sceptic stated by P #6 was that MMs could *'tranquillise'* people, leading to complacency and a lack of focus. There are several possible explanations for such scepticism. For example we find in the PM literature on MMs i.e. OPM3 and Berkley's (PM^2) the deconstruction of PM into certain best practices derived from e. g.: PMBOK®. In comparison to the domain of PM there is no combined body of knowledge captured in a book within LC. Further it is not surprising that a concept which is discussed in many industries is not without criticism. However, much of the literature on MMs criticises the flexibility of MMs in particular if organisations undertake unusual projects at the operational level (Kujala and Artto 2000). Furthermore it is argued that 'the main weakness of many MMs is the lack of a focus on the strategic management of project-based organisations' (Kujala and Artto 2000, pp. 47). Nevertheless some participants associated MMs as valuable. So was stated by P #7 that *'a MM will show the path of the LC evolution'*. Another key informant pointed out that they were not saying that *'MMs have no value'* whilst pointing further to Crosby's QMMG (Crosby 1979) as something that delivered positive and general boundaries within MMs.

The specific findings of both sub-themes are further presented and discussed in the next section.

Sub-theme 1: Conceptions of maturity in the context of LC

Four major topics regarding this first sub-theme were derived. As shown in *Figure 3* the topics are: (1) maturity is perceived as an *endpoint*, (2) maturity relates to a *set of tools* to measure it, (3) the term maturity could be replaced by a *substitute term* and (4) maturity in LC should be linked to the notion of *understanding and philosophy*. Figure 3 also shows example quotes from the participants in each of the topic areas i.e. P#4 is participant No.4. The next section discusses each of these four topic areas.

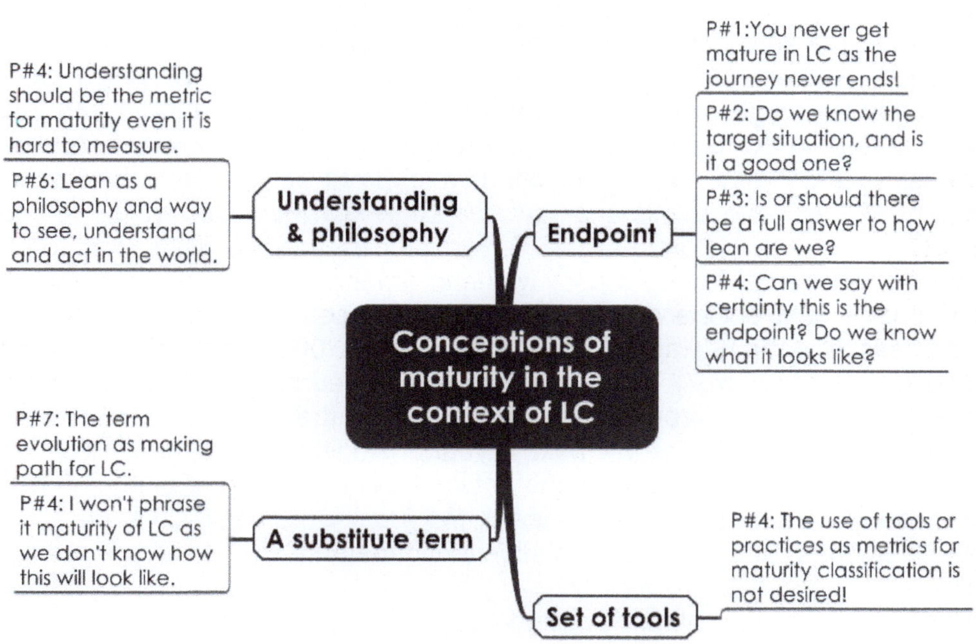

Figure 3 Conceptions of maturity

Topic 1: Endpoint

The first major topic 'endpoint' illustrates that the P #1, #2, #3 and #4 links the notion of maturity with a defined endpoint by using phrases like *'end of the journey'*, *'the target'*, *'target situation'* and *'full answer'*. This raises the issue of the usefulness of knowing or having such an endpoint in mind. Not only do these conceptions of the key informants indicating a potential lack of knowledge but also on answering the question: what is 'mature' in a LC context? Answering this introduces corollary questions: Is it useful to know what maturity in LC is? If yes, what does LC maturity look like? How would an organisation become more mature in LC? Is there really an identifiable endpoint in LC maturity? How Lean are we? Considering the large body of published work on maturity and maturity-related concepts, investigating the meaning of maturity in LC and trying to answer the above questions could potentially reveal new insights into both the field of MM and LC. Additionally there are already concepts of Lean maturity in other industries, such as aerospace, which are usefully applied to practice. However, the interview findings reveal a significant gap in knowledge as to what maturity in LC looks like. This is despite the fact that there has been work undertaken regarding the maturity in LC from industry policy makers such as the European Construction Institute (ECI) and the Highway Agency (HA) in UK. The latter of these two organisations re-contextualised the LESAT to a Highway Agency Lean Maturity Assessment Toolkit (HALMAT), with the aim being to determine the Leanness of the supply chain in relation to the objectives of the HA (Highways Agency 2010).

Given the emphasis by authors such as Caffyn (1999) on the importance of approaches to assess the organisational transformation and further consider the fact that implementing LC is a transformation (also often stated as journey) of an organisation (Mossman 2009, Sage et al. 2012, Zimina et al. 2012), the transformation can be seen as the movement towards maturity. However, as maturity is conceptualised as the movement towards an ideal point (Hogan and Roberts 2004) and an organisation is mature when it is perfectly conditioned to reach its objectives and goals continuously, in the real world it would be impossible to find a fully mature organisation (Andersen and Jessen 2003). This suggests that, contrary to the view of some of the participants, being mature is not a tangible end-point. Rather, it is more appropriate perhaps to illustrate the '*endpoint*' of being mature in LC as an idealised vision that one is constantly striving to achieve. This vision is similar then to the one used by what is arguably one of the most mature companies in terms of Lean: Toyota. Toyota strive to realise their vision of zero defects, 100 per cent value with the lowest costs and continuous flow production (Rother 2010). The vision of an ideal endpoint provides a clear direction towards maturity, although it is never actually reached (Andersen and Jessen 2003, Rother 2010) Being mature in LC can be seen as interchangeable with a vision or ideal as a goal of a Lean transformation within an organisation which is to be pursued (Zimina et al. 2012). The comment of P #1 reflects this analogy: *'the Lean journey never ends and you don't get mature in LC'*.

Another analogy is appropriate and that is with the concept of human maturity. The human maturity process is a circle without an end; regardless of where we go, we will get new experiences (Sams 1999, p. 7). So humans never stop learning and becoming more mature as they interact with their environment. As articulated by the P #2; *'LC is still developing and could involve something else in the next 5 years!'* Additional comments from P #2 made explicit the link between the concept of maturity in business and that of human maturity: *'We as humans recognise that the more we learn, the more we discover that we have to learn'*. So learning is an important process in relation to LC maturity.

The supporting data in the online discussion confirmed the perception of maturity as a final state or goal. With the additional consensus view that the term 'maturation' is appropriate to use in this context as one that describes the process of maturing towards a goal or vision. Yet all these views do not align with the current literature on the topic. The online discussion revealed further doubts as to whether there is currently a completely satisfactory answer to the question *'how mature are we?'*; which provides added impetus for the development of a MM.

Topic 2: Set of tools

The second major topic derived was that maturity in LC can be measured by a set of tools. This can be related to published PM MMs, such us the OPM3 and the Berkley (PM^2), where maturity is measured with best practices and tools from the PMBOK® (Cooke-Davies 2004). Nevertheless the counter-argument that the metrics for LC maturity should be based on a deeper understanding of the topic, rather than just the tools and practices, is a moot point which ought not to be overlooked. It must be recognised that there is a need for research to explore which metrics are more or less appropriate for LC and how LC can be conceptualised through these metrics.

Topic 3: A substitute term

The third topic, the use of a substitute term was neatly articulated by P #4, who stated: *'I wouldn't phrase it 'maturity' of LC'*. The theme was taken further by P #7, who suggested using the term 'evolution'. Taking into account the existing terminology of Lean 'maturity' in an aerospace context (LESAT), the work of the policy makers regarding LC 'maturity' (HALMAT-UK Highway Agency and LCTF of the ECI) and the increased volume of work using the term 'maturity' in the academic literature raises questions as to the desirability of adopting any substitute term for the domain of LC.

The appropriateness of the term '*maturity*' was also confirmed by a large part of the supporting data. Eleven out of the 19 individuals who participated in the online discussion regarded it as an appropriate term for LC – though 3 individuals were in agreement with the key informants in questioning the use of the term. Its appropriateness as a term was demonstrated by comments like: *'the concept of maturity is a commonly used in management terms'* and *'[the term maturity was] especially widely used in UK within management processes or Lean self-assessment kits*. It was also stressed that the term maturity has to be used *'in the interest of comprehension'* and that *'maturity is fully applicable to LC'*.

Topic 4: Understanding and philosophy

The final topic relates to the level of understanding of LC and whether a LC philosophy is followed; though in the words of P #4 this is something that is *'hard to measure'*.

An interesting attempt towards understanding LC maturity is evident from the supporting data derived from the LinkedIn discussions. Several ideas were put forward in terms of possible relationships with maturity i.e. it could be related to; experience, open considerations, concise decision-making, change in culture, efficiency, knowledge, productivity, learning, a state of mind, of a certain attitude or approach to business. An observation from one of the LinkedIn discussions that is worthy of further consideration was that, perhaps *'the start of LC being integrated with the existing project delivery methods is in itself a sign of maturity'*.

Sub-theme 2: Conceptions towards Maturity Models in the context of LC

The second sub-theme in the thematic framework was also broken down into four major topics, as shown in *Figure 4*. These are: (1) MMs are a kind of target situation; (2) a MM for LC requires a distinctive focus; (3) MMs involve a standard structure; and (4) MMs show the next level. The next section discusses each of these in turn.

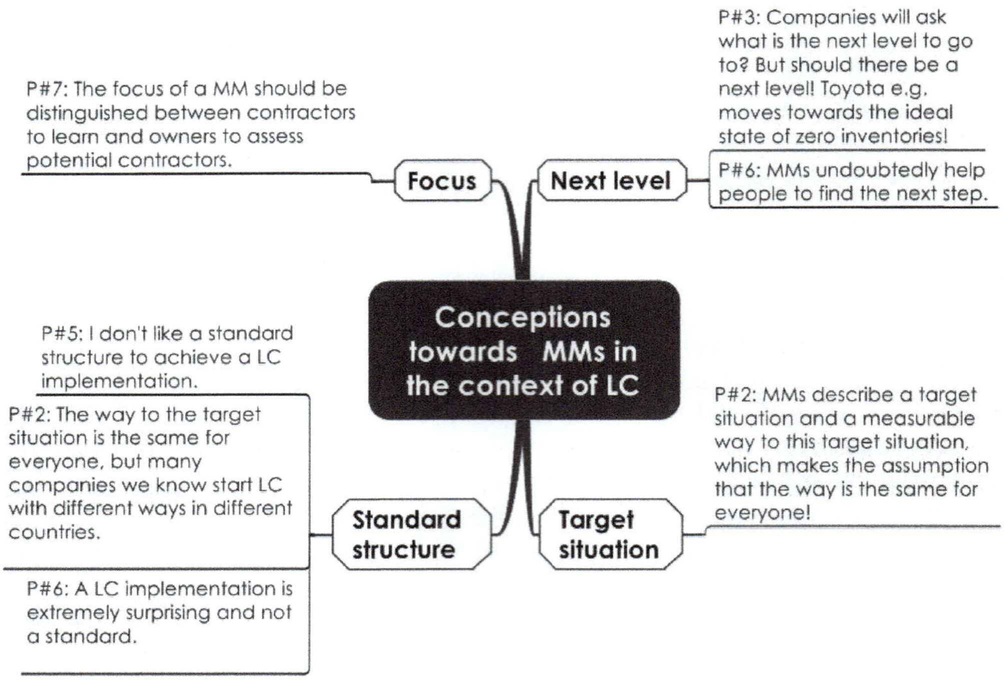

Figure 4 Concepts of maturity models

Topic 1: Target situation

The first area of discussion sees MMs as *'a kind of target situation and a way to that, which will be measured through the model'* (P #2). It was further implied that the route to this target situation in a MM is the same for everyone. This differs to the extant literature, for instance the UK Government report (OGC 2010), which argues that a MM supports organisations to implement their own unique change or improvement strategy in a managed way. Furthermore, MMs like the CMMI or the OPM3 does not prescribe how to improve specific areas, rather they point out the priority for improvement actions which may be different for each individual (CMMI Product Team 2010, Project Management Institute 2003).

Topic 2: Focus

One aspect of the data concerned what would be the focus of a possible MM in the context of LC. For example it was argued by P #7 that *'the focus of a MM within LC should be distinguished'.* The sense here is that it must be clear if the focus is on the clients – using it as a form of assessment tool to evaluate contractors, or used by contractors as an improvement tool within their organisations. It is worth noting in this respect that several MMs in other industries are used in both situations.

Topic 3: Standard structure

The attitude that MMs would deliver an undesired standardised structure for a LC implementation was articulated by P#5 who said *'I don't like a standard structure to achieve a LC implementation'.* This view was countered by the experiences of P #2 and #6 regarding the implementation process of LC. P #2 stated that *'many companies start with Last Planner® and companies in UK or Brazil start with different things'* and P #6 concurred; pointing out that they found the *'implementation of [LC] extremely varied and not a standard* However, it is worth considering more closely what kind of standard an MM would establish. MMs provide a common and shared language and understanding (Klimko 2001) i.e. in PM. Although somewhat dependent on the specific MM, in general they do not tell the organisation how and with what standard they should improve but, rather, point out where to improve; because most MMs are non-prescriptive.

Topic 4: Next level

Perceptions about a progression measured in maturity levels were made in the context of the Last Planner® System; with P #3 illustrating the point by stating: '*I know companies will ask what the next level to go to is, but should there be a next level! Since Toyota e.g. moves towards the ideal state of zero inventories.*' Hence a conception is that practitioners are interested in further steps to take within a LC transformation; however, a moot point is whether there is a next step which is the same for all? This links to the earlier discussion of a target situation (Topic: 1). Emphasising the power of MMs it was also stated (P #6) that MMs would undoubtedly help people to identify the next step. However, the scepticism about the ability to clearly identify boundaries, discussed in the previous section, stress the crucial task of defining maturity levels.

Conclusions

This paper has given an account of and the reasons for the widespread use of MMs in several industries. This study set out to determine the role of MMs in relation to LC, and to explore what we need to understand about maturity, and the conceptions of those closely involved in the construction industry about MMs in a LC context. A number of noteworthy issues have been identified that have not been discussed in the Lean literature. It has been shown that perceptions regarding the term 'maturity' of the key informants are diverse and could lead to potential misconceptions and concerns. Yet the findings suggest that maturity in LC can be generally perceived as the path of LC evolution. It was also revealed that metrics such as tools are regarded as inappropriate for assessing maturity in LC. It was further shown that the participants are rather sceptical about using the concept of maturity for LC. It is worth highlighting that, these perceptions are at odds with the literature in other management domains such as PM, and process management. Hence it is most likely that a lack of engagement with the concept of maturity by the key informants and the limited attention of MMs within the LC literature are the underlying reasons for such a finding. Despite the concerns and misconceptions about the term maturity and MMs there is certainly a growing body of literature on the subject of maturity; confirming that it is indeed a topic to which the research community could usefully paying increased attention to.

The results of this study indicate that there are no major constraints on MMs being contextualized to LC. Hence, there is the potential through a synergy of the concepts of MM and LC to deliver similar benefits as those reported when MMs are used in other management domains, for instance in providing effective support of ongoing transformations in organisations (Perkins et al. 2010). However as revealed in this study, it seems that the development of any MM for LC requires appropriate metrics which are coherent with Lean's underpinning philosophy, and which capture the behaviour, understanding and the outcomes of a typical organisation. The results also highlight that being mature is not an endpoint which will be achieved by everyone; rather it is a desired target that is forever moving.

It is necessary to note a limitation of this study, which is that the respondents might have a particular view on this investigation through their connection to the IGLC and other groups. An issue that was not addressed in this study was whether MM professionals are included in the sample. Therefore a larger and more random sample comprising of group discussion with key informants from both sides is suggested to generalise and validate the findings reported in this paper.

Recommendations for Further Research

As the concept of maturity has the potential to make a significant contribution to helping an organisation direct its change initiatives, further research on maturity in the context of LC is required. Linked to this is research into the appropriate metrics of LC maturity. Another avenue for further research is on whether the focus of LC MMs should be on the customers and clients, so that they can assess their contractor's maturity, or on the contractors and projects, in order to drive and prioritise internal improvement. Finally more work will need to be done to determine the relevance of the difference between MMs for LC within permanent operations-focused organisations and temporary (project) organisations.

References

Amaratunga, D., M., S. and D., B. (2002) 'Process improvement in facilities management: the SPICE approach', *Business Process Management Journa*, **8** (4), 318-37.

Andersen, E.S. and Jessen, S.A. (2003) 'Project maturity in organisations', *International Journal of Project Management*, **21** (6), 457-61.

Baiden, B.K., Price, A.D.F. and Dainty, A.R.J. (2006) 'The extent of team integration within construction projects', *International Journal of Project Management*, **24** (1), 13-23.

Caffyn, S. (1999) 'Development of a continuous improvement self-assessment tool', *International Journal of Operations & Production Management*, **19** (11), 1138-53.

CMMI Product Team (2010) *CMMI for services, version 1.3,* CMU/SEI-2010-TR-034, Pittsburgh: Software Engineering Institute.

Cooke-Davies, T.J. (2004) 'Measurement of organizational maturity', *In:* Slevin, D P, Cleland, D I and Pinto, J K (Ed.), *Innovations-Project Management Research 2004*, Newtown Square: PMI, 211-28.

Cooke-Davies, T.J. and Arzymanow, A. (2003) 'The maturity of project management in different industries: An investigation into variations between project management models', *International Journal of Project Management*, **21** (6), 471-8.

Cooke-Davies, T.J., Schlichter, J. and Bredillet, C. (2001) 'Beyond the PMBOK guide', In: Project Management Institute (Ed.), *Annual Project Management Institute Seminar & Symposium*, 7-10 November, Nashvill, USA. PMI, Vol. 32th.

Crosby, P.B. (1979) *Quality is free : The art of making quality certain*, New York: McGraw-Hill.

Curtis, B., Hefley, W.E. and Miller, S.A. (2001) *People Capability Maturity Model (P-CMM)*, Pittsburgh: Software Engineering Institute.

Curtis, B., Hefley, B. and Miller, S. (2009) *People Capability Maturity Model (P-CMM) Version 2.0*, Pittsburgh: Software Engineering Institute.

Eadie, R., Perera, S. and Heaney, G. (2012) 'Capturing maturity of ICT applications in construction processes', *Journal of Financial Management of Property & Construction*, **17** (2), 176-94.

Green, S.D., Harty, C., Elmualim, A.A., Larsen, G.D. and Kao, C.C. (2008) 'On the discourse of construction competitiveness', *Building Research & Information*, **36** (5), 426-35.

Hammer, M. (2007) 'The process audit', *Harvard Business Review*, **85** (1), 11-123.

Harmon, P. (2004) 'Evaluating an organisation's business process maturity', *Business Process Trends*, **2** (3), 1-11.

Highways Agency (2010) *Highways Agency Lean Maturity Assessment Toolkit (HALMAT)*, PR 145/10, Birmingham, UK: Highways Agency Publications.

Hogan, R. and Roberts, B.W. (2004) 'A socioanalytic model of maturity', *Journal of Career Assessment*, **12** (2), 207-17.

Huat, T.P. and Low, S.P. (1994) 'The fuzzy industry maturity grid (FIMG) and its application to the Singapore construction industry', *Construction Management and Economics*, **12** (2), 125-38.

Humphrey, W.S. (1989) *Managing the software process*, Boston: Addison-Wesley.

Ibbs, C.W. and Kwak, Y.H. (2000) 'Assessing project management maturity', *Project Management Journal*, **31** (1), 32-43.

Klimko, G. (Year) 'Knowledge management and maturity models: Building common understanding', *In:* Remenyi, D (Ed.), *Proceedings of the 2nd European Conference on Knowledge Management (ECKM)*, Bled, Slovenia. Bled School of Management, 269-78.

Koskela, L. and Ballard, G. (2012) 'Is production outside management?', *Building Research & Information*, **40** (6), 724-37.

Kujala, J. and Artto, K. (2000) 'Criteria for project performance in business context', *International Project Management Journal*, **6** (1), 46-53.

Kwak, Y.H. and Ibbs, C.W. (Year) 'The Berkeley project management process maturity model: Measuring the value of project management', In *Proceedings of the 2000 IEEE Engineering Management Society*, 13-15 August, Albuquerque, NM, USA. IEEE1-5.

Kwak, Y.H. and Ibbs, C.W. (2002) 'Project management process maturity (PM) model', *Journal of Management in Engineering*, **18** (1), 150-5.

Lockamy III, A. and McCormack, K. (2004) 'The development of a supply chain management process maturity model using the concepts of business process orientation', *Supply Chain Management: An International Journal*, **9** (4), 272-8.

Maier, A.M., Moultrie, J. and Clarkson, P.J. (2012) 'Assessing organizational capabilities: Reviewing and guiding the development of maturity grids', *IEEE Transactions on Engineering Management*, **59** (1), 138-59.

McGrath-Champ, S. and Rosewarne, S. (2009) 'Organizational change in Australian building and construction: Rethinking a unilinear 'leaning' discourse', *Construction Management and Economics*, **27** (11), 1111-28.

Mossman, A. (2009) 'Why isnt't the UK construction industiry going Lean with gusto?', *Lean Construction Journal*, **5** (1), 24-36.

NatCen Social Research (cited 2013) *The Framework method for qualitative data analysis*,

Nesensohn, C., Demir, S.T. and Bryde, D.J. (2013) 'Developing the True North route map as a navigational compass in a construction project management organisation', *Lean Construction Journal*, 2013 (1), 1-18.

OGC, O.o.G.C. (2010) 'Portfolio, programme and project management maturity model (P3M3®): Introduction and guide to P3M3®', In: Office of Government Commerce, Ed., Norwich, UK: Cabinet Office of Rosebery Court, 1-17.

Paulk, M.C., Bill, C., Chrissis, M.B. and Weber, C.V. (1993a) *Capability maturity model for software* 0471028959, Pittsburgh: Software Engineering Institute.

Paulk, M.C., Curtis, B., Chrissis, M.B. and Weber, C.V. (1993b) 'Capability maturity model, version 1.1', *Software, IEEE*, **10** (4), 18-27.

Pennypacker, J.S. (2005) *Project portfolio management maturity model*, Pennsylvania, USA: Center for Business Practices.

Perkins, L.N., Abdimomunova, L., Valerdi, R., Shields, T. and Nightingale, D. (2010) 'Insights from enterprise assessment: How to analyze LESAT results for enterprise transformation', *Information Knowledge Systems Management*, **9** (3/4), 153-74.

Phungula, M.G.P.G. (2008) *Review and Analysis of Organisational Project Management Maturity Of The South African Government Departments involved in Public Private Partnership (PPP) Projects*, University of South Africa.

Project Management Institute (2003) *Organizational Project Management Maturity Model OPM3 Knowledge Foundation*, 2nd ed,ed. Vol. second edition, Pennsylvania, USA: PMI.

Richardson, G.L. (2010) *Project management theory and practice*, Boston: Auerbach Pub./CRC Press.

Ritchie, J. and Spencer, L. (2002) 'Qualitative data analysis for applied policy research', In: Bryman, A and Burgess, R G (Ed.), *Analzing qualitative data*, London: Routledge, 173-94.

Ritchie, J., Spencer, L. and O'Connor, W. (2003) 'Carrying out qualitative analysis, In: Ritchie, J and Lewis, J (Ed.), *Qualitative research practice: A guide for social science students and researchers*, London: Sage, 219-62.

Rother, M. (2010) *TOYOTA KATA: Management people for improvement, adaptiveness, and superior results*, New York: McGraw-Hill.

Sage, D., Dainty, A. and Brookes, N. (2012) 'A 'strategy-as-practice' exploration of Lean construction strategizing', *Building Research & Information*, **40** (2), 221-30.

Sams, J. (1999) *Dancing the dream: The seven sacred paths of human transformation*, Vol. 1, San Francisco: HarperCollins.

Sarhan, S. and Fox, A. (2013) 'Performance measurement in the UK construction industry and its role in supporting the application of lean construction concepts', *Australasian Journal of Construction Economics and Building*, **13** (1), 23-35.

Sarshar, M., Haigh, R., Finnemore, M., Aouad, G., Barrett, P., Baldry, D. and Sexton, M. (1999) 'SPICE: Is a capability maturity model applicable in the construction industry', In: Lacasse, M A and Vanier, D J (Ed.), *International Conference on Durability of Building Materials and Components*, 30 May, Vancouver, Canada. National Research Council Canada, Vol. 8th, 2836-43.

Sarshar, M., Haigh, R., Finnemore, M., Aouad, G., Barrett, P., Baldry, D. and Sexton, M. (2000) 'SPICE: A business process diagnostics tool for construction projects', *Engineering Construction and Architectural Management*", **7** (3), 241-50.

Siriwardena, M.L., Kagioglou, M., Jeong, K.S., Haigh, R. and Amaratunga, D. (Year) 'SPICE 3: Facilitating organisational process improvement through good practice sharing', In *Salford Centre for Research and Innovation (SCRI) 2nd International Symposium*, University of Salford, Greater Manchester. University of Salford Manchester, 434-48.

Skulmoski, G. (2001) 'Project maturity and competence interface', *Cost Engineering*, **43** (6), 11-8.

Tapia, R.S., Daneva, M., van Eck, P. and Wieringa, R. (Year) 'Towards a business-IT aligned maturity model for collaborative networked organizations', In "*12th Enterprise Distributed Object Computing Conference Workshops*, 16-16 September, 276-87.

Van Looy, A., De Backer, M. and Poels, G. (2011) 'Defining business process maturity. A journey towards excellence', *Total Quality Management & Business Excellence*, **22** (11), 1119-37.

Vidal, R.V.V. (2006) 'Chuck Frey, power tips and strategies for mind mapping software, learn how to think better, improve your productivity and communicate with greater impact!', *European Journal of Operational Research*", **174** (2), 1348-9.

Wang, J., Xiao, J., Li, Q. and Li, K. (Year) 'Knowledge management maturity models: A systemic comparison', In *Proceedings - 2011 4th International Conference on Information Management, Innovation Management and Industrial Engineering, ICIII 2011*, 26 27 November, Shenzhen, China, Vol. 3, 606-9.

Wendler, R. (2012) 'The maturity of maturity model research: A systematic mapping study', *Information and software technology*, **54** (12), 1317-39.

Whittock, M. (2002) 'Women's experiences of non-traditional employment: Is gender equality in this area a possibility?', *Construction Management and Economics*, **20** (5), 449-56.

Zimina, D., Ballard, G. and Pasquire, C. (2012) 'Target value design: using collaboration and a lean approach to reduce construction cost', *Construction Management and Economics* **30** (5), 383-98.

ICT & Innovation: A Case of Integration in a Regional Construction Firm

8

Graham Brewer, (University of Newcastle, Australia)

Thayaparan Gajendran, (University of Newcastle, Australia)

Goran Runeson, (University of Western Sydney, Australia)

Abstract

Construction firms in regional areas face considerable challenges to their competitiveness. Logistics, availability of suitable personnel, unfamiliarity with trading partners, and the tyranny of distance conspire to obstruct profitable engagement beyond their local area. Electronic information and communication technology (ICT) such as web-based project management applications and other forms of collaborative data sharing have the potential to mitigate many of these obstacles, but have yet to gain widespread acceptance, particularly by smaller regional firms who question their cost effectiveness. The attitudes of decision-makers and the impact of their decisions on intra- and inter-firm culture lie at the heart of ICT-mediated innovation. This paper presents a case study of a specialist subcontractor located in regional New South Wales and its involvement in a New Zealand project, whose competitiveness arose from using ICT to integrate its own supply chain. It also provided additional value by triggering collaboration and integration in the broader project team. The case reveals that these outcomes were symptomatic of the organisational culture of the subcontractor, and were achieved through a combination of leadership, collaboration, flexibility and pragmatism, redolent of dynamic capabilities.

Keywords: ICT, Project team, Culture, Innovation, Regional

Introduction

Project team management and particularly the integration of construction project design and building phases has been a feature of human endeavour. As projects become more complex, so do the number and type of participant organisations. Whilst the project team objective must necessarily be to complete the project to specification, in the required time, and to the target cost the commercial imperatives of the participant firms may be at variance with these objectives. A measure of fragmentation and conflict is therefore an inevitable outcome of such conditions.

Successive industry reviews (e.g. Egan, 1998) have suggested the desirability of adopting production management strategies that have previously proved successful in the manufacturing sector. Examples include strategic alliances between trading partners (Sakal, 2005), lean construction, and supply chain integration (Khalfan, Asad, & McDermott, 2007). Whilst there have been reported instances of success following their use they have not been widely experienced, or copied. Various explanations have been advanced for this, and the absence of information integration is one such example (Atkin, 1999). It is therefore paradoxical that integrative ICT, which has the potential to assist firms in regional areas to compete with metropolitan firms especially where their innovative capacity relates to marketing a unique product or service (Rogers, 1998, p 11), is not more widely utilised.

Although such technologies exist and can be shown to function successfully at a local level, it is apparent that the construction industry as a whole lacks the technological capabilities and business process maturity to currently fully exploit them (Brewer & Runeson, 2009). Localised successes are therefore innovative in nature whether that be through product or service innovation (Teece, 2010) and the product of decision-making by one or more firms that is conducive to integrated, collaborative business processes. The research reported in this paper sought to surface the attitudes, decision-making behaviours and cultural consequences in relation to ICT-mediated innovation at the twin levels of the firm and project. A local Newcastle firm identified by previous research as demonstrating sustainable competitive advantage through innovation was found to be highly active internationally both in terms of managing its own supply chain relationships, and its involvement in internationally located construction projects. This case study is therefore a tale of two supply chains. One of them is regionally based and international in scope, successfully integrated by the ICT at the core of its specialist subcontracting business. The other is the project team of which it was a member, which accreted around a conventionally procured project, in which ICT use was largely absent.

Literature Review

The Decision to Innovate

At its most fundamental, innovation theory suggests that the decision to innovate is fundamentally driven by the search for sustainable competitive advantage, whether it be through the development of new products and services (Heertje, 2006; p 26) or through the adoption of products and services that are new to a particular firm (and therefore immitative from their perspective) (Manley & McFallon, 2003). It follows that the decision not to innovate – despite the presence of products or services that have the potential to deliver high levels of profit – will be made on the basis of an unfavourable risk/return on investment analysis (Heertje, 2006; p 11). Such decisions are often assumed to be made by *homo economicus*, on the basis of rational consideration of the economic alternatives in pursuit of maximal profits. In reality such business decisions are made by humans, often on the basis of incomplete knowledge of the "facts", influenced by prior experience and also frequently emotion, in a boundedly rational way (Simon, 1991). Business decisions made on the basis of bounded rationality are rarely optimal, instead resulting in "satisficing" solutions, outcomes that can be described as being "good enough".

A better understanding of the influences that shape the attitudes of potential innovators and adopters of innovation, and particularly the attitudinal traits (Venkatesh, Morris, Davis, & Davis, 2003) of strategic decision makers who sanction and dictate their deployment within construction organisations would help innovation levels in the construction industry. However, the human dimension has been largely overlooked by construction industry researchers and policy-makers.

Describing the attitudinal traits or attitudinal profile of a particular population is a technique that is familiar to both market and academic researchers, across multiple fields and disciplines. However in each case the starting point has to be a pre-existing framework of reference such as Ajzen's (1991) model, or multidimensional attitude profiling (Gann & Salter, 2000). Implicit in this is the understanding that a collection of decision-makers within a single project will necessarily develop a group culture, be that positive or negative.

Developing a Culture for Innovation

Organisational culture has been variously described as a "strong prescription for success" (Martin et al., 2004) and "an interpretation for better understanding" (Willmot, 2000). However in both cases the concept of "cultural analysis" has been mooted as an appropriate mechanism to allow its investigation.

Culture is an emergent feature of a group, which springs from the underlying assumptions and beliefs of its members about what they share in common, how the world operates and consequently, how they should relate to it. This shapes their attitudes and often their consequent behaviour. Schein (2004) defines culture as:

> *'a pattern of shared basic assumptions [beliefs] that was learned by a group as it solved its problems of external adoption and internal integration, that has worked well*

enough to be considered valid and, therefore, to be thought to new members as the correct way to perceive, think, and feel in relating to those problems' (p 17)

Group culture can manifest itself overtly through rituals and other behaviours, although often the most enduring cultural traits are embedded in the underlying beliefs held by its members (Schein, 2004; Rousseau, 1990). These beliefs can take one of two forms: espoused, and actual.

Espoused or claimed (beliefs) are usually those attributes that people want to be seen to possess, or believe they should demonstrate. By contrast actual beliefs are those made manifest through their unconscious behaviour. Literature (Schein, 2004) suggests that attempting to understand culture through surface level manifestations alone (e.g. overt behaviour) is unreliable. Schein (2004) recommends surfacing deeper psychological manifestations such as underlying assumptions or beliefs as a more reliable approach. Thereafter it is possible to observe "inconsistencies" or "conflicts" between overt behaviours indicative of claimed beliefs, and underlying beliefs.

The learning or transmission of cultural assumptions or beliefs from one person to another within a group is the way in which group cultures form, whether they form spontaneously (e.g. when a group is thrown together unexpectedly) or by design (e.g. through recruitment of like-minded staff). Various metaphors have been developed to explain the cultural development of groups, likening it to engineered control mechanisms (Archer, 2004), infection (Geertz, 1966), or evolution (Sperber, 1996): where the spread or promotion of cultural traits is not actively initiated or managed there has to be an alternative, naturalistic facilitation process, as in the latter two instances. In the context of this paper the development of project team culture and the cultural influence of the industry beyond the focal project are of central importance.

ICT as an Innovation

Many authorities regard the integration of ICT into construction industry supply chain activities as both desirable and imperative (e.g. Department of Trade and Industry, 2001). This runs counter to its cultural norms and project-centric structure. A conceptual supply chain model of the temporary organisations constructed to complete construction projects is therefore useful to analyse cultural influences on ICT use in this context. Such a model can be derived from concepts originated by the Industrial Networks School of supply chain management (Ford, 1997; Hakansson & Johanson, 1992). Thus each organisation within the project consists of a network of interacting actors, resources, and activities, all of which are necessary in order for it to function. Crucially this can be scaled to the level of the project, which itself becomes an *actant* in the network. The model therefore posits that a project team accretes around a project rather than an individual actor (e.g. the client).

There are a number of ways in which individual project organisations can be influenced by each other, and that these are not always contractual. Indeed they may not be formally recognised at all, rather being the product of a network of informal contacts and communications (Wasserman & Faust, 1994). It follows that whilst the adoption of ICT might facilitate the communication of ideas and information within a single project this is only one of a number of possible communication channels (Gajendran & Brewer, 2007). In the context of the current research it is clear that the characteristics thus conceptualised have the capacity to impinge upon the formation of culture in relation to the adoption and use of ICT within a project setting, both in a positive and a negative way.

Brewer & Runeson (2009) develops a model of innovation and attitude to explain the influences affecting the behaviour of decision-makers considering the innovative use of ICT across a temporary project organisation (TPO) in the construction industry. Using Ajzen's Theory of Planned Behaviour (1991) as the point of departure it acknowledges that the ideal decision ought to be to integrate ICT throughout the primary participant organisations in a particular TPO, with its ultimate expression in the adoption of web based communication combined with online access to a shared project information database.

The collective attitudes and behaviours of a group of people in large part serve to define their group culture (Schein, 2004). Culture has the ability to influence all areas and activities within organisations, and whilst these are usually thought of as being an individual firm, club or institution, cultural concepts can equally be applied to project teams. It has been observed that the culture of a construction project influences its level and quality of ICT uptake and integration.

Previous research identified the Critical Success Factors (CSFs) for integration of ICT as being: Organisational Commitment; Organisational Attitude to Communication; Rights and Duties of Organisations (in relation to ICT-mediated communications); Investment Drive, and; Risks related to ICT Usage (Brewer & Gajendran, 2006). Subsequent cultural analysis of these revealed that the cultural values espoused by the industry were analogous to the desired cultural values for an optimised project environment, which in turn ought to facilitate ICT integration. Unfortunately in practice it has been found that very few real life project cultures fully reflect these cultural ideals, resulting in a wide disparity between the levels of ICT integration experienced by participants in different projects and at different levels within the project supply chain.

It follows that a comparison of the actual culture of a project with the espoused values of the individual participants provides a basis upon which to identify the issues that lead to sub-optimal levels of ICT integration. The Cultural Analysis Framework for ICT integration proposed by Gajendran & Brewer (2007) maps the actual cultural characteristics of the environment into which ICT is deployed, comparing it to the cultural stereotypes first identified in the CRC-CI (2006) study.

ICT Innovation in a Regional Context

The concept of a regional culture is familiar in contexts as varied as the arts, sociology, or sport, and it is not surprising that regional distinctiveness may be detected in the business cultures of firms (Davidsson, 1995). Such cultures may develop endogenously, in response to competitive desires, or be fostered exogenously as a consequence of regional development policy initiatives. Indeed their regional location itself may be the trigger to innovate in order to survive (Boekema, Morgan, Bakkers & Rutten, 2000).

The competitiveness of regional firms is frequently theorised as leveraging technology to achieve innovation, where a business culture develops driven by 'capitalist market logic' (Moulaert & Sekia, 2003; p289). This observation suggests that more complex forces may be at play during the cultural development of innovative firms in regional locations; these are likely to be firm-specific.

ICT as a Facet of Dynamic Capability in a Regional Context

The strategic management literature has long recognised the importance of a firm's collection of capabilities as the basis upon which its performance is built. It is equally acknowledged within construction management literature that the overwhelming majority of firms in the architectural, engineering and construction (AEC) sector are SMEs, whose business objectives are to "make a living" in the short term, as they usually lack the time, resources and skills to formulate strategic plans. Their business capabilities can therefore be described as ordinary, or "zero-level" (Winter, 2002).

The concept of dynamic capabilities was initially defined as a firm's ability to "integrate, build, and reconfigure internal and external competences to address rapidly changing environments" (Teece, Pisano & Shuen, 1997; p 516). This approach emphasised the agility required to adapt to a dynamic operating environment, as a response to increasingly rapid changes in market tastes and requirements. Dynamic capabilities therefore reflected an organisation's ability to achieve new and innovative forms of competitive advantage given their current set of competencies, capabilities, and using their current market position as the point of departure in the planning process (Teece, Pisano & Shuen, 1997; p 516). It logically follows that dynamic capabilities are those which extend the firm's capabilities beyond a zero-level position.

Defining the specific attributes of a firm, which in turn combine to become parts of its capabilities is complex. Commentators are agreed that dynamic capability is rooted in managerial and organisational processes and routines (Teece, Pisano & Shuen, 1997; Teece & Pisano, 1994; Winter, 2002; Teece, 2011). Striking the right combination of reflective organisational learning, positional awareness, and robust business processes is central to a firm's dynamic capabilities: it is not simply enough to brilliantly improvise in times of stress or opportunity (Winter, 2002; p 3).

The durability of a competitive advantage derived from a firm's dynamic capabilities is largely determined by the length of time it takes a firm's competitors to replicate or imitate its innovation. This in turn reflects earlier choices made by the firm in response to its perceived market position and consequent options: dynamic capabilities are therefore highly paths-dependent, and it is this path dependency that may secure competitive advantage over the longer term, given appropriate agility (Teece & Pisano, 1994; p13).

Construction firms in remote and regional areas of Australia tend to be SMEs. Their physical distance from major metropolitan markets has traditionally imposed limits upon the size of the markets they could serve and therefore restricted their capacity to grow beyond a certain size. A key component in the latest dynamic capabilities research is the emphasis on technology, and particularly ICT as assets upon which capabilities can be built (Teece, 2010).

Reconceptualising businesses within the AEC sector as being essentially information-based could be critical in enabling regionally based firms to compete by providing new products and or services, and operating in non-traditional markets (Winter, 2000; p3). In order to achieve such outcomes firms would have to develop new dynamic capabilities based upon existing skill sets, and would likely require the development of a set of strategic business relationships based upon collaborative working with their supply chain partners (Teece, Pisano & Shuen, 1997; p 520).

Methodology

The design of this research was based upon four principles:

- Firms in the AEC sectors would consider their adoption of ICT to be an innovation;
- This decision would be made by key decision-makers within these firms;
- Although the decision-makers would consider their decisions rational, they would in fact be acting in a boundedly rational way;
- In the context of a TPO, the decision-making behaviour of individuals would contribute to the formation of a group culture, and that there was therefore a direct connection between the deeply held beliefs of individuals, their attitudes and consequent decision-making behaviours, and the culture of the TPO as evinced through its actions. (Brewer & Gajendran, 2012)

The design of research methods is shaped by a combination of theoretical and practical research considerations. These commence with consideration of ontology, what is believed to exist and its relationship to the existing body of knowledge. Thereafter epistemological considerations are addressed, by identifying the appropriate level of proof required to confirm or reject these beliefs. Methodology provides the framework within which to conceptualise a philosophically consistent method of collecting this proof, and the design of appropriate research methods become the tools with which this proof is generated (Livesey, 2003).

The foregoing considerations cannot be divorced from the researcher, which reflects their theoretical perspective (Livesey, 2003) and values (Burgess, 1982; Glaser and Strauss, 1967). They are also influenced by the reliability and validity of data collection (Strauss and Corbin, 1990), and the reliability/validity of data analysis/abstracts and methods (Morse and Richards, 2002). This study was concerned with *knowing*, by reconciling a wide range of individual experiences into a form that enabled readers to *know* the collective experience in a meaningful way.

Insofar as it is possible to truly *know* anything it was necessary to make pragmatic epistemological decisions to facilitate the process of knowing:

- Firstly, to identify the type of evidence that would generate an answer to the research question, and
- Secondly ensure this was sufficiently compelling to satisfy the intended audience.

On a practical level these considerations were reflected in the authors' commitment to multiple research methods, drawing on both qualitative and quantitative research paradigms.

The TPO was used as the unit of analysis. However its actions were observed through a phenomenological lens to reveal the decision-making behaviors of its individual members. This took the form of a case study protocol that collected three types of data.

- A quantitative survey instrument was administered to each case study participant: this was identical to the survey instrument used in the CRC for Construction Innovation project "Critical Success Factors for ICT Mediated Supply Chains" (Gajendran, Brewer & Chen, 2005), and its use was intended to reveal the espoused values of each respondent.
- A map of supply chain relationships was co-constructed with each respondent, identifying the connections and nature of their relationship with others in the TPO: These were then subsequently cross-referenced, a unified map of relationships produced, and validated by the respondents themselves.
- Interviews were conducted with the case study participants. These were semi-structured and in-depth, guided by thematic prompts whilst allowing respondents to raise new issues wherever necessary. The prompts themselves were designed to elicit the respondents opinions and experiences in relation to:
    - their own firms history and posture in relation to ICT adoption;
    - their own firms attitude to collaborative working with trading partners in project supply chains;
    - rights and responsibilities within collaborative ICT environments;
    - strategic expectations of ICT investments;
    - leadership and "championing" of ICT at the intra-and inter-firm levels;
    - their firms business processes, technologies, and agility to exploit opportunity.

The case study was completed over 14 months during the construction phase of the project, the qualitative survey instrument being administered at the outset, with the results being compared to the values obtained in the CRC project (Gajendran, Brewer & Chen, 2005).

Five representatives from primary stakeholders in the TPO (head contractor, client's in-house project manager, architect, steel fabricator, client's façade consultant) and three representatives from the façade supply chain (specialist subcontractor, steel fabricator, glass supplier) were interviewed during that period. The proprietor and engineering design manager of the facade specialist subcontractor were individually interviewed at the beginning of the case study, and once again at its conclusion.

Case Study

This project involved the design and construction of an iconic public building in Christchurch, New Zealand. The architect modelled the project in three dimensions using CAD. The main project supply network was defined by a conventional set of contracts/subcontracts, without the mandated use of collaborative ICT.

A key visual/physical component of the building was the "Sculpture Wall", a complex 3-D curved engineered facade. An international specialist engineered facade subcontractor was engaged, on the recommendation of the client's own engineer. The subcontractor's own supply network was distributed across many countries, including China, Australia, Dubai, and Germany. As a consequence it had developed agility based upon the use of ICT, particularly 3-D CAD, and had seen fit to "embed" members of its own staff within supplier organisations in order to ensure that they produced information that was compatible with the subcontractor's own design processes and quality assurance protocols.

The integration of the specialist subcontractor's designs and products with those of the base building had the potential to be problematic, since another subcontractor responsible for fabricating the support structure had no compatible ICT competence. Nevertheless the specialist subcontractor saw fit to provide documentation beyond their contractual obligation, and access to intellectual property and personnel, commercially sensitive nature, in order to facilitate prompt project completion. They received acknowledgement that this behaviour both resolved a potentially serious problem and created a positive change in the project team culture, but received no additional remuneration.

Results

Three sets of data were collected from the case study participants: questionnaire survey results; supply network relationship maps, and; interview transcripts.

The questionnaire administered to case study participants was identical in all respects to the one used for a national survey during the CRC-CI research project (2006), which identified six critical success factors for ICT innovation, differentiated by industry sector (clients, consultants, contractors and subcontractors). These were subsequently recognised to be analogous to the espoused values of ICT decision-makers in construction related professions, and consequently the questionnaire was used in this case study to reveal whether any of the participants espoused values that were at odds with their colleagues in the rest of the industry. It was found that all participants were found to hold views that were similar to others holding similar positions (i.e. within one standard deviation of the industry norms).

The second set of data was a series of supply network relationship maps that were constructed by each interviewee with the assistance of the interviewer. Each map detailed the interviewee's firm in relation to its immediate trading partners and those other project participants involved in the study. The nature of the relationships between these firms was detailed in terms of contractual, non-contractual, and monitored links, highlighting those that were considered by the mapmaker as being strategic in nature. The consolidated supply network relationships (subsequently reviewed and approved by all participants) are illustrated in Figure 1.

The final set of data was the interview transcripts. These were analysed using the techniques described previously in the methodological section of this paper. This resulted in a table of thematic codes, the highest level of which coincided with the six critical success factors arising out of the CRC-CI research. Immediately beneath these were a series of themes derived from the open coding process, each of which was subsequently illuminated where appropriate using axial coding techniques. 12 interview transcripts, containing a total of over 135,000 words were coded, resulting in 162 raw codes. Table 1 shows a representative extract from the consolidated coding table, where rationalisation of the coding has reduced the total number of codes to 93, and where alphabetisation has been performed in order to remove any connection to the original critical success factors.

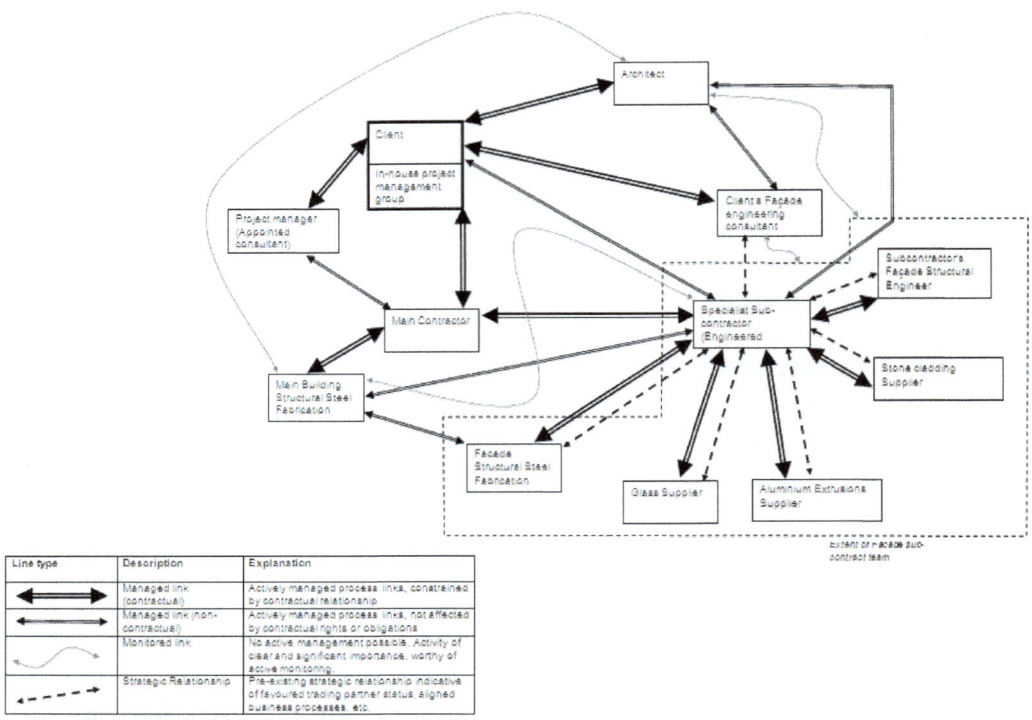

Figure 1 Consolidated supply network map

Given that the critical success factors were generalised to the entire industry, and that the detailed coding for the case study was intended to illustrate the specifics as to how these success factors found their expression in this particular case study project, it became apparent that there were in fact two stories to be told: one related to the TPO itself, as originally envisaged at the beginning of the research; the second story related to the specialist facade subcontractor and its upstream supply chain, which contrasted sharply with construction industry norms. Accordingly the data – and particularly the open/axial themes underlying the critical success factors – were reanalysed, this time placing the subcontractor (as opposed to the project) at the focus point, and using a Dynamic Capabilities lens. This enabled the researchers to drill down to the root causes of its success in innovating. It is important to note that whilst some of these were ICT-related, others were not, with issues such as leadership style, and organisational culture, agility, service innovation and process innovation all being found to contribute. It is also important to note that since no additional interview data was collected beyond that originally designed to investigate critical success factors, the dynamic capabilities analysis should be regarded as preliminary and exploratory, requiring further investigation to confirm or refute.

Open codes derived from first pass analysis	Explanation	*Axial codes derived from subsequent pass analysis*	Explanation
relationships (long term)	The notion that firms may enter into strategic business relationships that endure over time. In this case, ICT compatibility is seen as facilitating such relationships. Reflects such issues as technology compatibility, complimentary distinctive competences, niche markets, etc.	*relationships (quasi-strategic)*	A code group that acknowledges that enduring business relationships often reflect expediency on the part of one or more parties to the relationship.
revolutionary change	The antithesis of incremental change.		
reward	The expected outcome of using ICT.		
risk *risk (management)* *risk (minimisation)* *risk (prediction)*	Codes reflecting both the risks associated with adopting and using ICT within an individual firm, and within a project team for communication with other participants.		
ROI *ROI (difficulty of measuring, ICT investments)* *ROI (long term)* *ROI (short term)*	Codes reflecting the expectation of benefits accruing from the use of ICT, both in the short term and long term. Short-term has a specific meaning of ' within the current project', whilst long-term extends beyond a single project. Difficulties associated with specifying precisely what are and are not benefits, and how to measure the non-financial benefits are problematic.	*ROI (personal)*	A major and significant code group that recognises ROI existing on a number of different levels and over a number of different time spans. A significant addition is the recognition that individual users within a firm expect to derive a return on their own investment (usually of time and effort in learning new systems) in relation to ICT, and whilst the initial choice to adopt is usually determined from above, the choice to engage wholeheartedly is largely a personal one.

Table 1 Extract from consolidated code table

Discussion

The collaborative use of ICT across the TPO assembled to complete the design and construction of a construction project, is frequently held up as the future of project delivery. Its purported benefits include improved design coordination, lessened conflict, reduced communications costs and importantly, an accurate stream of information that can be used throughout the life of the constructed asset (Eastman, Teicholz, Sacks & Liston, 2008). However all of this is dependent upon the presence of participant firms that share compatible technologies, business processes, and cultures, led by people who hold attitudes and display behaviours conducive to collaboration (Gajendran & Brewer, 2007). In this case it was clearly apparent that the TPO did not possess these characteristics, whilst the specialist facade subcontractor's own supply chain did.

Experience has shown that it is unusual to find conditions conducive to ICT integration across an entire TPO, and this case was no exception. Nevertheless, it could be seen that where competitive conditions in the industry were suitable it was possible for stable supply chain relationships, based upon ICT infrastructure to develop and endure. Moreover it was observed that the presence of a project participant with experience in this regard enabled such collaborative behaviours to temporarily exist within the wider network of TPO members. The value of this phenomenon was not confined to the engineered facade either, "infecting" the operational culture of the project team, leading to a more collaborative working environment (Brewer & Gajendran, 2012; p 8).

From the client's perspective the nature of the trigger to positive project outcomes was of no significance, though the outcomes themselves were. From the architect's perspective the behaviour of the specialist subcontractor was both admirable and enviable. Yet although the architect possessed comparable technology they did not see fit to champion its use within the project network, citing low expectations of technological compatibility with the project network partners. Ultimately the success of the specialist subcontractor's own supply network was fuelled by commercial expedience based upon proven ability to provide the required performance in terms of time, cost, and quality. Nurturing and maintaining this network was regarded as essential to sustaining competitive advantage.

> *"You know, there are 6 billion people in the world, and it's surprising how difficult it is to find one of them who will reliably supply you with what you need on a regular basis..... When you find someone you can do business with, you cling onto them. They are pure gold. They become family."* (CEO, Specialist subcontractor).

This case study revealed a number of the paradoxes inherent in the industry at this time. Firstly, a number of the project participants had invested in BIM-capable technologies, and although each was satisfied with their use of it at an intra-firm level, the same could not be said for inter-firm information sharing. However none of these firms regarded this as an ICT failure, perhaps because none of them had genuinely BIM-ready business processes, which are inherently boundary-spanning, and require a degree of shared business processes. The specialist subcontractor came closest to this, however their supply network could not be said to be BIM-ready.

Secondly, for BIM integration to be successful it has to be championed by someone in the project supply network, ideally the client, or failing that the architect or head contractor. Their power position should allow them to mandate the use of a particular technology/protocol, thereafter requiring them to support its use, as they would be recipients of the benefits thus accrued. Disregarding the precise nature of BIM, the closest this condition came to existing within the case study project supply chain was where the specialist subcontractor championed the use of diverse ICT to facilitate a shared understanding of the technicalities of the Sculpture Wall. Moreover whilst this was confined to construction of a specific element of the structure, the collaborative effect it engendered pervaded other aspects of the project in a beneficial way, signposting the potential that this approach has when properly implemented.

Lastly, the beneficial effects of ICT-mediated collaboration on this project were realised in spite of the procurement mechanisms, not because of them. They arose because the specialist subcontractor was prepared to set aside the letter of the contract and pursue a more collaborative approach in order to achieve the project objective that was the sculpture wall. This required them to be exposed to what amounted to unregulated risk and unrewarded risk-taking, a behaviour that would be considered unusual in the construction industry. On the other hand, without such behaviour the project would almost certainly have taken a substantial turn for the worse. This highlights both the inadequacy of existing procurement mechanisms for suitability in the BIM age, and the necessity of developing new mechanisms to fill this gap, if BIM is to fulfil its potential.

It can be seen from the foregoing that the culture of the specialist subcontractor's firm performed a beneficial role by facilitating the effective completion of the Sculpture Wall, and this can be seen as a natural consequence of the pursuit of their own business interests. That they were able to see a clear business benefit in what could reasonably be seen as risky and apparently altruistic behaviour speaks to the nature of the firm itself. In particular it would seem to indicate a deep-seated confidence in the capabilities, both technological and interpersonal, of the firm to succeed in such an environment, which extended beyond brilliant improvisation in times of stress (Winter, 2002; p 3). As such it would seem reasonable to speculate that this behaviour was public manifestation of a dynamic capability that had been actively developed by the specialist subcontractor's management, and while exploring this concept was not originally an overt objective of this research the connection between deeply held beliefs and attitudes, subsequent decision-making behaviour, and the formation of project team culture certainly was. If one of the dynamic capabilities of the specialist subcontractor was perceived by its management to be to *integrate, build, and reconfigure internal and external competencies to address rapidly changing environments* (including project teams) (Teece, Pisano & Shuen, 1997; p516) then this certainly would deserve further investigation. Moreover it would be important to understand the influence that the firm's regionality had played in this behaviour (Boekema, Morgan, Bakkers & Rutten, 2000).

Conclusions

The focal company in this case study shows to innovate in a number of different ways: firstly it shows two reconceptualising its product in terms of the information and intellectual property that it generated, rather than the constructed artefacts that resulted from its implementation; secondly, it recognised the need to form and maintain a number of stable relationships with the suppliers in its supply chain; thirdly, it understood that these would only deliver value to it if their business processes, and specifically the ICT that facilitated them achieved a high level of integration. It was apparent, given the diverse international locations of the focal company's supply chain partners and projects that their location was a critical trigger to the foregoing strategic decisions. It is less clear whether their regional location was itself a trigger to seeking overseas supply chain partners.

The level of strategically driven reflection and reconceptualisation undertaken by the senior management of the specialist subcontractor could be thought of as atypical for SMEs in the construction industry. Their recognition of the need to fully exploit the capabilities of ICT in order to integrate their supply chain placed their relationship management abilities at the heart of their business model, requiring an investment in each trading partner in terms of ICT compatibility that to outsiders might seem like an act of faith. That they were able to successfully compete for, and profit from involvement in internationally located projects would appear to vindicate their decision. However closer examination revealed that this success was less founded on faith and more upon a careful matching of business opportunity to the capabilities of the firm.

Although this research was not originally designed to specifically probe the issue of dynamic capabilities it became apparent during data analysis that there were obvious links between the outcomes of senior management reflection upon its strategic planning and subsequent decision-making. Specifically they had connected: performance in previous projects (the need to produce and transmit more accurate information/the development of a zero-error culture within the firm); the recognition that ICT provided the tools to achieve this, but that it required the willing cooperation of its supply-side trading partners; the recognition that acceptable levels of supply chain integration would require a considerable investment of both time and money in order to train these trading partners to exchange information to an acceptable level; and finally, that this was ultimately an exercise in changing the culture of its entire supply-side chain. It seems reasonable to conclude that these outcomes were redolent of the exercise of the focal company's dynamic capabilities.

Finally, given the foregoing this research raises some intriguing questions. Firstly, did the firm's regional location trigger the development of its dynamic capabilities, or was due to the market in which it operated? Secondly, was the cultural change that the specialist subcontractor triggered within the project described in this research a manifestation of one of its dynamic capabilities (i.e. supply chain relationship management) or was it merely a fortuitous coincidence?

References

Ajzen, I. (1991) 'The theory of planned behaviour', *Organizational Behaviour and Human Decision Processes,* **50**, 179-211.

Archer, M.S. (2004) *Culture and Agency: The Place of Culture in Social Theory*, Revised Edition, Cambridge: Cambridge University Press.

Atkin, B. (1999) 'Measuring information integration in project teams', Paper presented at the CIB-W78 Workshop, Vancouver.

Boekema,F., Morgan, K. Bakkers, S. & Rutten, R. (eds) (2000) *Knowledge, innovation and economic growth: the theory and practice of learning regions,* Edward Elgar publishing. Cheltenham, UK.

Brewer, G., & Runeson, G. (2009) 'Innovation and attitude: mapping the profile of ICT decision-makers in architectural, engineering and construction firms', *International Journal of Managing Projects in Business*, (3).

Brewer, G. & Gajendran, J. (2012) 'Attitudes, behaviours and the transmission of cultural traits: impacts on ICT/BIM use in a project team', *The Journal of Construction Innovation, Information Process Management,* **12** (2)

Brewer, G. and T. Gajendran (2006) *Picturing success: critical success factors for ICT integration in the Australian construction industry,* Brisbane, Icon.Net Pty Ltd.

Brewer G., Gajendran T., & Beard, C. (2009) 'Inhibitors to the adoption of BPM/BIM: an Australian perspective', *CIBW78 Information Communication Technology in Construction Conference*, Istanbul Technical University, Turkey.

Burgess, R. (1982) *Field research: a source book and field manual*, London, Allen and Unwin.

Davidsson, P. (1995) Culture, structure and regional levels of entrepreneurship. Entrepreneurship & Regional Development: An International Journal Volume 7, Issue 1 pp41-62

Department of Trade and Industry (2001) *Foresight: constructing the future*, London, UK, DTI.

Eastman, C., Teicholz, P., Sacks, R. and Liston, K. (2008) *BIM Handbook: A Guide to Building Information Modelling for Owners, Managers, Designers, Engineers, and Contractors*, John Wiley & Sons, Inc., Hoboken, NJ, USA.

Egan, J. (1998) *Rethinking construction*, London: DETR.

Ford, D., Ed. (1997) *Understanding business markets: Interaction, relationships and networks,* London, Dryden Press.

Gajendran, T., Brewer, G., & Chen, S. (2005) 'Project teams and ICT: Surfacing the critical success factors', Paper presented at *the CIBW78 Information Communication Technology in Construction*, Dresden.

Gajendran, T. and G. Brewer (2007) 'Integration of information and communication technology - influence of the cultural environment', *Engineering, Construction and Architectural Management,* **14** (6), 532-549.

Gann, D. and A. Salter (2000) 'Innovation in project-based, service-enhanced firms: The construction of complex products and systems', *Research Policy,* **29**, 955-972.

Geertz, C. (1966) 'Religion as a Cultural System', in M. Banton (ed.), *Anthropological Approaches to the Study of Religion*, New York. Praeger, 1-46.

Glaser, B. and A. Strauss (1967) *Discovery of grounded theory: Strategies for qualitative research*, Mill Valley, CA, Sociology Press.

Hakansson, H. and J. Johanson (1992) 'A model of industrial networks', *Industrial networks: A new view of reality*, B. Axelsson and G. Easton. London, Routledge.

Heertje, A. (2006) *Schumpeter on the economics of innovation and the development of capitalism,* Cheltenham, UK: Edward Elgar Publishing.

Khalfan, M., Asad, S., & McDermott, P. (2007) 'Introducing innovative procurement methods for supply chain integration - implementing the challenge', Paper presented at *the CIB W92 Procurement Systems Conference.*

Livesey, C. (2003) 'A' Level sociology: A resource-based learning approach. Module One: Theory and methods: Unit M13: Research considerations', from www.sociology.org.uk.

Manley, K., & McFallan, S. (2003) 'Innovation adoption behaviour in the construction sector: The case of the Queensland road industry', Paper presented at *the 2nd International*

Conference on Innovation in Architecture, Engineering and Construction. Brisbane.

Martin, J., Frost, P. J., and O'Neill, O. A. (2004) 'Organizational culture: Beyond struggles for intellectual dominance', *Stamford Research Paper Series,* No 1864.

Morse, J. and L. Richards (2002) *Readme first: for a users guide to qualitative methods*, London, Sage.

Moulaert, F. & Sekia, F. (2003) 'Territorial Innovation Models: A Critical Survey', *Regional Studies,* **37** (3), 289-302.

Rogers, M (1998) *The Definition and Measurement of Innovation,* Melbourne Institute of Applied Economic and Social Research, The University of Melbourne, Melbourne.

Rousseau, D. M. (1990) 'Assessing Organizational Culture: The Case for Multiple Methods', In I. L. Goldstein (Ed.), *Frontiers of Industrial and Organizational Psychology.* California: Jossey-Bass Inc.

Sakal, M. (2005) 'Project alliancing: A relational contracting mechanism for dynamic projects', *Lean Construction Journal*, **2** (1), 67-79.

Schein, E. H. (2004) *Organizational culture and leadership*, San Francisco: Jossey-Bass.

Simon, H. (1991) 'Bounded rationality and organisational learning', *Organisation Science* **2** (1), 17-27.

Sperber, D. (1996) *Explaining culture: A naturalistic approach*, Oxford: Blackwell.

Strauss, A. and J. Corbin (1990) *Basics of qualitative research: Grounded theory procedures and techniques*, Newbury Park, Sage.

Teece, D. (2011) 'Dynamic capabilities: a guide for managers', *Ivey Business Journal*, Viewed online at https://www.iveycases.com/ProductView.aspx?id=48342 on 11/02/2012.

Teece, D. J. (2010) 'Technological Innovation and the Theory of the Firm: The Role of Enterprise-level Knowledge, Complementarities, and (Dynamic) Capabilities', In N. Rosenberg and B. Hall (eds.) *Handbook of the Economics of Innovation, Volume 1*. Amsterdam: North-Holland.

Teece, D., Pisano, G. & Shuen, A. (1997) 'Dynamic capabilities and strategic management', *Strategic Management Journal*, **18** (7) (Aug), 509-533.

Teece, D. & Pisano, G. (1994) The dynamic capabilities of firms: an introduction. Working paper WP-94-103 International Institute for Applied Systems Analysis A-2361 Laxenburg Austria.

Venkatesh, V., M. Morris, et al. (2003) 'User acceptance of information technology: Towards a unified view', *MIS Quarterly,* **27** (3), 425-478.

Wasserman, S. and K. Faust (1994) *Social network analysis: Methods and applications*, Cambridge, UK, Cambridge University Press.

Weisberg, S. (2001) i-collaboration - State of the industry. CADALYST Magazine Retrieved 21/9/2001, from http://209.208.199.147:85//features/0901icollab/0901icolab.htm

Willmott, R. (2000) 'The place of culture in organization theory: Introducing the morphogenetic approach', *Organization,* **7** (1), 95-128.

Winter, S.G.(2002) *Understanding Dynamic Capabilities*, Working Paper WP 2002-05. The Reginald H. Jones Center, The Wharton School, University of Pennsylvania.

Stimulating Learning with Integrated Assessments in Construction Education

9

Imriyas Kamardeen (University of New South Wales, Australia)

Abstract

Quality of learning students experience is heavily dependent on the effectiveness of course design. Assessments are a key component in course design and students determine their level of involvement in a learning activity based on whether it is assessed or not. Assessments are therefore a powerful tool that lecturers can utilise to drive learning. However, designing effective assessments to stimulate learning is challenging in the presence of disciplinary, contextual dimensions. A case study approach is adopted to demonstrate how effective integrated assessment schemes may be developed and implemented for construction education. The scheme in the case study amalgamated case-based learning, online quizzes and adaptive eTests to provide a variety of assessments, aligned with lecture topics and contemporary real-word scenarios. It was found that the presence of both formative and summative tasks in the assessment scheme complemented each other, kept students constantly motivated and engaged in learning, and resulted in a good learning experience for them. The study provide evidence, and valuable insights and tips for lecturers in similar degree programs as to how they could modify pedagogical styles in their courses for better learning experiences for students and improved teaching ratings for themselves.

Keywords: Pedagogy, Integrated assessment, Case-based learning, Online quiz, Adaptive eLearning

Introduction

Research shows that students regard assessments as more crucial than learning itself and they spend less than 10 percent of their time on non-assessed learning tasks. Students engage with a subject matter based on whether a task will be evaluated (Elton & Johnston 2002; Lombardi 2008). Kember and McNaught (2007) noted that assessments are critical to learning because students are largely driven and motivated by them. Hence, assessments need to be used as a means to achieve the ends: driving student learning and realising different learning outcomes. Brown (2004) reinforced that assessments are possibly the best tool that lecturers could utilise to drive student learning. Students could ignore a lecturer's teaching, but if they need a qualification they have to complete the assessment. Hence, it is worthwhile exploring how lecturers can achieve integration and constructive alignment to ensure that assessments help learning in their courses. The way lecturers assess will significantly influence how students learn.

It is a conventional practice in higher education that in-class quizzes, homework, class tests, laboratory practicals and other similar tasks are used to evaluate what students have learnt. Usually some form of credit, associated with feedback, is given to students' work. The feedback can be short or detailed, depending on the task and class size. In large classes, elaborative feedback is often limited due to resource constraints. Watkin (1998) criticised that while this form of summative assessment facilitates grading, it has failed in developing high-order learning outcomes as it makes students value rote learning and mark accumulation in preference to deep learning. Over the past decades, alternative approaches to assessment have been practised by a number of scholars in different disciplines in a bid to utilise assessments as a toolkit to promote student learning (Tang & Biggs 1998; Carless 2002). Although there are multiple examples of alternative assessment approaches, they cannot be merely repeated in other courses. Brown (2004) claimed that context is of critical importance in assessment; an assessment approach that works well for one subject/discipline may not work equally well in another subject/discipline. Thus, lecturers who endeavour to offer authentic learning experiences to students must formulate tailored solutions that are appropriate and meaningful to their subject/discipline (Lombardi 2008).

To this end, this study aims to demonstrate how effective integrated assessment schemes can be developed to improve the quality of learning and teaching in construction education. The paper first explores higher education literatures to understand key considerations for designing learning-oriented assessment instruments. Second, it discusses how the knowledge acquired from the literatures was operationlised in implementing an integrated assessment scheme in the context of Construction Management Degree program. Then students' views on the effectiveness of the assessment scheme for quality learning are expounded followed by a conclusion.

Theoretical Background

Assessments in Higher Education

Dietel, Herman and Knuth (1991) defined assessment as "any method used to better understand the current knowledge that a student possesses". Boud (1990) articulated that assessments serve two key purposes: improving the quality of learning and measuring student performance. Additional uses of assessments include: motivating students through the recognition of achievement and aiding administrative matters. Assessments are therefore a core element of instructional design.

Higher education literatures categorise assessments into two types: summative (traditional) assessments and formative (alternative) assessments. The summative approach aims to judge a student's competency/proficiency at the completion of an instructional unit or part of it. The most common summative assessment tools are: a mid-unit quiz/test, an end-of-unit quiz/test/exam, a final essay/paper/drawing, a final presentation, a final project, etc. Formative assessments instead inform students of their learning progress towards some expected learning outcomes, and offer help and guidance to strengthen their areas of weakness and thereby improving their performance. Examples of formative assessments include: problem-based learning, case studies, scenario-based learning, literature reviews, investigative project-based learning, portfolios, learning logs/journals, simulations/role plays, etc. Davies (2000) posited that formative assessments offer deep learning platforms as they promote active student engagement, as well as encourage them to take risk, and learn by doing and making mistakes. Additionally, formative assessments provide greater opportunities for lecturers to continually improve their teaching during an instructional period/unit. Formative checks of students' understanding of the subject unit during an instructional period can result in a lecturer changing the instructional strategy if the initially used method was ineffective.

Law and Eckes (1995) argued that summative assessments are an easy tactic for evaluating student performances as they are more objective and consistent. However, several disadvantages of summative assessments have been reported in the literature. Bailey (1998) critiqued traditional summative assessments as inauthentic and decontextualised, homogeneous, single-occasion tests, speed-centred and no meaningful feedback is provided to leaners. Law and Eckes (1995) added that this form of assessment can only measure learners' ability to perform a task at a given time and the test score is not a good indicator of student progression. Simonson et al. (2000) stated that summative assessments only evaluate learners' low-order cognition and thinking skills such as memorisation and recall. Summative assessments force students to demonstrate their knowledge in a predefined manner (Brualdi 1998).

Formative assessments, in contrast, evaluate higher-order cognition and thinking abilities as well as problem-solving skills (Reeves 2000). In this form of assessment, students have adequate opportunities to demonstrate their ability and therefore this tactic is more student growth focused. For example, if a student fails to perform in a task at a given time, s/he would still have the opportunity in a different scenario and time (Dikli 2003). Simonson et al. (2000) discussed many other benefits of formative assessments. Because formative assessments are designed by simulating real-world scenarios, they provide students with opportunities to practise authentic tasks, apply their knowledge and skills to real-life settings, and work collaboratively. Moreover, lecturers are able to conceive better insights of student learning. However, formative assessments demand more time and efforts from both the lecturer and students (Law & Ecke 1995), thus students may not favour formative assessment tasks.

In summary, while summative assessments help lecturers with grading students and administrative tasks, they promote rote learning and low-order cognition and thinking amongst students. Formative assessments, on the other hand, encourage deep learning with the aid of real-life contexts, and develop high-order cognition and thinking within students. Students are highly driven and motivated by assessments (Kember & McNaught 2007). Student will choose their learning style depending on the form of assessment; if an assessment rewards information retrieval, they will adopt rote learning and memorisation and if problem solving ability is rewarded, students will emphasise on solving problems. Assessments should be used as a means to achieve the ends, student learning. Moreover, in an educational setting grading students is as important as promoting learning. Hence, in order to obtain better outcomes for students and the lecturer, an integrated assessment scheme should be practised, which will allow attaining a balance between using assessment to drive learning and to grade students.

Key Considerations for Designing Learning-oriented Assessments

It is hypothesised above that the adoption of an integrated assessment scheme would simultaneously be a catalyst for student learning and a gauge for evaluating student performance. Several assessment instruments are suggested above that can possibly be combined to form an integrated assessment scheme. However, the selection of appropriate instruments is largely influenced by the context of the discipline and the course taught; what works well for one course/discipline may not work equally well in another course/discipline. Lecturers who desire to offer authentic learning experiences to their students should formulate meaningful instruments, tailored to their courses. Although this is a challenging task, closely following the key principles discussed below would facilitate a smooth process.

Keppell and Carless (2006) conceptualised three components for a learning-oriented assessment design:

- *Constructive alignment* – learning outcomes, objectives, contents, assessment items and instructional design should be aligned; assessment tasks should be designed such that time and efforts of students are spread evenly across the module study period rather than concentrating all at the end; establish connections between assessment tasks and real-life scenarios; and allow some degree of student choice to stimulate their motivation and engagement;

- *Student involvement* – tasks should encourage reflection, cooperation rather than competition, peer feedback, and self-evaluation. Gibbs and Simpson (2004) argued that for learning to be more effective assessments should encourage learners to spend ‘time on task’ in and out of class; and

- *Forward-looking feedback* – feedback has to be less final and conclusive (Boud 1995), more interactive and forward-feeding (Carless 2002), timely and with a possibility for remediation (Gibbs & Simpson 2004).

Seven further characteristics of assessments that help students to learn were postulated in the literature. These are:

- *Fit-for-purpose* – assessments need to be consistent with learning outcomes, i.e. designed to provide opportunities for students to acquire the desired learning outcomes by completing them. Different aims such as motivating students, encouraging activity, providing guidance and feedback for remediation, and grading will impact on the selection of assessment technique. Consider using a different/suitable method (i.e., portfolios, case studies, reflective commentaries, critical incident accounts, posters, reviews, role-plays or annotated bibliographies) to suit the purpose of assessment rather than over-using unseen time constraint exams, essays and reports (Brown & Knight 1994; Brown & Smith 1997);
- *Practice-oriented and authentic* – an effective assessment strategy would closely replicate the practice of the discipline and seek to measure how students apply to practice what they learnt (Kember & McNaught 2007);
- *Appropriately timed* – for assessments to be instrumental to learning, they should be scheduled appropriately to allow students to act upon the feedback given, and amend and remediate errors in their original work (Brown 2004);
- *Inclusive* – deploy a variety of assessment methods so that all students will have equal opportunities to demonstrate their skills and maximise their potentials, and the same students are not always disadvantaged due to a single format of assessment. Maintaining a variety is also crucial in aligning learning outcomes with assessments; that is, as courses have multiple learning outcomes, it is difficult to devise a single assessment item or type of assessment which achieves all the learning outcomes. In these cases, it is preferred to have an assessment package with different types of assessments (e.g. projects/case studies, exams, quizzes, games, participations etc.) (Brown 2004; Kember & McNaught 2007);
- *Transparent* – assessment criteria are clearly framed and communicated explicitly in simple language to students and are available from the onset (Thorpe 2000);
- *Reliable* – assessment criteria are clearly articulated to ensure inter-assessor reliability (i.e., different evaluators derive the same grade for similar work) and intra-assessor reliability (i.e., individual evaluators mark consistently) (Brown 2004); and
- *Efficient* – where possible, make assessments more efficient and valid by utilising technologies for creating more authentic tasks, enhancing the quality and timeliness of feedback, and enabling students to monitor their own learning (Brown, Rust & Gibbs 1994; Davies 2010).

The above literature synthesis provides an overarching guideline for developing integrated assessment schemes. However, their application is significantly influenced by the nature of the course/discipline, and therefore careful instructional planning and design is warranted. Considerations should also be given to aspects such as: resources - staff, budget and space requirements; ethics in circumstances where research dominates assessment tasks; and other logistical limitations.

Designing and Implementing an Integrated Assessment Scheme: a Case Study

This section demonstrates how the principles discussed above were operationalised in designing and implementing a learning-oriented, integrated assessment scheme for a course in Construction Management degree program. The case study could provide valuable insights for lecturers in similar degree programs as to how they might modify their pedagogical styles for better learning and teaching outcomes.

The Context

The study was conducted in the context of a first year core course, Construction and Property Economics, in the Bachelor of Construction Management and Property degree program at the author's university. The aim of the course is defined as "enriching students with competencies that are essential for performing economic analyses of construction projects and construction markets", and the course covers three key areas to achieve it, including: economics of building designs, project appraisal, and the construction industry and the national economy. The expected learning outcomes of the course are articulated as "at the conclusion of the course students will:

- be able to contribute significantly to decisions on the selection of appropriate design and construction methods in view of their cost and value effectiveness;
- have detailed knowledge of the commonly used project feasibility appraisal techniques and be able to apply them to construction projects; and
- be able to provide constructive suggestions for construction business planning in light of national and global economic trends".

It is noteworthy that this course has a student enrolment of up to 150. The author has been teaching the course for the last six years using the traditional model of delivery, involving lectures, tutorials, a major assignment based on literature review and a final class test. The author's experience of teaching the same course over the past years suggests some insights/reflections that are also crucial considerations. These are:

- Students are more grade-focused than learning-oriented. They appear interested in learning tasks and active participation in lectures or tutorials if these are directly related to an assessment or grade;
- Students favour the continual assessment system over the final exam system;
- Students prefer hands-on tasks over formal written submissions; and
- Flexibility, informality and simplicity are three features that students want in their studies.

The instructional strategy was modified in light of the assessment-driven learning/instructional paradigm and the insights above. The account below explains the newly adopted assessment and instructional approaches.

Aligned Assessments and Instructional Approaches

The theme of "contextualised teaching and assessment strategy" underpinned the design and implementation of the assessment and instructional approaches adopted in the course. Figure 1 portrays the integrated assessment scheme designed and implemented. Case-based assessments were positioned in the core of the scheme with regular online quizzes and an end-of-course class test supplementing them. The various assessment instruments were tightly aligned with the instructional mode to facilitate the achievement of the expected learning outcomes.

The selection of the above assessment methods for the course was informed by the literature on instructional design in higher education. Angelo and Boehrer (2002) argued that the case-based assessment method is a powerful student-centred instructional tactic in that students learn by solving complex real-world problems. It allows the application of theoretical concepts and encourages students to see them from an action perspective - bridging the gap between theory and practice. Daly (2002) added that the case-based approach develops several skills in students, viz.:

- Exploring a case entails researching and evaluating multiple data sources, which nurtures information literacy;
- Good time management and organisational skills are essential to work on a case;

- It enhances student aptitude in teamwork as well as written and oral communications; and
- The case-based learning approach is also effective for increasing real-world professional skills such as managing a meeting, negotiation, giving presentations, etc.

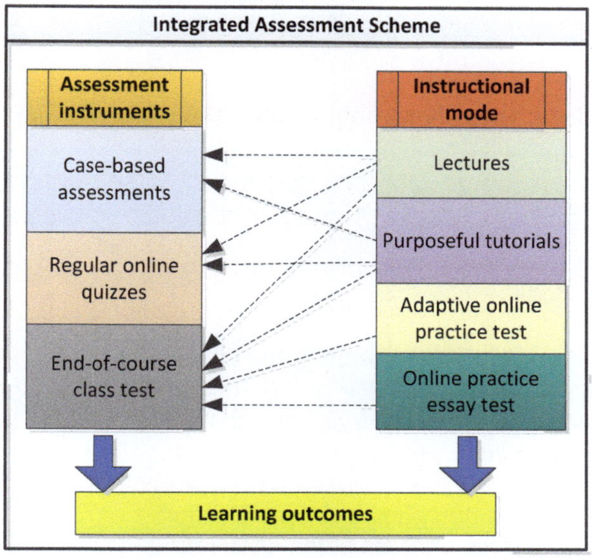

Figure 1: Integrated assessment scheme

Likewise, Salas-Morera, Arauzo-Azofra and García-Hernández (2012) reported that the integration of online quizzes with other instructional activities in a teaching strategy helped to keep students up with the subject, strengthened their involvement in other activities, and had a very positive impact on academic outcomes. Equally for end-of-unit tests, Law and Eckes (1995) claimed that they are objective, reliable and valid, and therefore an easy approach for measuring student performances for grading purposes.

Two case-based assessments were set to drive learning in the course and students were required to work in groups of five. The first case-based assessment dealt with the economics of building designs in that real building designs were given to student groups and they were required to make a class presentation on cost optimisation suggestions for the design provided. The assessment was introduced in week one just after the lecture that dealt with economics of building designs, which enabled students to deeply analyse and apply the theory to real-world contexts. During the analysis process, students were offered formative feedback to enable them to improve their work. When the student groups made final oral presentations of their findings, they were provided with both summative and formative feedback on their work. The entire process took place in the first two weeks of the session.

The second case-based assessment was related to project appraisal whereby the student groups were required to undertake a cost benefit analysis of a real-world infrastructure development project of their choice (for instance: roads, bridges, dams, railways, airports, tunnels, etc.). This assessment task was introduced just after the lecture that dealt with cost benefit analysis of public projects. Students were required to undertake a literature review to identify economic, social and environmental costs and benefits related to their chosen project type. For example, if a group chose a dam project for analysis, they were first required to undertake a literature review on economic, social and environmental costs and benefits of dam projects. Then the group would apply their knowledge and understanding in analysing the selected case. Extensive analyses, including benefit cost ratio calculations and sensitivity testing, constituted the task. The student groups were required to submit a full report of their study within eight weeks. Weekly meetings were held with the student groups to regularly assess their progress and provide formative feedback for remediation. Students were enthusiastically involved in the assessment as they were studying projects of their interests and from their localities. For both case-based assessments, detailed marking criteria were made available at the onset, which enabled the students to self-assess their progress from time to time. Table 1 shows the assessment criteria used for the cost benefit analysis assessment.

Table 1: Assessment criteria for cost benefit analysis

Specific Criteria	Rating Low….. ….High									
Effectiveness of executive summary/abstract: Provides a concise summary of the study, its findings, recommendations & implications.	1	2	3	4	5	6	7	8	9	10
Soundness of introduction: Soundness of study motivation, problem statement, the aim & the method.	1	2	3	4	5	6	7	8	9	10
Adequacy of literature review: Evidence of critical review of literature related to the case type.	1	2	3	4	5	6	7	8	9	10
Depth of case study: The case is analysed from multiple perspectives - identification and quantification of benefits and dis-benefits of the case from the perspectives of multiple stakeholders; accurate appraisals; and satisfactory sensitivity analyses.	1	2	3	4	5	6	7	8	9	10
Appropriateness of supporting evidence: The adequacy and accuracy of data that support arguments / analyses.	1	2	3	4	5	6	7	8	9	10
The intellectual coherence and quality of conclusion: The conclusion is supported by evidence and arguments – has original views and understandings (your own reflections of the topic studied, drawing from the investigation and findings).	1	2	3	4	5	6	7	8	9	10
Structure and organisation of the report: Clear and concise presentation - has all relevant sections and a clear structure; and accurately referenced.	1	2	3	4	5	6	7	8	9	10
Comments:										

Two online quizzes were also used for the topic of project appraisal in addition to the case-based assessment. The quizzes dealt with the calculation aspects of the topic. A third quiz was designed for topics related to the construction industry and the national economy. At the start of the lectures, students were informed that they would be required to complete an online quiz for the topic covered in the lecture on that day. This made students to be more attentive in the lecture and more diligent in their exercise in tutorial classes than previous years when online quizzes were not in place. The tutorials were avenues for applying the theories to solve problems, which essentially were simulated scenarios.

The end-of-session class test functioned as a summative evaluator/grader of the students. All the topics taught in the session were tested. In order to assist with test preparation, an adaptive eTest was set up on Moodle eLearning platform. The adaptive eTest had model questions that were application-focused. Students were able to practise the questions unlimited number of times and automated rapid feedback was provided for incorrect answers, which helped students to reflect on their mistakes and remediate them. Below shown is a sample question and corresponding feedback for different answers. The underlined option is the correct answer for the question. When a student selects wrong answers, the possible mistakes in his/her though process are highlighted, guiding towards correct understanding. When the selection is correct, the student is encouraged positively to boost confidence. Essay type model questions were also provided to encourage students to engage in discussions and brainstorming using those questions.

Question:

Which of the following will decrease the present value of the mixed cash flows for years 1 through 5 of $1,000; $4,000; $9,000; $5,000; and $2,000 respectively given a 10% discount rate?

a) *Decrease the discount rate by 2%*
b) *Switch cash flows for years 1 and 5 so that year 1 is $2,000 and year 5 is $1,000*
c) *Switch cash flows for years 2 and 4 so that year 2 is $5,000 and year 4 is $4,000*
d) <u>Switch cash flows for years 2 and 5 so that year 2 is $2,000 and year 5 is $4,000</u>

Feedback:

a) *This actually increases the PV of each individual cash flow and therefore the PV of the entire set of cash flows.*
b) *This will increase the PV of year 1 and decrease the PV of year 2. The net effect is that the PV of the entire cash flow sequence will increase.*
c) *This will increase the PV of year 2 and decrease the PV of year 4. The net effect is that the PV of the entire cash flow sequence will increase.*
d) *Well done!*

Measure of Efficacy of the Integrated Assessment Scheme for Improved Learning: Students' Perspectives

At the conclusion of the teaching session, a questionnaire survey was administered to ascertain how the integrated assessment scheme influenced students' approach to learning and their learning experience in the course. The questionnaire consisted of two sections. The first section evaluated the effectiveness of the integrated assessment scheme with four subsections, including: overall integrated assessment scheme, effectiveness of the case-based assessment approach, effectiveness of the summative assessment approach, and overall learning experience. Altogether there were twenty-six questions belonging to these subsections. Participants' responses to these questions were collected on a 5-point Likert scale. The last section of the questionnaire received descriptive comments on two questions: what were the best features of the assessment scheme used in the course; and what aspect of the assessment scheme needs improvements and how. All students in the class were invited to participate in the survey, however only 64 out of 112 students responded to it, making a response rate of 57%. The survey data was analysed to explore students' perspectives. Descriptive statistical measures of the quantitative responses to the questionnaire were computed as shown in Tables 2 - 4. Qualitative comments made by the students were also analysed separately.

Table 2 summarises the responses of the students on the questions that assessed how well the assessment scheme satisfied the key characteristics of learning-oriented assessment, which are: constructive alignment, practice-orientation, variety, appropriate distribution, and assisting progressive learning through feedback. The findings suggest that the integrated assessment scheme has satisfied all these well; the mean ratings of responses to all relevant questions are above 4.00 out of 5.00.

Table 2: Efficacy of the assessment scheme

Ref	Survey question	Mean rating	Standard deviation
A1	The assessment scheme deployed in the course was well aligned with the course contents and expected learning outcomes.	4.28	0.55
A2	The assessment scheme in the course was well distributed in terms of content coverage, time and efforts.	4.08	0.72
A3	The assessment scheme in the course maintained a variety to provide you with opportunities to demonstrate and maximise your potentials.	4.31	0.73
A4	The assessment scheme deployed in the course acted as catalyst for your learning (motivated you to be actively involved in learning).	4.00	0.80
A5	The assessment scheme in the course better prepared you for a professional role that involves economic analyses of construction.	4.17	0.72

Table 3: Comparison of assessment methods

Ref	**Survey question**	**Case-based assessments**		**Online quizzes & adaptive eTest**	
		Mean rating	***Standard deviation***	***Mean rating***	***Standard deviation***
B1	The assessment tasks used in the course were practice-oriented, and intellectually challenging and stimulating.	4.08	0.63	4.03	0.69
B2	The assessment tasks used in the course engaged you in self-directed, deep learning in and out of class.	3.91	0.73	3.95	0.76
B3	The assessment tasks offered opportunities for receiving timely and forward-looking feedback on your progress that can be acted upon to remediate your mistakes.	4.06	0.79	4.06	0.85
B4	Completing the assessment tasks fostered your information literacy (ability to research and evaluate information from multiple sources for a given purpose).	3.95	0.72	4.00	0.84
B5	Working on the assessment tasks in the course improved your organisational and time management skills.	3.98	0.83	3.98	0.79
B6	Working on the assessment tasks in the course increased your proficiency with oral and written communications.	3.75	0.85	3.81	0.83
B7	Completing the assessment tasks in the course developed your teamwork and collaboration skills.	4.22	0.73	3.86	0.83
B8	Working on the assessment tasks in the course enhanced your analytical and critical thinking abilities required for problem solving.	4.03	0.78	4.13	0.77
B9	The assessment tasks were effective for advancing your ability to engage in independent and reflective learning.	3.92	0.78	4.05	0.81
B10	The assessment tasks promoted active engagement in learning activities.	4.05	0.74	4.16	0.86

Table 3 compares the performances of the case-based approach and the summative approach (only quizzes and the adaptive eTest) in driving student learning. The wordings of the questions in this table have been modified to make them common to both assessment tasks so that they can be presented and compared in a single table. However, in the exact questions references were made to appropriate assessment tasks and questions were presented separately for each type. The evidence suggests that the case-based tasks and online quizzes have performed equally well in:

- Intellectually challenging and stimulating students;
- Engaging students in self-directed, deep learning;
- Offering opportunities to remediate progress/mistakes based on feedback;
- Improving organisational and time management skills of students;
- Developing information literacy and communication proficiencies;
- Enhancing critical thinking and analytical capabilities for problem solving; and
- Engaging in learning activities.

Moreover, the case-based assessment tasks have been better for developing teamwork and collaboration skills whilst the online quizzes have been better for independent and reflective learning. Overall, they have been complementing each other and as a result learning has been driven quite effectively as evidenced by the high mean ratings for all the variables.

Table 4 confirms that the quality of learning that the students experienced with the integrated assessment scheme was ranked to be between "very good" and "excellent" as evidenced by the mean rating of 4.29 out of 5.00 in a continuum wherein 1= poor quality, 2 = average quality, 3 = good quality, 4 = very good quality and 5 = excellent quality.

Table 4: Overall learning quality

Ref	Survey question	Mean rating	Standard Deviation
L	Overall, how would you rate the quality of learning you experienced in the course, which was driven by the assessment scheme deployed?	4.29	0.59

A thematic analysis was conducted on the qualitative responses of students to identify the aspects of the assessment scheme that resulted in a better learning experience. The analysis discovered five themes from the textual descriptions, which are: online quizzes, group work, assessment variety, alignment of assessment, and relevance to industry. Online quizzes, including the adaptive eTest, were quoted the most (43%) while alignment and relevance to industry were quoted the least (5%) in the descriptions. Students stated several reasons for favouring the online quizzes in the assessment scheme. Some of the direct quotes with those reasons are as follows:

> *The quizzes based on the lectures provided me with more understanding of each topic and what I need to concentrate on more to understand better in the future.*

> *Really enjoyed the online quizzes, they were challenging and supplied me with immediate feedback, thus allowing me to work on my mistakes.*

> *The online quizzes relating to each tutorial exercise were very helpful in learning skills throughout the progression of the course. They provided a good diversity in the course assessment scheme.*

> *The content of the quiz was challenging and helped to promote consistent learning of all topics covered.*

> *Efficient online quizzes published after the lectures improved the understanding of the knowledge.*
>
> *In class exercises and adaptive e-test gave us chances to practise what we have learnt.*

Similarly, students liked the case-based assessments because they featured group work, interactions and research. Some direct quotes of students on these are as follows:

> *Interacting with others in a group environment enhanced the formulation of ideas.*
>
> *The course had more group work rather than individual work; therefore, it improved my communication skills dramatically.*
>
> *Encouraged further research and deeper thinking, and group tasks fostered further knowledge.*
>
> *Allowed engaged learning and participation in helpful and detailed activities.*

The presence of a variety in assessment tasks, aligning tasks closely with lectures and tutorials, and relating assignments to the real-world were other aspects that students liked in the assessment scheme as quoted below.

> *The variety of assignments.*
>
> *Variety of assessment work spread throughout the semester rather than a heavily weighted final exam.*
>
> *I believe the assessments in alignment with the topic enriched my ability to learn.*
>
> *They followed the course very well, easy to access.*
>
> *Relevant to real-world scenarios.*

A similar thematic analysis discovered two themes for potential improvements to the scheme: online quizzes and case-based tasks. Students suggested the following for future improvements:

For online quizzes:

> *Not only the answer but the working out showing how the answer was derived should have been shown with the results of the online quizzes.*
>
> *Answers of all online quizzes should be published after everyone has finished them.*
>
> *More online quizzes.*

For case-based tasks:

> *More assessment tasks based around promoting communication skills; this is important as students enter the workforce.*
>
> *The case-based assessment needs to be guided through more regularly in each tutorial.*
>
> *The major assessment should be marked in parts, ie. Draft worth 10%...., ...?%, ...?%, and be handed in over a series of weeks.*

Summary and Conclusion

Students regard assessment as the most central element of their university learning and decide the importance of a learning activity based on whether it is assessed or not. Hence, assessment is probably the most important element in course design that lecturers can utilise to drive student learning. Assessments are of two types: summative and formative. While summative assessments help lecturers with grading, they promote rote learning and low-order thinking amongst students. Formative assessments on the other hand encourage deep learning and develop high-order thinking within students, but helps minimally with grading. In an educational setting, grading students is as important as promoting student learning. It has been proven in this study that an integrated assessment scheme is essential for driving student learning. In developing such a strategy in a course, the following key principles should be absorbed into:

- Closely align lectures with assessment tasks both in content coverage and timing;
- Utilise online quizzes to reinforce the understanding of lectures;
- Relate assessments to real-world scenarios through approaches like case-based learning;
- Offer ample opportunities for discussions and teamwork through assessments;
- Break the monotony in learning through a variety of assessment tasks;
- Provide rapid, detailed feedback to enable students to reflect on their mistakes and remediate them before it is too late; and
- Schedule regular progress reviews and marking for case-based or similar assessments, which involve significant efforts and time.

Despites the valuable findings, the study suffers from a limitation. It was conducted in the context of only a single course in a single episode. Thus, it is hard to generalise the findings to university learning as a whole. Further studies in other courses with similar assessment schemes may be conducted to ascertain similarities and differences in findings. Based on this, general conclusions can be drawn. Nonetheless, the study could be an exemplary demonstration and provide useful insights for future efforts by other lecturers.

References

Angelo, T & Boehrer, J. 2002, 'Case learning: How does it work? Why is it effective?', Case Method Website: How to Teach with Cases, University of California, Santa Barbara. URL: http://www.soc.ucsb.edu/projects/casemethod/teaching.html. [1 Oct 2013]

Bailey, K. M. 1998, Learning about language assessment: dilemmas, decisions, and directions. Heinle & Heinle, US.

Boud, D. 1990, 'Assessment and the promotion of academic values', *Studies in Higher Education*, 15 (1), 101-111.

Boud, D. 1995, 'Assessment and learning: contradictory or complementary', In: P. Knight (ed.) Assessment and learning in higher education, Kogan Page, London, 35-48.

Brown, S & Smith, B. 1997, Getting to Grips with Assessment, SEDA Publications, Birmingham.

Brown, S. & Knight, P. 1994, Assessing Learners in Higher Education, Kogan Page, London.

Brown, S. 2004, 'Assessment for learning', Learning and Teaching in Higher Education, 2004-05 (1), 81-89.

Brown, S., Rust, C. & Gibbs, G. 1994, Strategies for Diversifying Assessment, Oxford Centre for Staff Development, Oxford.

Brualdi, A. 1996, 'Implementing performance assessment in the classroom', Practical Assessment, Research & Evaluation, 6(2). URL: http://ericae.net/pare/getvn.asp?v=6&n=2. [1 Oct 2013]

Carless, D. 2002, 'The 'mini-viva' as a tool to enhance assessment for learning', Assessment & Evaluation in Higher Education, 27(4), 353–363.

Daly, P. 2002, 'Methodology for using case studies in the business English language classroom', Internet TESL Journal. 8(11), 1-7. URL: http://Daly, 2002/Techniques/Daly-CaseStudies/. [2 Oct 2013]

Davies, A. 2000, Making Classroom Assessment Work, Connections Publishing, Merville, BC.

Davies, S. 2010, Effective Assessment in a Digital Age. URL: http://www.jisc.ac.uk/media/documents/programmes/elearning/digiassass_eada.pdf. [15 Oct 2013]

Dietel, R. J., Herman, J. L., & Knuth, R. A. 1991, What does research say about assessment? NCREL, Oak Brook. Available online: http://www.ncrel.org/sdrs/areas/stw_esys/4assess.htm. [1 Oct 2013]

Dikli, S. 2003, 'Assessment at a distance: Traditional vs. Alternative Assessments', The Turkish Online Journal of Educational Technology, 2(3), 13-19.

Elton, L & Johnston, B. 2002, Assessment in Universities. URL: http://eprints.soton.ac.uk/59244/1/59244.pdf. [5 Oct 2013]

Gibbs, G. & Simpson, C. 2004, 'Conditions under which assessment supports students' learning', Learning & Teaching in Higher Education, 2004-05 (1), 3–31. URL: www.glos.ac.uk/departments/clt/lathe/issue1/index.cfm [19 Oct 2013].

Kember, D. & McNaught, C. 2007, Enhancing University Teaching: Lessons from Research into Award-Winning Teachers, Routledge, Oxon.

Keppell, M & Carless, D. 2006, 'Learning-oriented assessment: a technology-based case study', Assessment in Education, 13 (2), 179-191.

Law, B. & Eckes, M. 1995, Assessment and ESL, Peguis publishers, Manitoba.

Lombardi, M. M. 2008, Making the Grade: the role of assessment in authentic learning. URL: http://net.educause.edu/ir/library/pdf/eli3019.pdf. [8 Oct 2013]

Reeves, T. C. 2000, 'Alternative assessment approaches for online learning environments in higher education', Educational Computing Research, 3(1), 101-111.

Salas-Morera,L, Arauzo-Azofra, A & García-Hernández, L. 2012, 'Analysis of online quizzes as a teaching and assessment tool', Journal of Technology and Science Education 2(1), 39-45.

Simonson M., Smaldino, S, Albright, M. & Zvacek, S. 2000, Teaching and Learning at a Distance: Foundations of Distance Education, Prentice-Hall, Upper Saddle River, NJ.

Tang, C. & Biggs, J. 1998, 'Assessment by portfolio', In D. Watkins, C. Tang, J. Biggs & R. Kuisma (eds), Assessment of university students in Hong Kong: how and why, assessment portfolio, students' grading, City University, Hong Kong.

Thorpe, M. 2000, 'Encouraging students to reflect as part of the assignment process: student responses and tutor feedback', Active Learning in Higher Education, 1(1), 79-92.

Watkins, D. 1998, 'Assessing university students in Hong Kong: how and why', In: D. Watkins, C. Tang, J. Biggs & R. Kuisma (eds), Assessment of university students in Hong Kong: how and why, assessment portfolio, students' grading, City University, Hong Kong.

Sustainability: Its Adaptation and Relevance in Remote Area Housing

10

Rosemarie Rusch and Rick Best, (Bond University, Australia)

Abstract

Little consideration has been given to the context of housing in remote areas. It is important for the economic survival of many remote communities that appropriate and sustainable housing solutions are decided and implemented. This report examines housing at St Pauls, Moa Island in the Torres Strait, using site information, historical research and a review of cultural and geo-political factors to compare the current model with similar studies in self-build housing undertaken in the region between 1986 and 1992. It not only demonstrates tangible economic benefits, but also evaluates the environmental and social improvements which can be achieved with a re-examination of the existing model. It is important to consider the value of investing in policies and practices of sustainable development that can play a pivotal role in potential capacity building within such communities. Current land tenure policy means that families wishing to own their own homes must leave St Pauls, or build illegally. Economically they cannot establish businesses so must leave the island or remain on welfare. The result of this is an exponential increase in the cost of providing community housing and the associated increase in social and health stresses. Acknowledgement at a policy level of the links between social and emotional well-being, and 'Closing the Gap' initiatives, have the potential to offer a wide range of funding opportunities and innovative approaches to solving the housing crisis in remote Australia, if they can be implemented in an open and effective manner.

Introduction

There is a burgeoning desire in remote Australia for housing projects that involve self-build or individually directed outcomes in tune with residents' needs and aspirations, which are not only cost effective, but culturally and demographically appropriate. While many studies have been undertaken with regard to publicly funded community housing, little consideration has been given to the context of housing in remote areas, nor the whole of system approach, not just for indigenous communities but their non-indigenous neighbours. If future housing in these areas is to be appropriate and sustainable it is imperative that it is understood that housing in remote areas does not in any way resemble urban housing and there should not be a replication of urban models. This report examines remote area housing at the grass roots level and uses the village of St Pauls on Moa Island in the Torres Strait as a focus for discussion of the link between housing and social well-being, economic certainty and, ultimately, sustainability. It compares the current model with a similar study in self-build housing undertaken at Moa between 1986 and 1992.

Overview

Foundations of Sustainability

Housing for Health[1] identifies a link between a healthy living environment, health and wellbeing, education and training, good governance, resources and continuing development review. The sustainability continuum is enhanced by the resilience of the community and their courage to pursue follow-on enterprise through knowledge and economic security.

---

[1] Housing for Health are projects undertaken by Healthabitat which are based on principles which include safety and healthy living practices.

Figure 1 Foundations of stable, sustainable human settlement and development in Australian remote areas

Sustainability requires reconciliation of environmental, social and economic demands to harness the potential for long-term maintenance of human well-being. This is not as simple as providing a house. It is about a process of empowerment through inclusion and capacity building, which results in social transformation.

Sustainable Design

Sustainable design should relate to regional climate, soil conditions and cost effective building materials and technologies. Cultural aspects are also important as this affects the way people use buildings and their incorporated systems. The short lifecycle of housing in remote areas has proved to be a direct result of not considering the context of culture and environment in which they are constructed.

Evidence-based data collected by Housing for Health (Healthabitat, 2013) since 1987 to improve housing in Indigenous Australia initially showed that 80% of all problems were due to faulty construction, lack of supervision and lack of maintenance, dispelling the stereotypical view that vandalism was the cause. Well considered housing should draw upon community resources. Implementing a program of education and training releases the responsibility for maintenance of existing housing to the community, providing not only employment but also a sense of self-reliance while reducing the cost of uninterested fly-in-fly-out contractors.

Remote Area Costs and Sustainability

There is no doubt that distance directly affects the cost of construction. Apart from transportation of materials, the current practice of bringing in external labour at premium rates and the lack of competition in tender processes impose cost pressures which impact heavily on the number of houses that can be built within the allocated budget.

There may also be a reduced work season, due to inaccessibility during the wet season and high summer temperatures. These inhospitable conditions are not always reconcilable with the standard of housing constructed. Corrosion caused by heavy rainfall and high humidity in the tropics and sand and heat in the arid areas of Australia, coupled with overcrowding resulting from low housing availability, mean that sustainable design needs to address the robustness of materials if houses are to last in these environments. Meeting the triple bottom line (people, planet and profit) may mean initial increased expenditure in construction but substantially less expense in maintenance or replacement over time.

The triple bottom line (TBL) has been generally accepted as a measure of corporate social responsibility and aims to measure corporations' social and environmental, as well as financial performance over time. It works on the premise that only when a company produces a TBL is it taking account of the full cost of doing business. Arguably a succession of governments has not achieved a TBL balance when viewed in the context of health, happiness and employability that comes from self-help housing, particularly as demonstrated at St Pauls.

The National Partnership Agreement on Remote Indigenous Housing (NPARIH) (2009) was to establish a strategy to improve the poor standard of housing in remote Australia and address the high levels of overcrowding and homelessness, with the first phase intended to bring existing houses to a safe and habitable standard. The Agreement was also purported to ensure ongoing maintenance, infrastructure and essential support services as well as economic development opportunities through local training and employment in construction and housing management. It was hoped the Agreement would use housing as a platform for achieving targets in health, education and employment through improved living standards, and contribute to developing and maintaining sustainable communities in remote Australia. Whether this percolated through to the Torres Strait and Moa Island in particular, is questionable.

Inclusive, integrated planning is required in remote areas to improve construction processes. The current government policies are non-inclusive. Planning, design and funding are decided in boardrooms thousands of miles away with very little consultation and even less qualitative evaluation. To fully understand the implications one needs to be aware of just how remote some of these areas are. Moa Island is north of Bamaga, the furthest northern community on the tip of Queensland and is in the 'Very Remote' category in the Banks Channel of Torres Strait.

The topography is mountainous, with a near pristine biodiversity including rainforests, mangroves, reefs and plains. Moa Island has two villages, Kubin and St Pauls. In 2011 the population of Moa was estimated at 421 with approximately half living at Kubin and half at St Pauls. Of that population 95% identify as Aboriginal or Torres Strait Islander.

Logistics

Construction costs on Moa are at least double those in Brisbane (Rawlinsons 2012). Transportation alone[2] could justify this, however the cost of community housing in remote areas is considerably higher than this and patently unsustainable, not just in fiscal terms but also in terms of social and economic degradation. Established patterns of malaise across health, justice and incarceration, overcrowding and lack of employability or training associated with the limited options in community housing, make it imperative to reconsider alternatives in self-help.

---

[2] While there is air transport for smaller items, most provisions come by barge from Cairns once a week. If some of the locally available building materials could be used, not only would transportation costs be reduced but the economy of the island would be lifted and, with the resultant availability of jobs and community infrastructure, more economically sustainable.

Pholeros (2013a) estimates the cost of a single community house on the similarly remote Tiwi Islands north of Darwin to be in the $1 million range, and that "*money is simply not an issue*"[3]. He confirms that most new housing in remote areas is public housing, is a result of policies that are the product of often chaotic processes and is 'delivered' with little involvement of tenants or communities. He suggests that this lack of interaction is common for social housing and has been more evident since the introduction of the 2009 NPARIH, contrary to its own output benchmarks.[4]

If money is not an issue at present, then accountability for its mismanagement will surely become an issue in the future as a key indicator of unsustainable government practice. For the Australian legal and bureaucratic framework for remote area housing to become a worthwhile proposition it needs a coordinated effort on the part of Local, State and Federal governments to manage the system as a whole, rather than making decisions and funding projects independently within their own portfolios. One could contend that such participation is professionally threatening. Finding common ground means sharing authority and being accountable.

Remoteness Scope

Remote Australia is frequently portrayed as a series of dysfunctional Indigenous settlements and seen therefore as an 'Indigenous' issue. However it can be claimed this is a result of remoteness rather than cultural or racial divisions. Predominantly non-indigenous settlements in remote Australia suffer from similar issues. While this report is centered on Moa Island, it does so with reference to remote Australian housing as a whole and an ineffective and disengaged government response despite years of agreements with, and supplication from pastoralists, indigenous communities, health workers, educators and others who have an interest in working jointly to bring about the systemic changes needed.

Walker (2012) cites Fred Chaney[5] who suggested that there is a social, ecological and economic crisis facing remote Australia as Australia's political and economic system serves the large number of people residing in urban Australia at the expense of remote Australians. This, Chaney suggested, will have significant consequences for the whole of Australia although most urban Australians are oblivious to it. Statistics (Walker, et al., 2012) show that 85% of our land mass contains only 11% of our population but produces more than 60% of our national wealth through mining, primary industry and tourism. Surely this should entitle people in these regions to better, fairer housing options and an opportunity to remain economically viable.

As Healthabitat (2013) has demonstrated, programs that focus on delivery of a physical product alone, such as housing, will not address issues of poverty and chronic illness in remote areas. Achieving change requires much more than the design of structure; it is about recognizing the potential in people to control and manage their own affairs and removing the obstacles and constraints of legislation which prevent them realising that potential.

---

[3] Pholeros is referring to money available under the National Partnership Agreement. His frustration is with the way the money is allocated and how little is directed where it was intended.

[4] This Agreement between the Commonwealth Government and individual State Governments and Territories (Council of Australian Governments) aims to facilitate reform with a ten year ($5.5bn) funding strategy to provide, among other initiatives, housing for Indigenous people in remote communities and to address overcrowding, homelessness, poor housing condition and severe housing shortage in remote Indigenous communities.

[5] Frederick Michael (Fred) Chaney, AO is a former Australian politician who, until April 2007, held the position of deputy chairman of the Australian Native Title Tribunal and is Chair of Desert Knowledge Australia and on the Board of Directors of Reconciliation Australia.

The Context of Place-making

Regardless of the diversity of locations, climate and conditions in Australia, housing delivery is homogenous; arguably lacking in relevance or any connection to the concept of place-making.

Government and external professionals, perhaps well intentioned, provide recipients with little opportunity to obtain housing that is appropriate and meaningful to them.

> I think architects, engineers and administrators are taught to be clever with materials and technology, to make statements on art and society, and to solve and colonise the unknown. Whilst I like to think this preoccupation at times has its place, it has less relevance when working with Aboriginal and Torres Strait Islander peoples. Here we must have faith in and make real space for the creativity and lifestyle aspirations of dynamic communities of Indigenous Australians. They are not needy subjects of White benevolence but equal partners in the challenging times we have ahead. (Haar, 2000)

Architect Paul Haar was involved in a number of self-help housing projects during the 1980s and early 1990s, one of which was on Moa Island. What he found was that aspirations, logistics and processes varied enormously, not only between communities and regions but also between individuals and families within the Moa community.

Haar is considered a specialist in self-build remote area housing and his work on Moa over 20 years ago has provided a benchmark against which contemporary housing policy in remote communities can be compared.

Although the St Pauls community is regarded as a discrete indigenous community, this report is not about indigenous housing or anthropological or racial issues, rather it comments on tangible housing outcomes that might benefit and enhance remote communities. Research has shown that there is a common link between pastoralists, Aboriginal people and anyone else living ‘out bush’, expressed in their concern for their country, the land and the sea which sustain them (Walker, et al., 2012).

The cattle industry would not have succeeded without the active involvement of Aboriginal stockmen and their families. Sugar cane, a major export, would not have been possible without Islander and other immigrant labour and missionaries would not have survived without the support of the remote communities in which they were stationed. St Pauls men have served in wars, serviced remote lighthouses and since 1968 have held the record for railway track laying on the Port Hedland to Mt Newman railroad construction project. Islanders were described as tremendous workers, strong, willing, fast and efficient; men who were commended for their skill in handling the heavy rail, their fitness, their rhythmic team work and pride in their work (Newie, 2012). These qualities were evident in the people Haar assisted in the 1980s and is still apparent today. Yet they struggle to bring economic security to their island, for the right to self-determination and even to put a roof over their heads.

Housing procurement is not static. It is subject to the vagaries of globalisation, cultural intermingling and environmental and economic relationships. Within this dynamic framework, why then has remote area housing languished as is evident on Moa?

There is no place here for a ‘one size fits all’ approach which disregards the relationship between people and place-making. St Pauls is an evolving community and as such housing delivery needs to be divorced from mainstream architectural and construction theory and red tape. Instead of continuing to provide social housing at huge expense and encouraging a co-dependent relationship between government providers and passive remote area recipients, it may be prudent to explore more fully a policy of interdependence and participation, resilience and productivity as facilitated by Paul Haar on Moa more than 20 years ago.

From that allocated to the Torres Strait region out of the Government's $5.5bn (over 10 years) commitment to housing delivery, only one new house has been built at Moa in recent years (TSIRC, 2013). Based on Pholeros' estimation of $1m per house this could be the full allocation for Moa for the foreseeable future. More definitive construction figures have been difficult to obtain from TSIRC as there has been no response to requests for information or clarification. This budget only takes into account 'provided' housing; it does not consider opportunities for home ownership which has in the past been restricted by:

- underlying community land tenure arrangements
- the high cost of construction in remote areas
- the absence of commercial lending and financial services
- low incomes
- a lack of education about home ownership

Despite the 20 year gap since Haar's work at St Pauls, this situation prevails.

Settlement History

Population Statistics

The St Pauls population is almost entirely descended from South Sea Islanders who were permitted to settle there after the forced repatriation of Pacific Islander labourers, following the Federal Government's introduction of The White Australia Policy in 1904. Permission to stay was based on age, the length of time they had been in the Torres Strait, marriage to Torres Strait Islanders (thought to have settled on Moa 1000-1500 years ago) or property ownership.

Settlement Timeline

The pearl rush which began around 1879 brought foreign seamen from the Pacific Islands, the Philippines, Europe, Indonesia, Malaya and Japan, and merchants from China and Sri Lanka. In 1872 islands within 60 miles of mainland Australia were legally annexed to Queensland. The remaining islands were annexed in 1879.

In 1885 the Government Resident Magistrate, the Hon John Douglas, stationed at Thursday Island, instituted a system of indigenous self-government which was revoked after his death in 1904 when the Queensland government placed Islanders under the Protector of Aboriginals, essentially an administrative policy of control over their lives and earnings.

Despite this, by 1911 the ethos at St Pauls was one of hard work and industry. The community was self-sufficient with individually owned homes, established gardens, a church and community boats. Island taxes paid for all island infrastructures. (Bell, 2012) The residents of St Pauls were not subject to the Torres Strait Island Act because missions were exempt from control. They held full citizenship rights including the right to vote. The community had a viable, successful and sustainable social and economic base.

By the 1930's St Pauls residents had become indigenous under legislation. They were required to contribute a percentage of their wages to their Island Welfare Fund, some of which disappeared in unexplained charges and taxes. A nightly curfew was instituted and travel passes were required for any movement off the island unless at the pleasure of the government. Community boats (and their associated profits) became government boats.

In early 1936, a general strike by Islanders working the boats was instigated to demand freedom of movement, the right to run their own affairs, choice over where and with whom they could work, and better pay rates. It is believed to be the first organised Islander challenge to European authority, but gained little in real terms. Government still retained control of boats and wages. Nevertheless it was a further verification of the Islanders initiative and determination to work toward a better future.

In 1947 the first islanders were allowed to move to the mainland for work, beginning the post-war diaspora which halved the population and threatened island viability. It is debateable whether St Pauls has ever fully recovered from these policies either socially or economically. (Bell, 2012)

By the early 1980's when several extended families returned to Moa seeking a more traditional lifestyle to that on mainland Australia, St Pauls was officially classified as a Torres Strait Community. Politically this was seen as a step in the right direction. Unfortunately they were subject to the vicissitudes of unpredictable political manoeuvring where land tenure and self-determination were concerned.

Land Tenure

Brief History

St Pauls land leases, although locally registered since 1905, never existed officially and were replaced in the early 1970 by Permits to Occupy, which did not confer exclusive possession and were a means of producing revenue for the Lands Department, authorising (for a small fee) informal leaseholders to occupy land upon which they would otherwise be trespassing!

By 1984 the land status changed to a Deed of Grant in Trust (DOGIT)[6] which was amended again in 1999 under the Native Title Determination to an Indigenous Land Use Agreement. This meant the people of St Pauls were now under a Native Title Prescribed Body Corporate with no individual ownership possible. (Bell, 2013b)

In 2008, with the creation of the Torres Strait Island Regional Council, under the Queensland Government amalgamation policy, jurisdiction of land has become a political 'hot potato'. Buying or building a home on communal land is simply not an option unless land tenure is secure.

Moa Land Tenure Status 2013

Legal information obtained in February 2012, qualified as being neither advice nor legal opinion, (Bell, 2013a) acknowledged that Historic Title exists in St Pauls, but dealing with it is problematic as it has no legal status under any legislation. It appears that individual land holdings, leases and grants have evaporated. Currently owners who have built their own houses don't own the land or have leases, and can neither raise finance for mortgages nor get insurance. While the Australian Government's Indigenous Business Australia offers loans to buy or build on communal land, the policy is somewhat restrictive. To be eligible, the community must first have been assessed and accepted as a location in which the program is to be made available and then, only if land tenure and administrative

---

[6] A Deed of Grant in Trust (or DOGIT) is the name for a system of community-level land trust established in Queensland allowing community councils to be created and to own and administer former reserves or missions. These Incorporated Aboriginal Councils have the power to pass by-laws, appoint police for the community, and are responsible for maintaining housing and infrastructure, running the Community Development Employment Program and issuing hunting, fishing and camping permits. They work much like local government, but are different in that they own the land they administer on behalf of the community.

arrangements are in place (IBA, 2013). This clearly does not apply to St Pauls and as at March 2013 the IBA has not extended its program to the Torres Strait region.
The dearth of good integrated (and hence sustainable) decision making within the public service, without considering cross-disciplinary impacts, has undoubtedly contributed to the mayhem surrounding land tenure in remote Australia.

In terms of housing it is imperative to progress land tenure policy because there is more at stake. Council of Australian Governments (COAG), National Affordable Housing Agreement (NAHA) and National Partnership Agreement on Remote Indigenous Housing (NPARIH) need this resolution in order to facilitate home ownership opportunities, thereby generating a means of transferring the responsibility for housing to the community and allowing governments to be seen to be facilitating social wellbeing while concurrently meeting their own economic efficiency goals. As policy currently stands, finance for home ownership is only available in economically sustainable communities, leaving the onus on governments to continue funding community housing provision and maintenance for a much longer term.
If Fred Chaney's predictions are correct, home ownership will never be possible; communities already struggling will no longer be economically viable. In summary, on Moa:

- There is no land available to build on which is not Crown land
- Where there is land title it is a DOGIT therefore home loans are difficult to obtain

One of the main objectives for economic development in the Torres Strait region was to ensure resolution of land tenure issues so that there is sufficient land available for economic development opportunities and that this will mean housing options for both affordable ownership and community rental are addressed. Regrettably this target has not been met.

Architectural History

The Early Years

The roots of self-build housing began with early settlement at Moa. Like many of the other settlements in Australia, Moa homes were constructed of logs, bark, grass, leaves and bamboo, depending on the available materials in a particular area, which were adopted and assimilated to construct substantial thatched roof stilt houses. Mangrove and Bloodwood were also used, particularly with mission buildings. The Islander's homes emphasised external living with social interaction for activities such as cooking and eating, while environmental factors influenced the design and setting. The houses were sited, designed and built by the occupants.

The New Order

Ecological degradation and the resultant species protection policy make it unlikely that timber could now be legally harvested for construction on Moa, although this resource was still available during the 1980s. Haar (Haar, 1992) recommended the establishment of a forestry program and possible sawmill to provide building timber for future generations. No indication of implementation is evident, although there is sufficient land available and this could still be a viable and sustainable industry with venture capital and policy review. There is already a CDEP Ranger Program in place for environmental management of this resource.

Moa Islanders, like many others in remote Australia, are a dynamic and self-reliant community in a continual process of change. Their homes are not merely shelters but also a means of self-expression, reflecting the needs of family, tradition and social patterns. As Haar discovered, contemporary remote area housing aspirations at Moa are not necessarily based on the Melanesian or Pacific Islander culture from which they originated. Traditional materials such as bamboo and woven palm were by the 1980's considered inadequate and *'proper solid'* housing such as mud brick and stone was favoured (Haar, 2000).

Through a process of political and religious change, as well as the migrations to the mainland and back, both imposed and voluntary, many people now lean towards a more 'western' housing ideal. This should not necessarily be construed as wanting community provided housing.

Social Cohesion and the Self Help Model

Similar to the situation described for Cape York, in the words of Noel Pearson (Pearson, 2009) *"they want a hand up, not a hand out".*

Haar recognized that housing in the Torres Strait was dependent on external forces with little input from Island people. Disregarding the cost of imported building materials and labour, he expressed concern at the long term social costs, believing the process contributes to a loss of interest and skills for expression shown by their grandparents as members of a self-sustaining community (Haar, 2013). Interviews and observations carried out in 2012 showed this to be a continuing and valid concern. Many of the following generation are content to accept community housing while their parents and elders struggle to understand the loss of status of their Historic Title in St Pauls community and the resultant difficulty in securing home ownership.

Haar presented self-build housing as one alternative for many communities whose access to a fragmented and failing housing delivery system was limited at best. He does not suggest that it is suitable for all and discusses both the advantages and disadvantages in detail (Haar, 1992). It is interesting to note that his time at Moa and other remote communities was during a period of groundswell movement in Australian self-build housing. Individuals across the country, following the ideals of Alistair Knox, GF Middleton and others, started to consider alternative building techniques, particularly mud brick. While for Knox this had been philosophically driven, for most it was a resistance and determination driven by financial necessity or an anti-establishment agenda. In remote pockets, away from official reach, self-build houses were established using whatever materials were available with scant regard for building permits. In Kuranda (North Queensland) for example, the practice was so widespread that by the 1990's the Mareeba Shire Council was forced to offer amnesty, in an attempt to track down and 'approve' these homes. While this was a beneficial and desirable outcome for the owners, it also provided Council with a dramatic increase in rates collection without the need to include additional services or infrastructure.

Haar's experience at St Pauls was similar and sadly the situation has not improved in the intervening years. The self-build housing he facilitated was undertaken by families with little choice but much determination and hard 'yakka'. Unrelenting resistance from QBuild, the government building contractors, their financial position and low priority in the housing delivery line meant they initially built their houses with no permits.

All levels of Australian government must recognise the importance of ensuring remote area communities such as St Pauls can access the necessary resources to meet their basic needs, as well as being given opportunities to use, develop and retain their resources in ways that will improve their quality of life and their cultural, social and economic development, as promised in the plethora of policy reforms on housing and homelessness in remote areas. Progress has been made however none come to mind in recent years at St Pauls that encourage community participation, nor adapting the construction methodology to suit the workforce skills or passing those skills on to the next generation. The mediocrity of public service decision making, without consideration of social and locational logistics has only exacerbated the problems of remote area housing, both community and self-help.

Available Resources

Building Materials

A number of building materials are readily available at Moa, but for a variety of reasons they are not being utilised. It is not the purpose of this report to describe building materials in detail, merely to confirm that there are abundant resources available which need to be explored in an attempt to offer alternative pathways to housing procurement, whether self-help or community housing. Potential materials include:

- Granite (abundantly available and currently used for decorative purposes)
- Bloodwood, Mangrove
- Bamboo
- Concrete masonry block
- Recycled plastic from ocean borne debris
- Clay (Clay pits previously used by Haar and others for mud brick making)

While it is reasonable to conclude that some materials are finite in supply or impractical for intensive harvesting, several others could be beneficial in the long term.

Timber

Certainly there is no suggestion that timber stocks should be depleted, but there is room for continual improvement in sourcing and studying available construction materials and building technologies as part of the procurement process, while examining affordable alternatives. Despite Haar's recommendation for a forest management program in the late 1980's, little progress has been made to implement a forest industry and the situation has been further complicated by overlapping government policy under the 'Caring for Country' and 'Conservation Management' banners.

Mangrove is further protected because of its importance to fisheries and the marine ecosystem. The Mangrove Watch Program (Mangrove Watch Ltd, 2013) monitored and analysed by James Cook University, is not restricted to monitoring of mangroves but encompasses the whole shoreline and promotes better management and conservation of tidal and freshwater wetlands. If Moa is to maintain the existing near pristine marine bio-diversity these studies are extremely important. Anecdotal evidence suggests however that when money is scarce for building materials the islanders still harvest from the back of Moa *"where no-one can see"*. Frustrated by lack of progress in reducing overcrowding and homelessness, it seems that desperate times lead to desperate measures, which will not be sustainable in the long term. Without research into the viability of a forest harvesting program, conflict between housing needs and conservation will remain.

The Bamboo Factor

Bamboo forests are extensive on Moa but are severely overgrown and there is uncertainty about this species' suitability for structural use. While not advocating that this resource is a major building material, with the current trend towards renewable resources and innovative material science studies into bamboo uses, it is unquestionably worth future consideration.

The current impediment to bamboo use in relation to construction at St Pauls is that the plantations need to be re-invigorated and studies undertaken into the efficacy of the species and its uses. Aided by better road access, the study could be incorporated into the existing Ranger or CDEP programs and provide subsequent regional employment. Should research show this species not to be of a structural quality there are many other uses for this renewable resource which could contribute to the economy of St Pauls in the long term. The plantation in its current state is of little benefit to the community.

The Torres Strait islands have many advantages over other remote communities. The sewer system was upgraded in 2009 and St Pauls is well serviced with reticulated water, water tanks and power. In keeping with the objectives of TSIRC commitment to economic development, CDEP workers were up-skilled during the sewer system upgrade and transitioned to full time employment. However other projects have not benefitted the community in the same way.

Concrete Block

A concrete masonry block plant once supplied Moa and most of the outer islands but has been closed since the amalgamation of Councils in 2009. With willing local government agreement and an injection of venture capital this plant could provide jobs and reduce the cost of importing building materials to some degree.

Granted, concrete block is not viewed by architects as appropriate for tropical design, where lightweight, low mass materials are recommended, however if supply of building materials remains a major issue (and it has proven to be so) then there is some merit in pursuing the commercial viability of this resource.

When considering key design responses in remote communities as suggested by the Australian Housing and Urban Research Institute (Fien, 2008), concrete block certainly doesn't meet all of the criteria. It is probably not culturally appropriate or eco-efficient however it is possible that if it could be made readily available it may fulfil the criteria for employment opportunities, life cycle costing and procurement.

Simple construction materials such as blockwork may be suited to the community workforce and their attraction towards solid construction is not necessarily a bad thing if it brings home ownership within the reach of families through better affordability. Blockwork is also durable and cyclone resistant. Individual owners are perhaps best placed to decide the homes they want.

Figure 2 Blocks from abandoned plant, a forlorn and neglected resource

Figure 3 Abandoned block plant has fallen into disrepair

Ocean Borne Debris

In the case of ocean borne debris, a potentially unlimited renewable resource could be explored for future economic benefit to the community, with the dual benefit of protecting the environment and valuable bio-diversity of the Torres Strait region. Observation has shown that most of the debris washed up daily on St Pauls' beaches is plastics from foreign shipping, is of a large quantity and is becoming an increasing problem for the local community in terms of refuse collection and disposal.

Muddy Waters

Mud bricks are the ideal building material from an environmental perspective and are a forgiving construction medium. They are also cheap. The use of earth construction is well established in energy efficient housing.

There don't appear to be any restrictions on the use of mud on Moa other than the hard graft required in production and the fact that there is little incentive for the next generation of young people to bother. Most were babies during the time Haar was on Moa. Many have been away to Thursday Island or mainland Australia for higher education and have loftier expectations. They have no desire or motivation to build because they can see no actual benefit to themselves in doing so therefore the requirement for manual labour or 'sweat equity' is of absolutely no interest to them.

The innovative possibilities for self-help housing promulgated by their parents are anathema to the next generation. It is far easier to take social housing with no responsibility for initial outlay or future maintenance if there is no certainty of ownership in any case. Thus the status quo prevails and desire for action is cauterised.

American economist Milton Friedman suggested there were four types of cash, the spending results of which were dependent on the source and user of the cash. I would suggest that his third type of cash applies in regard to the attitude toward community housing on Moa at present:

> I can spend somebody else's money on myself. And if I spend somebody else's money on myself, then I'm sure going to have a good lunch! (The American Spectator, 2011)

Cost Planning

Prime Ministers Report

The Federal Government claims to have exceeded targets for home refurbishment (FaHCSIA, 2013). While the reported targets are Australia wide, with regard to the Torres Strait and Moa Island in particular, no figures have been made available against which to measure success. As previously stated only one house has been constructed on Moa in recent years. There are no immediate plans for more and no mention was made of refurbishment. This is not to suggest that nothing is being done, merely that there is little evidence to support the government's claim, and no way of measuring outcomes.

As early as 2007 Healthabitat (Pholeros, 2013) queried inconsistencies between expenditure and outcomes in remote area housing. Pholeros casts doubt on the gap between the stated budget and the program scope since the inception of the NPARIH. Alarmingly he suggests that there is no information about where the money is going and that the program is proceeding with scant regard for evidence based solutions. His response to the publicly stated budget is detailed on the Healthabit website and can be summarised as follows:

- If administration and overhead costs were 8% (as originallty stated) the budget should have been $3,478,680,000
- The published budget was $5,475,700,000 suggesting that administration and overhead costs were actually 70% (Healthabitat, 2013)

Pholeros questions whether it is possible that fees and charges have been raised to 70%, when the original agreed cap was 8% and why there seems no correlation with the national programmed budget. Success is contingent on transparency and accountability or we will have yet another vexing housing delivery model. The above figures are food for thought. Pholeros does qualify his calculations as having insufficient information but one would have to ask – why the secrecy? In this instance perhaps Friedman's fourth type of cash applies.

> I can spend somebody else's money on somebody else. And if I spend somebody else's money on somebody else, I'm not concerned about how much it is, and I'm not concerned about what I get. (The American Spectator, 2011)

Regional Employment

Pholeros is also cautious about the employment figures claimed. Clearly if only 30% of the construction workforce is indigenous (FaHCSIA, 2013) this is still well below the Healthabitat recommended figure of 78%. He is far less optimistic and contends that there are places where undertaking a safety course was counted as employment (Pholeros, 2013).

The Torres Strait Regional Authority asserts that it follows these employment objectives (TSRA, 2013) however discussion with the CDEP workers at Moa suggest that they are rarely included in any construction work or transitional training.

Lack of due process will only contribute to incremental dependence-inducing behaviour and displacement from community. Unless some tangible economic benefits can be gained in terms of mainstream employment, migration of adolescents and families to the mainland for education and jobs will once again change the demographic of St Pauls. Only the ageing will remain, wealth and income will go to the mainland, and Centrelink will continue to be the dominant economy on the island. This renders home ownership and self-builds all but impossible regardless of incentives.

Housing Alternatives

Self-help

During Haar's time at Moa the self-help housing approach brought home ownership within the reach of families who were prepared to work hard. He introduced simple construction methods suited to the community workforce and respected their attraction towards a more local and traditional building culture using bush materials that were readily available.

An effective ownership strategy means allowing people who can be involved to believe that what is being offered has the potential to make a difference in their lives. Progress in the intervening years has meant that the self-build model has foundered, in the main due to land tenure and funding restrictions. But the desire for a self-help model has intensified and fanned the remaining coals of persistence.

Reflections on the issues confronting St Pauls residents in terms of a habitable home show that despite the competing imperatives of economic survival and social cohesion, pockets of resourceful independence simmer.

Several islanders, tired of waiting and confronted by antagonistic public servant interventions have *"just built"* regardless of land tenure, building permissions or finance. Unimpeded by perceived legal barriers and knowing that there would be scant assistance from government at all levels they were faced with the choice of leaving the island to pursue work and a home of their own or go down the self-procurement path by whatever means possible. While researchers and governments talk of design frameworks and culturally appropriate housing, such lofty ideals are a long way off for the people of St Pauls. Any modest shelter is better than none at all. Tired of the rhetoric and hollow promises, they have chosen to get on with unfettered implementation.

Having already passed retirement age, the chances of gaining employment off the island was not an option for this group and in any case, family ties mean they want a place where their grandchildren can return to during school holidays or *"if they get into trouble"*.

New Technologies

There have been many innovative advances in technology, in particular in regard to construction, which are markedly more cost effective than the methods government is currently employing on community housing. Mining camps and cattle stations for example face similar issues of remoteness and yet they can construct in record time and (because they have to be competitive and accountable) for much less cost. It appears that taking the word 'Indigenous' out of the remote area housing equation is synonymous with more affordable housing. Mining companies in Darwin are purchasing fully fitted out transportable three-bedroom houses for about $200,000 (Armbrust, 2012). While the total cost may be more than that for housing in urban areas it is reasonable considering the context in which it is required and far less than the $1 million per house the government is currently spending.

There are innumerable building groups that have developed innovative construction technology which delivers housing by significantly faster, more economical and more sustainable methods. Pre-fabricated building solutions, suitable for a wide variety of remote housing construction projects, are able to be delivered in up to half the time of traditional methods of construction with significant cost savings and higher quality finished products. Competitive private enterprise means that most systems can adapt to the specific design features of any given project, as well as flat pack delivery suitable for community labour to erect. This flexibility allows collaborative and adaptive solutions, which is the key to underpinning partnerships and cultivating a sense of belonging and responsibility. A million dollars spent supporting private housing would save many more millions on community housing expenditure and give islanders accountability for their own objectives.

Innovative and intensive approaches are needed to bring about the major shift in motivation that might energise self-help housing. One could conclude that the main impediment seems to be the relentless interventions of successive adversarial government departments who seem unable to 'close the deal'.

Conclusion

There is a quirky expression in remote areas which encompasses a myriad of meanings. 'Which way?' can mean 'how are you' or 'what's up' or 'where are you going' but is also a combination of all of the above; a way of confirming that everything is OK. So, 'which way' for remote area housing policy?

Acknowledgement at a policy level of the links between social and emotional well-being, and 'Closing the Gap' initiatives, have the potential to offer a wide range of funding opportunities and innovative approaches to solving the housing crisis in remote Australia. Breaking down departmental boundaries and giving existing community housing tenants, and those that have self-built the option to take ownership of the homes they live in, would contribute to ending welfare dependence if bureaucracies moved away from their own agendas for 'social' housing. Supporting private housing also supports economic stimulation of communities, which are more than capable of making their own decisions. It is evident that there are only three obstructions to achieving this:

- lack of secure land tenure
- lack of economic security
- difficulty of marshalling political will

The land tenure impediments at St Pauls inhibit the residents' ability to be independent and earn a living. There is sufficient evidence to suggest that they legally purchased their land as early as 1905. Through a series of governmental policies this ownership was erased. It's time the inequity was redressed and land tenure was granted, preferably at no cost. Where family lands are mapped, as has already been proven in St Pauls, there should be no deterrents to transferring the leases out of the DOGIT and into private ownership along traditional lines and inheritance agreements that already exist. Community housing tenants should be given the choice of ownership of the houses they live in at no cost (and of course the responsibility for future maintenance), or a sliding scale of purchase, particularly if they have been paying rent on those houses for longer than five years. The majority of the community houses at St Pauls are old and 'tired' and many of them have long passed their life cycle replacement value.

The intractable problems of land ownership at Moa are the result of legislative and administrative arrangements that should have played no part in the lives of these Islanders. It was their land to start with and should now be returned.

Currently, families wishing to own their own homes must leave St Pauls, or build illegally. Economically, for similar land tenure reasons, they cannot establish businesses so must leave the island or remain on Centrelink. The thrust of this is an exponential increase in the cost of providing community housing and the associated increase in social and health stresses. Encumbrances such as these contribute to the malfunction of place and a loss of social cohesion and self-esteem.

The St Pauls community has spirit and grit. Prior to what amounted to little more than theft of their land and their freedoms in the early 1900s, St Pauls was a thriving, self-reliant community and it could be again, with a modicum of competent leadership from politicians. Is there any point in investing huge amounts of money on community housing that people can feasibly build better and more cheaply for themselves? Give them the building blocks of

land tenure, skills training, self-build opportunity and access to affordable finance. These basic requirements are no different to what is readily available in urban areas of Australia. With a population of just over two hundred at St Pauls, and existing historical maps of family allotments, this should not be difficult to implement.
With all that is wrong, what is the direction for a solution, and what might that look like? Or is it an insoluble dilemma?

> After years of lobbying, meetings, summits and conferences, the Australian Government's Department of Families, Housing, Community Services and Indigenous Affairs in March 2013 endorsed a further paper to 'explore' available information and 'identify' opportunities in housing and broader policy (FaHCSIA, 2013)

What part of the previous thirty years or more of research, reports, reviews and gnashing of teeth have they not understood?

References

ABS, 2013. *National Regional Profile Torres Strait.*

[Accessed 20 July 2013].

AIQS, 2003. *Australian Cost Management* Manual. *Evaluation of Sustainable Development.* 1st ed. Canberra: AIQS.

Armbrust, N., 2012. *AUSCO Homes, Personal Interview* [Interview] (November 2012).

Bell, H. R., 2012. *St Paul's Story.* Smythesdale: Unpublished.

Bell, H. R., 2013a. *Notes from conversation between HRB and Chris McLaughlin.* St Paul's: Unpublished.

Bell, H. R., 2013b. *Personal Communication* [Interview] (January - August 2013b).

FaHCSIA, 2013. *Closing the Gap Prime Minister's Report*, Canberra: Department of Families, Housing, Community Services and Indigenous Affair.

Fien, J. e. a., 2008. *Towards a Design Framework for Remote Indigenous Housing*, Melbourne: Australian Housing and Urban Research Institute.

Haar, P., 1992. *Housing in the Torres Strait Region Towards a Self-Help Approach*, Thornbury, Victoria: Paul Haar.

Haar, P., 2000. A Self-help Approach to Remote Area Housing, St Paul's Village, Moa Island, Torres Strait. 1986–92, In: P. Read, ed. *Settlement: A History of Australian Indigenous Housing.* Canberra: Aboriginal Studies Press, pp. 221-233.

Haar, P., 2003. Community Building and Housing Process Context for Self-Help Housing. In: P. Memmott, ed. *Take 2 - Housing Design in Indigenous Australia.* s.l.:Royal Australian Institute of Architects.

Haar, P., 2013. *Personal communication* [Interview] (June-August 2013).

Healthhabitat, 2013. *Healthabitat.* [Online] Available at:

IBA, 2013. *Buying a Home on Communal Land.*

MacKellar, D., 1908. *My Country. London*: Official Dorothea MacKellar Website.

Mangrove Watch Ltd, 2013. *Mangrove Watch Torres Strait Regional Authority.*

Newie, G., 2012. *Moa Island elder, Mt Newman railway track worker - Personal Communication* [Interview] (February 2012).

Newie, M., 2012. *Moa Island elder - Personal Communication* [Interview] (February 2012).

Pearson, N., 2009. *Up from the Mission*. Brisbane: Black Inc.

Pholeros, P., 2013. *Personal communication* [Interview] (19 June 2013).

Pholeros, P., 2013. *Personal Communication* [Interview] (2 August 2013).

Rawlinsons, 2012. *Australian Construction Handbook*. 30 ed. s.l.:Rawlinsons Publishing.

The American Spectator, 2011. *Friedman's Four Ways*.

TSIRC, 2013. *Annual Report 2011-2012*, Thursday Island: Torres Strait Island Regional Council.

TSRA, 2013. *Torres Strait Development Plan 2009-2013*, Thursday Island: Torres Strait Regional Authority.

Walker, B. W., Porter, D. J. & Marsh, I., 2012. *Fixing the Hole in Australia's Heartland: How Government needs to work in remote Australia*, Alice Springs: Desert Knowledge Australia.

A Review of Productivity Analysis of the New Zealand Construction Industry

11

Malcolm Abbott, (Swinburne University of Technology, Melbourne, Australia)

Chris Carson, (UNITEC, Auckland, New Zealand)

Abstract

In recent years there has been a rising interest in the level of productivity and efficiency of the New Zealand construction industry. This interest has meant that there has been an increased use of statistical techniques to determine the productivity and efficiency of the overall industry. The purpose of this paper is to review the various measures that have been used to gauge the levels of productivity and efficiency in the New Zealand construction industry; as well as analyse some of the results to date. Finally, it considers potential areas for future research, including possible improvements to measurements techniques that maybe used.

Keywords: Productivity, Construction NZ

Introduction

In recent years a number of attempts have been made to identify the reasons for the lack of growth in productivity and income in New Zealand. One area where this lack of growth of productivity is most marked is in the case of the construction industry. Indeed, official statistics seem to indicate that the construction industry in New Zealand has a productivity level roughly comparable to what it was in the late 1970s (Statistics New Zealand, Industry productivity statistics). This has raised considerable concern in the industry itself, and more generally, about the degree to which the industry is operating below its full potential (New Zealand, Department of Building and Housing, Productivity Taskforce 2009; Van Dai Tran 2010; New Zealand Productivity Commission 2011).The possible reasons behind this low level of productivity in New Zealand are various and could include such things as low economies of scale in the industry, a lack of competition, regulatory impediments, faulty innovation and management practice, poor investment quality and a low levels of skills (Davis 2007).

The purpose of this paper is to review the various measures that have been used to gauge the levels of productivity and efficiency in the New Zealand construction industry and to make suggestions to how these measures might be improved. In meeting this objective the results of the past studies will be analysed in order to identify some of the issues surrounding the apparent stagnation of growth in productivity in the New Zealand construction industry. The manner in which this is approached is to compare the different approaches used and to then identify the pros and cons of the different approaches. The final task is to then suggest improved ways to determine productivity change of the New Zealand construction industry that would take account of the difficulties that were identified in the previous approaches used.

More improved measures of productivity of the construction industry would be useful as the construction industry constitutes a substantial part of the New Zealand economy. Furthermore in recent years there has been a concern that this sector has been a relatively stagnant one. More improved measures would assist policy makers and industry leaders to determine ways to improve the performance of this industry.

The paper is structured as follows. In the first section a broad description of the condition of the New Zealand construction industry is given. This is followed by a section that examines the various ways in which productivity can be assessed and some of the issues involved in doing so. A section is provided that examines past studies of productivity in the construction industry in New Zealand and in the final section some conclusions are made.

The Construction Industry in New Zealand

The construction industry constitutes a significant part of the New Zealand economy. In 2010 the industry made up over four per cent of New Zealand's GDP and employed over eight per cent of the employed workforce (Table 1). This high level of employment compared to output reflects the relative labour intensity of the industry.

	GDP	GDP	Employment	Employment
	Constant 1995/96 $m	%	000	%
Agriculture, fishing, forestry and mining	9,980	7.7	150.2	6.9
Manufacturing	16,395	12.7	247.8	11.5
Electricity, gas and water	2,634	2.0	14.4	0.7
Construction	5,521	4.3	178.8	8.3
Wholesale trade	9,374	7.2	103.7	4.8
Retail accommodation and restaurants	9,901	7.6	333.7	15.4
Transport and Communication	13,891	10.7	135.4	6.3
Finance, business &personal and community services.	55,061	42.5	869.3	40.2
Government administration and defence	6,754	5.2	121.2	5.6
Gross Domestic Product	129,511	100.0	2163.4	100.0

Table 1 Gross Domestic Product and employment by industry, 2010, Annual (March years)
Source: Statistics New Zealand, *Gross Domestic Product; Household Labour Force Survey*

Growth of the construction industry was solid from the late 1990s and through most of the 2000s, but tapered off in the last few years of the 2000s (Figure 1). Over the longer term the construction sector has grown at a similar rate to the rest of the New Zealand economy, driven by rising incomes and population growth, although there is evidence that the growth of the industry is more volatile than that of the economy in general (Figure 1).

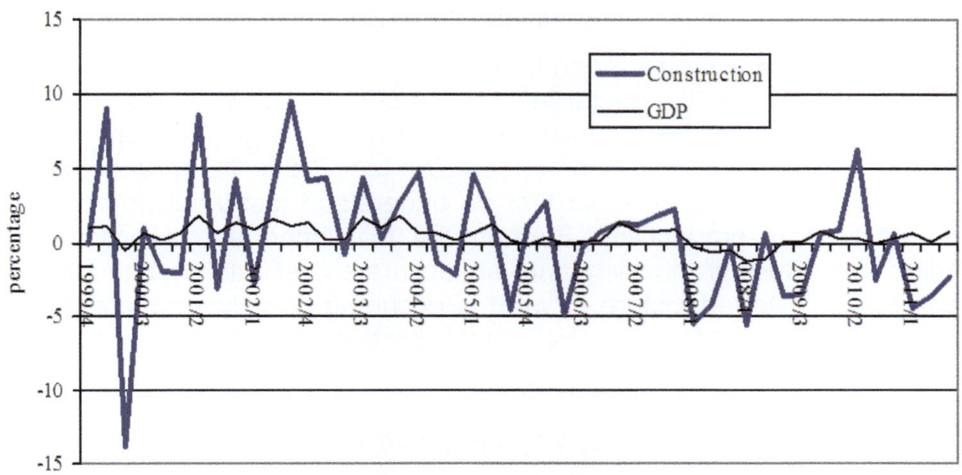

Figure 1 Quarterly growth of GDP and construction in New Zealand
Source: Statistics New Zealand, *Gross Domestic Product*

The construction sector is broadly speaking divided into three sub-sectors:

- Residential building (houses, units, flats);
- Non-residential building (offices, shops, hotels);
- Engineering construction (roads, bridges, water sewerage).

The private sector operates predominately in the construction of residential and non residential building, whereas the public sector operates primarily in engineering construction. Some major engineering projects are delivered through public-private partnerships, which in its simplest form is a contractual agreement between the government and private sector for the delivery of public infrastructure.

Despite the growth of the construction industry and its importance to the generation of employment in New Zealand, there has been a growing concern that the sector has been a stagnant one in recent times in terms of growth of productivity. Indeed some studies have found that the sector's level of productivity is no higher than it was in the late 1970s (Figure 2). A Building and Construction Sector Productivity Taskforce, for example, was established in 2008 as a joint measure of the industry and government in New Zealand to investigate the limited productivity gains in the sector in recent years and reported a number of reasons why this had been the case (New Zealand, Department of Housing Productivity Taskforce 2009).

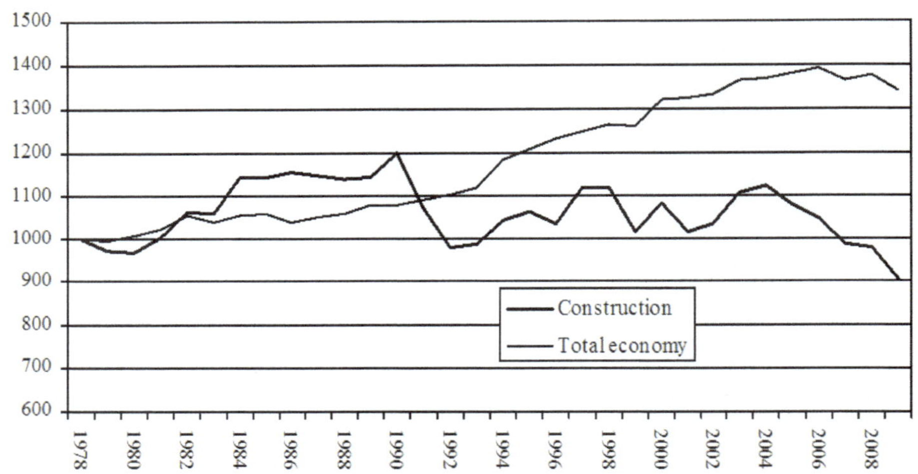

Figure 2 Productivity indexes of construction and the New Zealand economy, 1978 to 2009

Source: Statistics New Zealand, *Industry Productivity Statistics*

In the past there has been some research undertaken that looks at the growth of productivity of the construction industry at the sector level. Most of this past work has been undertaken as part of larger work looking at productivity growth of the larger economy, but in recent times some more focused work has been undertaken (see Table 2). To understand the past work that has been undertaken it must first be understood just what approaches can be made to the analysis of productivity. Furthermore it is important to policy makers and industry leaders that future productivity analysis accurately reflects changes in the industry and does not give a distorted picture of what is occurring. Finally accurate measures can help to raise our understanding or what drives productivity change in the industry, or conversely impedes it.

Traditional Efficiency and Productivity Measures

To understand how productivity analysis has been used in the past to determine change in the New Zealand construction over time it is important to understand the different ways in which this can be estimated. The way in which the productivity and efficiency of the construction industry has been analysed in the past has been influenced by the way in which productivity analysis has developed more generally over the years. Structural reform of the economy by various governments has encouraged the study of the productivity and efficiency performance of a variety of industries and government service providers. In undertaking this research a range of productivity and efficiency techniques have been used.

To begin with, efficiency can be defined as being the degree to which resources are used in an optimal way to produce outputs of a given amount. At the level of an industry there are three main aspects of economic efficiency: allocative, technical, and scale (Farrell 1957). Allocative efficiency indicates the factor input proportions given relative input prices. An allocatively efficient industry produces using the lowest cost combination of inputs. Technical efficiency takes place when the maximum output possible is delivered using given inputs (labour, capital, and other inputs). Scale efficiency refers to optimal size of an industry (and firms within it). It could be that an industry may be allocatively and technically efficient, but may still be producing too little or too much output. Economic efficiency is a combination of allocative and technical and scale efficiency. It helps to identify whether cost reductions can be made.

Productivity change differs from efficiency in that it is a dynamic indicator of the change in the ratio of outputs to inputs over time. Productivity growth reflects changes in allocative and technical efficiency, technological and managerial improvements and changes in the external environment in which the production of an industry occurs. When an industry improves its efficiency, it moves towards a best practice frontier, and in doing so also increases its productivity. When the frontier also shifts outwards, due say to technological change, then productivity also rises.

Over the years a range of different measures have been developed to quantify how efficiency and productivity levels change over time. These measures include the calculation of indexes, the estimation of cost and production functions, data envelopment analysis and stochastic frontier measures. Most of these types of measures have been used in the past to analyse changes in the New Zealand construction industry.

Partial and Total Factor Productivity Index Measures

The initial way in which productivity changes over time were measured was by using an 'index' approach. The index approach involves the estimation of index numbers, which can be used to indicate the partial or total factor productivity change of an industry over time. Partial productivity measures generally show the ratio of an index of an industry's output to an index of a single input factor. In the case of the construction industry, for example, an index of the volume of construction activity per employee is a labour-based partial productivity measure.

Partial productivity indicators have the advantage of being easy to compute, require only limited data and are intuitively easy to understand. This has meant that they are used very commonly in a range of different industries. Unfortunately they can also be misleading when looking at the change in productivity of an industry over time. It might, for instance, be possible for an industry to raise productivity with respect to one input at the expense of reducing the productivity of other inputs. Partial indices of output to labour, for instance, often tend to overstate the growth of total productivity because over time the use of capital as an input substitutes for labour. A further problem is that capital productivity measures are difficult to calculate given the difficulty that often occurs in measuring capital inputs, and the often very long life of some assets.

Total factor productivity indexes were developed to overcome the problem of the substitutability of inputs. Total factor productivity measures are the ratio of a total aggregate output quantity index to a total aggregate input quantity index. The index approaches used to combine outputs and inputs that might be used include the Laspeyres, Paasche, or Fisher approaches. A Tornqvist index approach has been used in many total factor productivity studies. Total factor productivity growth, therefore, is the difference between the growth of the output and input indices.

Although there are a few studies of industry level total factor productivity in the construction industry the literature is not as extensive as it is in many other industries. Most research on construction industry productivity tends to be concerned more with site level labour productivity, which has a more direct relevance to industry management (Ganesan 1984; Lowe 1987; Maloney 1983; Allen 1985; Thomas, Maloney, Horner, Smith, Handa and Sanders 1990; Thomas and Sakarcan 1994). Despite this some measures of the total factor productivity of construction were developed using the index approach. On the whole this occurred as part of efforts to establish the growth of productivity for the entire economy and other industry sectors, rather than exclusively for the construction sector.

Total factor productivity indexes were first developed at the National Bureau of Economic Research in the United States in the late 1940s. Initially studies were undertaken of labour productivity for a range of different industries (including construction). This was followed by the estimation of a range of other non-labour partial productivity measurements such as for capital and in the case of agriculture land and livestock. A range of studies were then undertaken that attempted to combine the various types of inputs together in a similar fashion to the way in which earlier researchers like Simon Kuznets and Colin Clark had for output in determining levels of Gross National Product. In undertaking this work the research by Stigler (1947; 1961), Barton and Cooper (1948); Kendrick and Jones (1951), Schmookler (1952), and Fabricant (1954) were important. Starting with agriculture this work spread to other sectors of the economy as well as national wide studies. By the mid-1950s the National Bureau of Economic Research had published a great deal of work using total factor productivity indexes for a range of industries (Kendrick 1956a, 1956b, 1961, 1973). As part of this work total factor productivity indexes were also determined for the construction industry. Over the years studies have also been conducted in a range of other countries besides the United States using the index approach (see for instance Hong Kong; Chau and Walker 1988; Chau 1990 and Singapore: Tan 2000).

In the case of the New Zealand construction industry examples of partial productivity indexes include those by Marks (1984), Davis (2007), Janssen and McLoughlin (2008) and the New Zealand Institute of Economic Research (2011), who all estimated New Zealand labour productivity figures. In terms of total factor productivity indexes there have been a number of main works undertaken on the New Zealand construction industry as part of economy wide productivity estimates. These are by Diewert and Lawrence (1999), Black, Guy and McLellan (2003), and Statistics New Zealand, Industry productivity statistics (2011).

Econometric Measures – Cost and Production Functions

The index approach to determining total factor productivity assumes that the ratio of inputs used and products produced are kept constant over time. As this seems unrealistic other methods were developed. Econometric methods involve the estimation of cost or production functions. The estimated function can then be used to identify changes in productivity or efficiency. The estimation of cost functions has been the most commonly used method of determining the levels of efficiency in the industry, although a number of techniques have been used in estimating cost functions. This approach is also often referred to as the Growth Accounting Approach.

Dacy (1965) was perhaps the first to use an estimated production function for the United States construction industry (1947 to 1963) and found increasing levels of productivity. Later examples of estimated production functions include those by Allen (1985); Stokes (1981) Goodrum and Haas (2004), Schiver and Bowlby (1985), Mao, Goh, Wang and Ofori (2003) and Kau and Sirmans (1983).Examples of the use of cost functions estimations include those by Chau (1993) and Chau (2009) who in both cases found rising productivity levels in Hong Kong; and Schriver and Bowlby (1985).

In the New Zealand case a number of researchers used the growth accounting approach to estimate productivity over time by estimating production functions. These include the studies of Orr (1989), Chapple (1994), Philpott (1995), and Mason and Osborne (2007). In each case they made estimates at the national level and then broke down productivity into various industry sectors including that of construction.

One difficulty that arises in using either the index approaches or econometric approaches is that it relies upon there being reliable, market determined prices for both inputs and outputs. In the case of industries such as the utilities where market power distorts prices, and in government service delivery where output prices are often subsidised or non-existent there arose a need to develop productivity measures that was not so reliant on price data. This led to the development of Data Envelopment Analysis (DEA).

Data Envelopment Analysis (DEA) and Stochastic Frontier Measures

DEA was pioneered by Charnes et al (1978) based on the work by Farrell (1957) and there are now many texts offering a detailed discussion on DEA, including the algorithms used (see, for example, Lovell and Schmidt (1988), Färe, Grosskopf and Lovell (1985) and Coelli, Rao and Battese (1998)). Data envelopment analysis (DEA) is a linear programming technique which estimates organisational efficiency by measuring the ratio of total inputs employed to total output produced for each unit. This ratio is then compared to others in the sample group to derive an estimate of relative efficiency. DEA identifies the most efficient providers of a good or service by their ability to produce a given level of output using the least number of inputs. Other units in the sample group receive an efficiency score determined by the variance in their ratio of inputs employed to outputs produced relative to the most efficient unit in the sample group. DEA is therefore a measure of relative efficiency against the sample group's benchmark best practice. The advantage is that it can be used without input or output prices, which is useful where data of this sort is not available. As well as being used to benchmark units against one another it is possible to use DEA to estimate changes in productivity of individual units, or the sample as a group, over time.

DEA has been used extensively in a number of industries to assess productivity and efficiency levels (especially utilities). In the case of the construction industry only a fewer number have been undertaken, The DEA approach was used in the case of Hong Kong construction by Chau and Wang (2005), Wang (1998); Wang and Chau (1997, 2001). Each study found a rising level of productivity there. In the case of New Zealand a single study by Fare, Grosskopf and Margaritis (1996) has been undertaken.

Stochastic frontier measures involve the use of parametric techniques to estimate technical efficiency by constructing first the production frontier derived from the best practice firms and then comparing the actual output of firms relative to the best practice firms. Studies of this sort in the construction industry are rare and in the case of the New Zealand industry have not to date been undertaken.

Although markets for construction tend to be fairly competitive and more market driven than the case of the utilities and government service provision there have been concerns that prices in these markets do face some distortions. For this reason the studies previously mentioned made use of the DEA approach at determining changes to productivity over time.

Issues

In addition to the need to use total factor indicators to determine productivity change over time rather than partial indicators and the possible need to use indicators that don't rely so much on price data a number of other issues have arisen. Most studies that look at the productivity of the construction industry are centred on the industry in the United States. One element some researchers found was that there appeared to be a degree of stagnation or decline in productivity in the United States construction industry during the 1970s and 1980s. These include work by Stokes (1981), Allen (1985) and Schriver and Bowlby (1985). These researchers had some difficulty explaining why this might have occurred, given that it was a period of some technological change and innovation in the industry. Tan (2000) found a similar decline in productivity in the Singapore construction industry over the period 1980 to 1996, and Mao, Goh, Wang and Ofori (2003) found Singapore had a slow rate of growth between 1984 and 1998. The explanations of this perceived stagnation are of some interest to New Zealand researchers as a similar stagnation has been detected in this country.

In the American context some thought that this decline could be explained by the use of inappropriate price indexes (Dacy 1965; Stokes 1981). In addition a more influential view expressed at the time in the United States was that construction was largely a labour intensive industry and the introduction of new equipment and materials could only increase outputs with a given level of labour and capital. Rosefielde and Mills (1979), in contrast, argued that official data did not fully account for changes in quality and so tried to adjust for changes in the scale and structure of the quality of output. They argued that as construction is immobile and designed to achieve some particular function, at some specific site, a diverse and changing type off output is produced by the industry. This coupled with a changing composition of materials makes it difficult to measure quality improvements over time, which in turn leads to an underestimation of output levels, and therefore productivity improvements. In using quality adjusted figures for output Rosefielde and Mills relied upon the work of Cassimatis who used the number of rooms per dwelling and square footage of buildings as an indicator of the quality of output.

In the New Zealand case past studies have all used some constant dollar valuation of the industry's production as an in indication of output, which assumes that this value reflects any changes that are occurring in quality and size, whether it be in terms of area or other form of quality. This might, perhaps, be a fair assumption but nowhere has it been justified. Indeed over the periods looked at by analysts the average size of both residential and commercial properties has risen considerably. It is notable that unlike countries like Australia, quality adjustments for the residential building industry are not included in the New Zealand estimation of construction value added (Bollard and Barrow 2012). As most New Zealand studies on productivity use some estimation of value added of the construction industry as the indicator of output it is possible that these studies significantly underestimate New Zealand productivity growth.

Schriver and Bowlby also found that much of the fall in productivity could be attributed to a change in the output mix from high to low productivity building projects. One problem in the New Zealand case is that the past studies have all bundled construction together as a single industry and do not attempt to break it up into its various components (residential, large engineering projects and commercial property). Therefore no attempt has been made to account for either changes in quality of changes in the composition in the output mix.

In light of the issues that have arisen in determining productivity change of the construction industry more generally a number of measures can be undertaken. First all a total factor rather than partial approach should be undertaken. Secondly it is necessary to use an approach like DEA or the stochastic approach along with any index or econometric approach to see if there are any price distortions that are important. Finally any approach should try to make use of physical values of output, along with monetary ones, in order to ensure that distortions to the estimates of productivity change are not being made. Bearing in mind these issues it is possible to observe how productivity analysis has been conducted in the New Zealand case.

Past Measures of Productivity in the New Zealand Construction Industry

Since the 1980s a number of economists in New Zealand have made estimates of New Zealand's productivity change over the longer term. Notable examples include Marks (1984), Orr (1989), Smith and Grimes (1990), Chapple (1994), Philpott (1991, 1993, 1995), Sarel (1996), Grubb, Jackman and Layard (1982),[1] Fare, Grosskopf and Margaritis (1996) and Conway and Hunt (1998). In some of these cases sectoral productivity was also identified, including that of the construction sector (i.e. Chapple; Philpott; Fare, Grosskopf and Margaritis). Further work was initiated from 1999 onwards by government agencies as well (Diewert and Lawrence 1999; Black, Guy & McLellan; and Janssen & McLoughlin (2008).

It should be remembered that all the studies, with one exception (Davis 2007), undertook a study of construction productivity as part of broader works on productivity of the New Zealand economy as a whole. In none of these cases, therefore, was any attempt made to adapt the methods used to account for any special conditions in the construction industry (such as quality adjustments). The separate studies on the New Zealand construction industry are listed in Table 2.

The first work undertaken was by Marks (1984), which used an index approach to determine labour productivity growth in New Zealand between 1965/66 and 1980/81. Marks' work was published in a work by the New Zealand Institute of Economic Research (NZIER) on the Labour Market, funded by the Social Science Research Fund Committee of the New Zealand Government. Marks was concerned with analysing the slowdown in labour productivity in all industries that had occurred in New Zealand from the early 1970s. As well as national figures, however, she also looked at a number of sectors of the New Zealand economy, including construction. In the case of the construction industry labour productivity was found to have increased at a, steady rate through the 1960s and 1970s.

Marks's work was followed by three related works, which all used a growth accounting approach to calculate total factor productivity change along with labour and capital productivity changes (Orr 1989; Chapple 1994; Philpott 1995). These studies were undertaken as part of a joint project between the NZIER and the Institute of Policy Studies of the Victoria University, Wellington (for the Ministry of Commerce). Orr's work was the first of these and his work and was later followed by work by Chapple (1994). Both Orr and Chapple worked for the NZIER which was to retain an interest in productivity analysis and as late as 2011 published work on this theme (NZIER 2011).

Orr's work ranged from 1961 to 1986 and found that total factor productivity rose over the period at an annual average of 1.75 percent (Labour productivity: 2.30 percent, Capital productivity: 1.02 percent). Chapple's work ranged from 1971 to 1991 and showed a rise in annual average productivity of 0.7 per cent over the entire period (1.0 per cent for labour productivity and 0.2 per cent for capital productivity). Both estimated a Cobb-Douglas production function and equated one output to that of the inputs of labour hours and net capital stock, using capital stook figures developed by Philpot. Orr's work confirmed Marks' earlier findings that productivity rose steadily during the 1960s and early 1970s. Chapple also found that productivity rose in the early 1970s. Both Orr and Chapple also found that productivity tended to peak in the years of construction booms in 1975/76 and then again in 1986/87. At much the same time Philpott worked on similar figures for the period 1960 to 1995 (Chapple 1991, 1993, 1995). He looked at twenty separate industries including construction. His results were similar to that of Chapple and Orr, in that he found that productivity growth was steady up until 1975, stagnated somewhat after that and surged during periods of construction boom and flagged during recessions.

[1]Sarel (1995) and Grubb, Jackman and Layard (1982) both undertook studies on New Zealand productivity growth as part of wider studies of growth of the OECD countries.

Figure 3 provides an overview of these works on productivity of the construction industry. As can be seen productivity tended to rise through the 1960s and early years of the 1970s before peaking around 1976 and then declining. A recovery in productivity took place in the early 1980s. Despite the decline in productivity in the late 1970s the situation in the late 1980s still seemed to be a relatively good one. The productivity of the industry tended to rise over the twenty-five year period (1960 to 1985) and this is reflected in the works of Marks, Orr, Chapple and Philpott who all recorded increases in total factor productivity of the construction industry (Table 2).

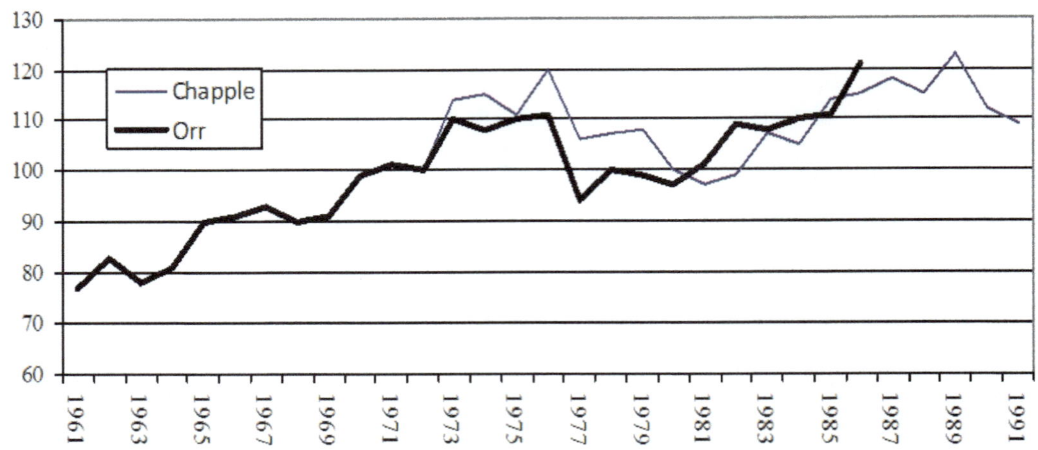

Figure 3 Total factor productivity in New Zealand construction, 1972=100

Source: Orr 1989; Chapple 1994

The first study to come to more pessimistic conclusions was that of Fare Grosskopf and Margaritis (1996). These economists used a different approach and undertook a study on a national basis and looked at twenty industries (including construction). They used the DEA approach and equated value added as the output to labour hours and capital using the perpetual inventory method as the input for the years 1973 to 1994. They found for construction that although productivity rose slightly between 1973 to 1984 (annual average of 0.3 per cent) it declined over the period 1984 to 1994 (-1.2 per cent). For the entire economy between 1973 and 1974 productivity rose by 1.46 per cent, mainly due to technological change rather than efficiency changes. Because their study concentrated more on the period from the mid 1970s onwards their results overall showed stagnation in the growth of construction productivity.

In 1999 Diewert and Lawrence - in work commissioned by the Department of Labour, the Reserve Bank of New Zealand and the Treasury - used labour hours and net capital stock (valued using a user cost formula which accounted for depreciation, a rate of return and capital gain) as inputs and an index of total national output to compute labour productivity, capital productivity and a total factor productivity between the years 1972 and 1998. In doing so it used the index number approach. In this work the construction sector showed a worse performance compared to other sectors, although it enjoyed a trend annual total factor productivity growth of 0.6 per cent although it was the subject of cyclical downturns, especially in 1980 and 1993 (Diewert and Lawrence 1999, p. 74).

Statistics New Zealand received funding from the Growth and Innovation Framework in 2003 to develop official productivity statistics for New Zealand. In doing so they built on the earlier work undertaken by Diewert and Lawrence (1999) and Black, Guy and McLellan (2003).

The results of the work by Diewert and Lawrence, Black, Guy and McLennan and Statistics New Zealand are all summarised in Figure 4. In each case productivity does seem to have stagnated across the whole of the period covered with especially low levels of productivity encountered during the recessions of the late 1970s and early 1990s. There does, therefore, seem to some evidence that productivity in the construction industry correlates to business cycles.

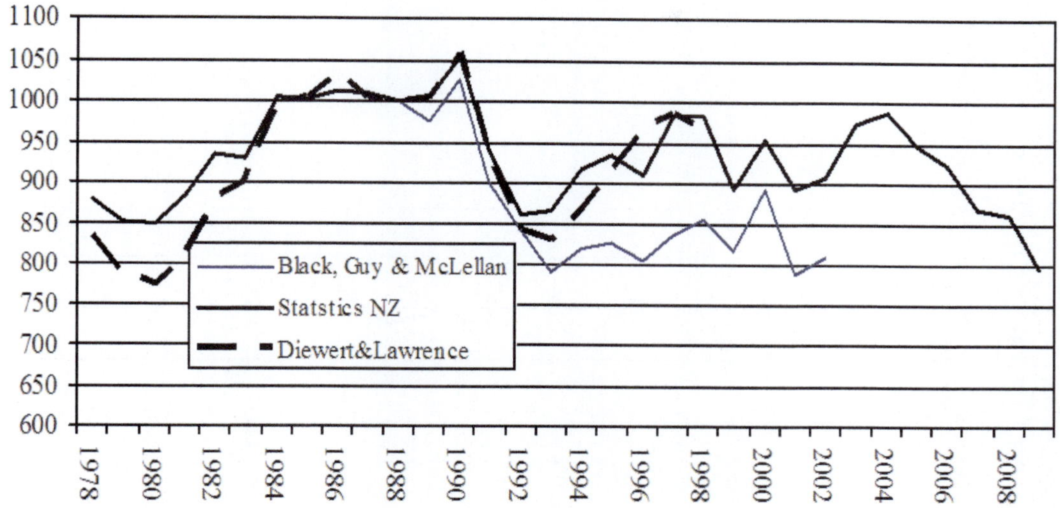

Figure 4 Total factor productivity of the New Zealand construction, 1978=1000 construction industry

Source: Statistics New Zealand 2011; Diewert & Lawrence 1999, Black, Guy & McLellan 2003

Additional studies have been undertaken by Janssen and McLoughlin (2008), Davis (2007), and Mason and Osborne (2007). Each of these studies has tended to concentrate on the periods from the mid-1990s to the end of the 2000s. Given that these studies miss the downturns in productivity of the early 1990s and late 2000s they tend to provide more positive results.

Regardless of the periods selected it appears from the various studies that although productivity increased to some degree up until the mid-1970s, since then the results have been mixed. This appears to be the case even though there have certainly been a range of technological changes that have occurred in the industry since then. New tools and equipment have been introduced that have tended to be labour savings. This has meant the introduction of hand held powered tools (nail drivers, sanders, saws and drills), and improved lifting and moving machinery (cranes, loaders, earth movers, graders, forklifts). New materials and processes have also been introduced along with a greater use of pre-fabricated materials.

The introduction and greater use of new equipment, materials and processes has meant that it is hard to explain why growth in the productivity of the New Zealand construction industry has been sluggish. It is possible that the various studies of productivity growth may have underestimated the growth in output of the industry, and therefore productivity, for the reasons explained by Rosefielde & Mills, and Schriver & Bowlby; namely the lack of quality adjustments in value added data and the shift of construction from high productivity to low productivity activities. In the future, therefore, it would be useful if studies made greater use of output measures that have some quality adjustment, and that greater use is made of techniques that rely less on pricing.

AUTHOR(S)	DATE	MEASURES	PERIOD STUDIED	APPROACH, VARIABLES & RESULTS
Marks	1984	Labour productivity	1965/6 to 1980/1	Index approach Output: Sectorial real GDP. Input: Labour employed LP: 1.0% per annum
Orr	1989	Partial and total factor productivity	1961 to 1986	Growth Accounting - Production function Output: gross domestic product. Inputs: employed, net capital stock TFP: 1.75% per annum (LP: 2.30%; KP: 1.02%)
Chapple	1994	Partial and total factor productivity	1972 to 1991	Growth Accounting Production function Output: value added. Inputs: labour hours, net capital stock TFP: 0.7% per annum (LP: 1.0%; KP: 0.2%)
Philpott	1995	Total factor productivity	1960to1994	Growth Accounting Production function Output: Sectorial real GDP. Inputs: employed, average of gross &net capital stock. TFP: 0.3% per annum (LP: 1.0%; KP: -1.4%)
Fare, Grosskopf & Mararitis	1996	Total factor productivity	1973 to 1994	DEA approach Output-value added. Inputs-labour hours, net capital TFP: -0.4% per annum
Diewert & Lawrence	1999	Partial and total factor productivity	1978 to 1998	Index approach Inputs: labour hours and net capital stock. TFP: -0.4% per annum
Black, Guy & McLellan	2003	Partial and total factor productivity	1988 to 2002	Index approach Output: sectorial real GDP. Inputs: labour hours, net capital stock TFP: -1.51% per annum
Janssen & McLoughlin	2008	Labour productivity	1989 to 2007	Index approach Output: sectorial real GDP. Inputs: labour hours. LP: -3.1% per annum
Davis	2007	Labour productivity	1997 to 2006	Index approach Output: sectorial real GDP. Input: labour hours LP: 0.3% per annum
Mason & Osborne	2007	Labour and total factor productivity	1995 to 2004	Growth Accounting Production function Output-Real value added. Inputs: labour hours and net capital stock TFP: 1.12% per annum (LP: 0.84% per annum)
NZIER	2011	Labour productivity	1989 to 2006	Index approach Output: value added. Input: labour employed, net capital stock TFP
Statistics New Zealand	2011	Partial and total factor productivity	1978 to 2009	Index approach Output: Sectorial real GDP. Inputs: labour hours, flow of capital services generated by capital TFP: -0.2% per annum (LP: 0.4%;KP: -1.9%)

Table 2 Past studies on the productivity of the New Zealand construction industry

Summary

Over the years there has been an interest in the performance of the construction industry in New Zealand and its contribution to national output. In this paper we have reviewed the various measures that have been used to gauge the levels of productivity growth in the industry and some of the key findings, as well as made recommendations on improved measures.

From the past research that has been undertaken to date on the productivity growth of the New Zealand construction industry a few conclusions can be made. The first of these is that in all probability growth in productivity in the industry was higher in the 1960s and early 1970s then it was in subsequent decades. The apparent stagnation that occurred since the mid 1970s, however, might not have been as great as it first appears given that some form of value added, rather than quality adjusted measure of output, has always been used. Even given the difficulties associated with determining levels of output, productivity growth in the industry probably was greater in the period up until 1975.

The second factor that can be identified from past studies is the strong cyclical nature of productivity in the industry. Strong periods of growth in productivity in the industry are associated with strong periods of growth in demand. Sharp falls in productivity are also associated with slumps in building activity and recessions. This is especially true during the recessions in New Zealand of the early 1980s and early 1990s. Typically in industries productivity slumps during recessions because companies have to regard some of labour and most of their capital as fixed and so output slumps relative to inputs. During boom periods the opposite occurred. One associated problem with this phenomenon is that those studies that cover only a limited number of years are generally dominated by the predominate business cycle (i.e. boom or recession). Any study of productivity change in the construction industry in New Zealand really has to cover a long period of time if it is to net out the impact of cyclical fluctuations.

Finally no attempt has been made in the New Zealand context to determine if there is any reallocation within the construction industry of activity from high to low productivity areas of activity which might be affecting estimates. It appears that research stills needs to be undertaken in these areas in the future if we are going to be able to make informed inferences about the performance of the industry in terms of productivity. In particular greater use of quality adjusted outputs and less reliance on techniques that require reliable data on prices would be useful.

References

Allen, S.G. (1985) 'Why construction industry productivity is declining', *Review of economic statistics,* **117** (4), 661-665

Alterman, J. and E.E. Jacobs (1961) 'Estimates of real product in the United States by industrial sector, 1947-1955, *Output, input and productivity measurement*, National Bureau of Economic Research, Princeton University press

Barton, G.T., and M.R. Cooper (1948) 'Relation or agricultural production to inputs', *Review of economics and statistics*, **30** (7), 117-126

Black, M., Guy, M. and N. McLellan (2003*) Productivity in New Zealand 1988 to 2002*, Treasury Working paper 03/06, Wellington, The Treasury

Bollard, A. and R. Barrow (2012) 'Could we be better off than we think?' Reserve Bank of New Zealand, Speech to a meeting of the Trans-Tasman Business Circle, Auckland, 17 February

Chapple, S. (1994) 'Searching for the Heffalump?', An exploration into sectoral productivity and growth in New Zealand', New Zealand Institute of Economic Research Working paper 99/10, Wellington. May

Charnes, A., W.W. Cooper and E. Rhodes (1978) 'Measuring the efficiency of decision making units, *European journal of operational research*, **2** (6), 429–444

Chau, K.W. (1990) *Total factor productivity of the building industry of Hong Kong*, unpublished PhD thesis, University of Hong Kong, Hong Kong

Chau, K.W. (1993) 'Estimating industry-level productivity trends in the building industry from building cost and price data', *Construction management and economics*, **11** (4), 370-383

Chau, K.W. (2009) 'Explaining total facto productivity trend in building construction: empirical evidence from Hong Kong', *International journal of construction management*, 45-54

Chau, K.W. and A. Walker (1988) 'The measurement of total factor productivity of the Hong Kong construction industry', *Construction management and economics*, **6** (3), 209-224

Chau, K.W. and Y.S. Wang (2005) 'An analysis of productivity growth in the construction industry: a non-parametric approach', in Khosrowshahi, F. ed. 21st Annual ARCOM conference, 7-9 September 2005, SOAS, University of London, Association of Researchers in Construction Management, **1**, 159-69

Coelli, T., D.S. PrasadaRao and G. E. Battese (1998) *An introduction to efficiency and productivity Analysis*, Boston, Kluwer

Conway, P. and B. Hunt (1998) 'Productivity growth in New Zealand: economic reform and the convergence hypothesis', Reserve Bank of New Zealand, internal paper, Economics Department, February

Dacy, D.C. (1965) Productivity and price trends in construction since 1947', *Review of economics and statistics*, **47** (4), 406-411

Davis, N. (2007) *Construction sector productivity scoping report*, Wellington, Martin Jenkins

Diewert, E. and D. Lawrence (1999) *Measuring New Zealand's productivity*, Treasury working paper 99/5, Wellington, Treasury

Fabricant, S. (1954) *Economic progress and economic change*, Washington, DC, National Bureau of Economic Research

Fare, R., S. Grosskopf and D. Margaritis (1996) 'Productivity growth' in *A study of economic reform: the case of New Zealand*, B. Silverstone, A. Bollard, and R. Lattimore, eds., Elsevier Science, New York

Farrell, M. (1957) 'The measurement of productive efficiency', *Journal of the Royal Statistical Society*, Series A, CXX, 253-81

Ganesan, S. (1984) 'Construction productivity', *Habitat international*, **8**, 3-4

Mao, Z., B. H. Goh, S. Q. Wang and G. Ofori (2003) 'Total factor productivity growth accounting in the construction industry in Singapore', *Construction Management and Economics*, **21** (7), 707-718

Goodrum, P.M. and C.T. Haas (2004) 'Long-term impact of equipment technology on labor productivity in the U.S. construction industry at the activity level', *Journal of construction engineering and management*, **130** (1), 124-13

Grubb, D., Jackman, R., and R. Layard (1982) 'Causes of current stagflation', *Review of economic studies,* **44**, 707-730

Haber, W. and H. Levinson (1956) *Labor relations and productivity in the building trades*, Ann Arbor, University of Michigan Press

Janssen, J. and S. McLoughlin (2008) *New Zealand's productivity performance*, New Zealand Treasury Productivity Paper 08/02, Wellington, the Treasury

Kau, J.B. and Sirmans, C.F. (1983) 'Technological change and economic growth in housing', *Journal or urban economics*, **13**, 283-95

Kendrick, J.W. (1956a) *Productivity trends: capital and labor*, Occasional Paper 53, Washington, DC, National Bureau of Economic Research

Kendrick, J.W. (1956b) *The meaning and measurement of national productivity*, PhD thesis. George Washington University, Washington D.C

Kendrick, J.W. (1961) *Productivity trends in the United States*, New York, NY, National Bureau of Economic Research/Princeton University Press

Kendrick, J.W. (1973) *Post-war productivity trends in the United States 1948–1966,* New York, NY, National Bureau of Economic Research

Kendrick, K.W. and C.E. Jones (1951) 'Gross National Farm product in constant dollars, 1910-1950', *Survey of current business*, **31**, September, 12-19

Lowe, J.G. (1987) 'The measurement of productivity in the construction industry', *Construction management and economics*, **5**, 115-21

Lovell, C.A. Knox and P. Schmidt (1988) 'A comparison of alternative approaches to the measurement of productive efficiency', in *Applications of modern production theory: efficiency and productivity*. eds. A. Dogramaci and R. Färe, Boston, Kluwer, Boston

Maloney, W. (1983) 'Productivity improvement: the influence of labor', *Journal of construction engineering management,* **109** (3), 321-334

Mason, G. and M. Osborne (2007) *Productivity, capital-intensity and labour quality at sector level in New Zealand and the UK,* New Zealand Treasury Working Paper 07/01, Wellington

Marks, P. (1984) 'The slowdown in labour productivity growth rates in New Zealand in the 1970s', in B. Easton, ed., *Studies in the labour market*, Wellington, NZIER

New Zealand, Department of Building and Housing, Productivity Taskforce (2009) *Report of the Building and Construction Sector Productivity Taskforce: a modern efficient and productive New Zealand built infrastructure industry*, Wellington, N.Z.: Dept. of Building and Housing

New Zealand, Productivity Commission (2011) *Housing affordability inquiry: draft report,* Wellington, Productivity Commission

New Zealand Institute of Economic Research (2011) *Industry productivity and the Australia-New Zealand income gap*, Wellington, Working paper 2011/3, NZIER

Orr, A. (1989) *Productivity trends in New Zealand: a sectorial and cyclical analysis 1961-1987,* Wellington, NZIER

Philpott, B. (1991) 'Economic growth in New Zealand models and Experience', in Long run perspectives on the New Zealand economy, *Proceedings of the Sesquicentennial Conference of the New Zealand Association of Economists*, New Zealand Association of Economists

Philpott, B. (1993) 'Data for sectorial productivity analysis and some preliminary results for 1978-93', RPEP Paper 256

Philpott, B. (1995) *New Zealand's aggregate and sectoral productivity growth 1960-1995*, Research project on economic planning paper 274, Victoria University of Wellington, October

Rosenfielde, S. and D.Q. Mills (1979) 'Is construction technologically stagnant?', in J.E. Lange and D.Q. Mills, eds., *The construction industry*, Lexington MA, D.C. heath and Company

Sarel, M. (1996) 'Growth and productivity in New Zealand', IMF internal paper 14, November

Schmookler, J. (1952) 'The changing efficiency of the American economy, 1869-1938',*Review of economics and statistics*, **34** (3), 214-231

Schriver, W.R. and R.L. Bowlby (1985) 'Changes in productivity and composition of output in building construction, 1972-1982', *Review of economics and statistics*, **67** (2), 318-322

Schultz, C.L. (1959) *Prices, costs and output for the post ware decade: 1947-1957*, New York, Committee or Economic Development

Statistics New Zealand (2011) *Gross domestic product*, Wellington, Statistics New Zealand

Statistics New Zealand (2011) *Household labour force survey,* Wellington, Statistics New Zealand

Statistics New Zealand (2011) *Industry productivity statistics, 1978-2009,* Wellington, Statistics New Zealand

Stokes, H.K. (1981) 'An examination of the productivity decline in the construction industry', *Review of economics and statistics*, **63** (4), 495-502

Smith, R. and A. Grimes (1990) 'Sources of economic growth', *Reserve Bank of New Zealand Bulletin*, **53** (2), June, 140-48

Stigler, G.J. (1947) *Trends in output and employment*, National Bureau of Economic Research, University of Princeton Press

Stigler, G.J. (1961) 'Economic problems in measuring changes in productivity', In *Output, input and productivity measurement,* Conference on Research in Income and Wealth, UMI

Tan, W. (2000) 'Total factor productivity in Singapore construction', *Engineering, construction and architectural management*, **7** (2), 154-158

Thomas, H. and A. Sakarcan (1994) 'Forecasting labor productivity using factor model', *Journal of construction engineering management*, **120** (1), 228-239

'Thomas, H., Maloney, W., Horner, R., Smith, G., Handa, V., and Sanders, S. (1990) Modelling construction labor productivity'*, Journal of construction engineering management*, **116** (4), 705-726

Van Dai Tran (2010) *Exploring construction productivity statistics in New Zealand*, Master of Engineering thesis, Auckland University of Technology, Auckland

Wang, Y.S. (1998) *An analysis of the technical efficiency in Hong Kong's construction industry*, unpublished PhD thesis, University of Hong Kong, Hong Kong

Wang, Y.S. and K.W. Chau (1997) 'An evaluation of the technical efficiency of construction industry in Hong Kong using the DEA approach', Proceedings of ARCOM97 Conference, Cambridge, 690-701

Wang, Y.S. and K.W. Chau (2001) 'An assessment of the technical efficiency of construction firms in Hong Kong', *International journal of construction management*, **29**,105-122

Econometric Modelling of Risk Adverse Behaviours of Entrepreneurs in the Provision of House Fittings in China

12

Rita Yi Man Li, (Hong Kong Shue Yan University, Hong Kong)

Abstract

Entrepreneurs have always born the risk of running their business. They reap a profit in return for their risk taking and work. Housing developers are no different. In many countries, such as Australia, the United Kingdom and the United States, they interpret the tastes of the buyers and provide the dwellings they develop with basic fittings such as floor and wall coverings, bathroom fittings and kitchen cupboards. In mainland China, however, in most of the developments, units or houses are sold without floor or wall coverings, kitchen or bathroom fittings. What is the motive behind this choice? This paper analyses the factors affecting housing developers' decisions to provide fittings based on 1701 housing developments in Hangzhou, Chongqing and Hangzhou using a Probit model. The results show that developers build a higher proportion of bare units in mainland China when: 1) there is shortage of housing; 2) land costs are high so that the comparative costs of providing fittings become relatively low.

Keywords: Imperfect information, Risk, Entrepreneur, Fittings, Probit model, Housing.

Introduction

In many places, such as Australia, the United Kingdom and the United States, housing developers provide floor and wall coverings, window frames, cupboards, and electrical fittings as standard equipment. In some cases, developers also provide heated floors and wine storage. There is, however, a completely different norm in mainland China. For instance, the majority of developers in Beijing and Nanjing sell bare dwellings to home buyers. There are no floor coverings, kitchen cupboards or basic bathroom fittings (Li, 2009). Table 1 shows the percentage of bare flats in Nanjing. Among all the residential developments available for sale in 2004 - 2007 in Nanjing, 90% were bare units. Table 2 displays a similar phenomenon. It shows the percentage of bare flats in Beijing from 1997 - 2008. Of all the first hand residential units sold in Beijing districts, more than 70% were sold as bare units. Similarly, bare units can also be found in Shanghai and Hangzhou. This paper aims at finding out the major drivers behind these housing entrepreneurs to build bare residential units in these cities.

Hypotheses

Purchasing bare units without any fittings in mainland China implies that buyers need to spend time in identifying relevant parties to decorate their housing units before they move in. Why do busy people, often trapped in crammed schedules accept such a time consuming building activity? There has been very limited study of this phenomenon. This paper aims to remedy this and test potential reasons behind this by testing theories of brand name, risk adverse behaviours and information costs etc. Specifically, the author proposes four hypotheses on the emergence of bare flats:

1. Developers tend to build housing units with fittings in high land price areas.
2. The larger the proportion of low income residents the higher the proportion of bare flats.
3. Developers with well renowned reputations (brand names) build more furnished flats.
4. The greater the shortage of residential units, the higher the proportion of bare flats.

District	Total residential developments available for occupation in 2004 - 2007[1]	Percentage of bare flats
Bai Xia	32	92
Da Han	4	50
Gao Chun	3	100
Gu Lou	65	94
Jian	71	93
Jiang Zhu	133	98
Li Shui	11	91
Lu He	12	92
Pu Kou	60	97
Qi Xia	50	100
Qian Huai	42	100
Xia Guan	24	92
Xuan Wu	35	92

Table 1 Percentage of bare residential units in Nanjing (Li, 2010)

District	Residential projects which build bare flats	Total	Percentage of bare flats
Chao Yong	618	853	72
Chong Ping	160	182	88
Chong Wen	62	74	84
Da Xing	140	165	85
Dong Cheng	59	79	75
Fang Shan	83	90	92
Feng Toi	260	305	85
Hai Dian	379	470	81
Huai Rou	37	44	84
Mi Yun	43	52	83
Shi Jing Shan	48	54	89
Shun Yi	102	126	81
Tong Zhou	194	219	89
Xi Cheng	62	82	76
Xuan Wu	88	114	77
Yan Qing	17	24	71
Others	92	110	84

Table 2 Percentage of bare flats in Beijing from 1997 - 2008 (Li, 2009)

[1]Each of the residential development refers to one residential project. Some of the projects consist of up to, or more than 1000 units. The districts refer to small areas inside the city. This applies also to Table 2.

The first proposition: *Developers tend to build housing units with fittings in areas with high land prices.*

While fittings in high land price areas only make up a small proportion in construction costs, fittings in low land price areas constitute a relatively large share. The costs of providing fittings which do not suit the taste of customers are relatively lower in areas with high land prices compared to the costs of providing wrong fittings in areas with low land prices as home purchasers might choose not to buy the flats when developers provide unsuitable fittings. As developers are risk adverse, it is natural that they provide fewer fittings in those areas to avoid the relatively high risks:

$$P(f)/L_H < P(f)/ L_L$$

where P (f) refers to the price of the fittings, L_H refers to high land price area and L_L refers to low land price area.

The second proposition: *The larger the proportion of low income residents, the higher the proportion of bare flats.*

Developers in affluent countries such as United Kingdom, Singapore, United States offer more fittings as compared to those in developing countries, such as China, Ghana, Indonesia etc. Buying favourite fittings implies that the buyers are investing the time and money in the act of searching information. A man who earns $4 per hour has a low discretionary income but may be happy to search by himself rather than pay the contractor to do all on behalf of him, while a man who earns $ 400 finds it not worthwhile to do so. Searching costs him $400 per hour which is more than the cost of letting the contractor do it. In view of this, high income areas theoretically should have a smaller portion of bare flats. Developers tend to build more well-furnished flats with fittings to suit the needs of customers.

The third proposition: *Developers with well developed reputations build more well-furnished flats and vice versa.*

Bare dwellings may avoid losses but do not help the developers to maximize gains from value added. Hence, well regarded developers may tend to build more housing units with good fittings to enhance their reputation for quality. For others, it may be better to provide no fittings as the provision of poor fittings may set back reputations which have taken years to build up.

The fourth proposition: *The supply is so limited that the huge demand lowers developers' motivations to provide well-equipped units.*

Previous literature shows that shortages may lead to use of unqualified staff and poor materials. Compared to many overseas countries, China experiences a shortage of housing by private developers. With substantial excess demand, developers do not need any gimmicks to compete for potential buyers. Provisions of kitchen and bathroom fittings are unnecessary.

Literature Review

Assumptions, Functions and Factors affecting the Supply of Entrepreneurs

Another important role played by entrepreneurs is risk-taking. Entrepreneurs are risk-takers (Blanchflower and Oswald, 1998) who accept the firms' risks (Carland et al., 1984) when running a business.

Information, Imperfect Information and Asymmetric Information

One major problem that the entrepreneurs face is imperfect information. Imperfect information affects market power and prices and leads to a dis-equilibrium where firms can charge monopoly prices (Granlund and Rudholm, 2011). Hence asymmetric information is a market failure which undermines the efficiency of product transactions. It is not unusual that consumers are uninformed about risks which affect their ability to choose terms which reflect their preferences correctly. Firms often exploit this ignorance by degrading contract quality intentionally and will then have little incentive to offer better deals as these will not increase sales (Bechern, 2008).

Risk-averse Human Behaviour

Because of imperfect information, individuals and entrepreneurs make their decisions under risk. Risk has been identified as the potential for threat, damage, injury, or other loss (Zou et al., 2007, Jin and Doloi, 2008). It may also be conceptualized as variance in outcome in any project (Das and Teng, 2001, Fellner and Maciejovsky, 2007). Risk can be managed by catastrophe planning, easing, insurance, control, identification, quantification and shifting it to other agents (Zou et al., 2007, Jin and Doloi, 2008).

People's attitude to risk affect their behaviour and is important in decision making (Richard, 1975, Xiao and Yang, 2008). People's attitude towards risks is not consistent. Some people are more risk adverse than the others.

Buyers cannot obtain information on housing quality from the previous owners when they buy new housing. Many of them only know they have bought a poor housing when they open the doors of their units (Gwin and Ong, 2000). Therefore, home buyers often rely on developers' reputation and previous residential project to make decisions so as to minimise their risk in home purchases.

Loss-averse Human Behaviour

Apart from risk aversion, many entrepreneurs are loss-averse which refers to a situation when decision-makers are more sensitive to losses than to gains. This phenomenon represents a discontinuity in their utility function (Berkelaar et al., 2004), graphically expressed as an abrupt change in the slope of the utility function at the reference point (Wang et al., 2009) which distinguishes gains from losses It also signifies that the utility function is steeper for losses than for gains, i.e. the disutility that one experiences in losing money is greater than the utility associated with gaining the same amount. Loss aversion has become an important tool to explain all sorts of phenomena which are not explained by traditional theory, such as the endowment effect (Tovar, 2009).

Branded Product

A brand name is a firm's most important asset when it comes to evaluating companies against each another (Laforet, 2011). Entrepreneurs spent a lot of effort on boosting the name of their companies. Branded products have valuable merits to consumers as well as sellers (Table 3) as they convey quality information to consumer. As early as 1960s, the American Marketing Association defined "brand" as:

> "a name, term, sign symbol or design, or a combination of these, intended to identify the goods or services of one seller or group of sellers and to differentiate them from those of competitors" (Zilg, 2011, p.284).

With asymmetric information in China's housing industry, buyers use brand names to assess product quality. Reputation is an effective signal which provides quality information because it is firm-specific – gradually built up from the quality of the projects built by a developer in the past. In case where developers have a history of working in a community, home buyers can obtain information on quality of the units by observing previous projects.

Merits of brand name	Examples
Allows firms to escape from the confines of generic prices	Jensen and Drozdenko (2008); LeBel and Cooke (2008); Rotfeld (2004); Vukasovič (2009); Souiden and Pons (2009)
Easier adoption of a new product	Rotfeld (2004), Vukasovič (2009)
Price premium	Corkindale and Belder (1985); ; Olson (2008); Chen (2007) ; Li et al. (2009) disagree)
Enhance perceptions of product or service quality	Corkindale and Belder (1985); Horppu et al. (2008); Shannon and Mandhachitara (2008); Vukasovič (2009);
Facilitate promotional effectiveness	Corkindale and Belder (1985)
The "personality" of brand product add value to the consumer	Charters (2011)
Increase market share	Vukasovič (2009)
Decrease risk	Vukasovič (2009); Matzler et al. (2008)
Barrier to entry for competitors	Omar et al. (2009)
Provides information to the consumer	Baltas and Saridakis (2009); Pechtl (2008); Pitta and Franzak (2008); Souiden and Pons (2009)
Product differentiation	Zilg (2011)
To attract a group of loyal customers Not likely for housing	Burnett and Bruce (2007); Horppu et al. (2008)

Table 3 Merits of brand name

Shortage and Sellers' Goods Quality

Market forces not only affect prices, they also determine suppliers' incentives to provide 'extra' value which aims at being perceived as quality improvement. In case of shortage, suppliers do not need to compete with other sellers. As a result, they may minimize production costs by providing only limited accessories to their buyers (Hawthorne and Birrell, 2002, Ingersoll and Smith, 2003).

Housing investment in the planned economy was limited before 1976. The government preferred to spend money on investments other than housing (Wang and Murie, 1996). Private housing investment was allowed after the death of Mao and rapid growth in population and subsequent alterations in national urban policy have brought about a rapid urbanization in China. Registered urban population has been increasing at a rate of 4% annually since 1980 (Wu, 1999). Residential unit supply never met demand. Housing shortages problems had became serious by the end of the Cultural Revolution in 1976 when there was only 3 m^2 of floor space per person on average. Despite the large quantity of housing that had been built since 1978, 4 million urban households still live with 4 m^2 per family member in 1994 and 400 000 had an average of 2.5 m m^2 living space per person (Wang and Murie, 1996).

In view of the above, does it mean that the popularity of bare units in China is the results of a shortage of housing, risk adverse behavior by developers or something else?

Research Method – Probit Model

To find out the reasons behind the sale of bare units in China, the Probit model will be used. Covariates are measured from errors in regression model estimates. In binary dependent variables, the error term can also be dichotomous in nature, hence normal distribution assumption of error term in ordinary least square is not essential. Therefore, OLS regression which is suitable for continuous dependent variables may not be the best choice in case of dichotomous dependent variables to estimate an equation (Gujarati, 2006).

In a binary response model, interest lies primary in the response probability

$$Yt^* = Xt$$

$$P(y=1|x) = P(y1|x_1,x_2,\ldots,x_k)$$

Consider an equation with binary response (0,1) :

$$P(y=1|x)= Y(\eth_0 + \eth_1 x_1 + \eth_2 x_2 + \ldots + \eth_k x_k) = Y(\eth_0 + x\,\eth)$$

Where x denotes a complete set of explanatory variables, Y is a function which takes on values between 0 and 1, that is $0<Y(z)<1$ where t represents all the real numbers $x\,\eth = \eth_1 b_1 + \eth_2 b_2 + \ldots + \eth_k b_k$ (Wooldridge, 2003).

Data Analysis

Data of provincial GDP per capita, construction costs for each year, annual supply of newly built residential units and change in number of household, background information of the residential developments such as developers which built the dwellings were collected to test the four hypothesis listed at the beginning of the paper.

Year	Real GDP per capita	Construc-tion costs (m^2)	Number of develop-ment	Average Estimated Land Cost (per m^2)	Change in No of Households (ten thousand)	Area of newly built dwelling (ten thousand m^2)	Percentage of bare deve-lopment (%)
Hangzhou (Hangzhou Statistics Bureau, 2008, Hangzhou Soufun, 2009)							
2003	32667	2275	26	9350	2.85	516	82
2004	38593	2023	61	11435	3.40	528	83
2005	44555	2459	62	11535	2.90	662	83
2006	51650	2340	51	10484	2.49	580	81
2007	52590	2953	59	6616	2.08	675	81
Chongqing (Chongqing, 2008, Hangzhou Soufun, 2009)							
2003	7261	827	47	3242	15.32	1232	85
2004	7633	929	80	3281	11.58	1228	90
2005	7745	1074	106	3694	21.82	1714	93
2006	7689	1319	146	2800	20.25	1700	93
2007	7854	1452	151	3023	26.31	1769	93
Shanghai (Shanghai Statistics Bureau, 2008, Shanghai Soufun, 2009)							
2003	39128	2987	149	6294	4.29	2 281	88
2004	46338	2958	180	8107	4.52	3 270	83
2005	51529	2960	304	7204	6.11	2 819	84
2006	57695	2963	151	8491	2.85	2 747	80
2007	66367	3014	128	9319	3.75	2 844	80

Table 4 Economic and housing data

2003 - 2007 GDP per capita data of the three places - deflated by GDP deflator - were obtained from Statistics Bureau in each place (Shanghai Statistics Bureau, 2008, Chongqing,

2008, Hangzhou Statistics Bureau, 2008). Land price was a proxy obtained by subtracting the average selling price of residential units (Shanghai Soufun, 2009, Wooldridge, 2003, Soufun, 2009) by construction costs available from the respective Statistical Year Books of (Shanghai Statistics Bureau, 2008, Chongqing, 2008, Hangzhou Statistics Bureau, 2008).

The table below summarizes the abovementioned information in Hangzhou. In this five year' period, about 80% of the dwellings were bare units. The number of developments ready for sale has been fairly constant at about 50 - 60 per year. The real GDP per capita[2] rose from 32667 to 52590 in 2003 and 2007 respectively. Construction costs have increased since 2004. Estimated land costs have risen from 9350 in 2003 to 10484 in 2006 but dropped to 6616 in 2007. In Hangzhou, during this five year period GDP per capita has increased from 32667 in 2003 to 52590 in 2007. Construction costs per m^2 and number of residential developments rose from 827 to 1452 and 47 to 151 during the five years' observation period respectively. In Shanghai, the percentage of bare units dropped from 88 to 80 between 2003 and 2007. Number of housing developments rose from 149 to 304 in 2003 and 2005 but decreased afterwards and reached a trough of 128 in 2007. Construction costs oscillated around 2900 in this 5 year period. Annual change in number of household rose from 42900 to 61100 between 2003 and 2005, dropped to 28500 in 2006 and rose to 37500 in 2007. Area of newly built dwellings climbed from 22,810,000 m^2 in 2003 and reached a maximum at 32,700,000 m^2 in 2004 (Table 4).

Research Results and Analysis

To test the hypothesis, five variables LDMV, BUILDONCE, BUILDONCE*LOG(CCMV), A_CHANGEHSE, and GDP were included (Table 5). Table 6 Results of Probit regression

shows the results of the Probit regression.

Variables	Explanation
LDMV	Land price divided by open market value of the dwellings that is, selling price of the units.
BUILDONCE	It represents developers which built once only during the study period. As branded developers are characterized by repeated sales, dummy "1" is assigned if the housing development was built by a developer which had built once only and "0" if they had engaged in repeated sales activities from 2004 - 2007.
A_CHANGEHSE	To test the impact of housing shortage on provision of fittings, dwelling space was divided by the change in number of household (Powell and Ansic, 1997) and it is represented by A_CHANGEHSE.
BUILDONCE*LOG(CCMV)	CCMV denotes the proportion of construction cost in dwelling value.

Table 5 Variables in Probit model

The Probit regression models confirm the hypothesis that there is a significant negative relationship between bare units and land costs: there is a higher probability that developers build bare units when the land costs are low. Costs of fittings provisions become relatively high in low land costs areas as compared to the whole construction costs. On top of that, buyers may choose not to purchase units if the fittings do not suit their tastes. China is a newly emerging real estate market as there were no private companies prior to 1978 due to the planned economy. Hence many housing entrepreneurs are newcomers to the industry and lack information for predicting buyers' tastes in house fittings. In sharp contrast, the well established markets in places, such as Hong Kong, Europe, and Australia have enabled

[2] All costs and prices are expressed in RMB unless otherwise specified. Currently 1USD equals 6.23 RMB

local housing entrepreneurs in these locations to learn about the preferences of buyers. Housing developers are risk averse. They all aim to maximise their profits without too much risk. To reduce their risk, these entrepreneurs choose not to provide any fittings. Furthermore, traditional entrepreneurs relay on many sources of external financing: debt, venture capital, private equity and public stock offerings. Investors who commit funds to business start-ups expect to receive back their investment along with a handsome return. Entrepreneurs must ask themselves if their goals are congruent with those of possible investors (Williams et al., 2006). If housing entrepreneurs make an incorrect decision, such as putting undesirable fittings into their units, investors will be affected and develop negative opinions about these entrepreneurs. Potential joint ventures may also become futile if the housing entrepreneurs have a history of incorrect decisions. Under the postulate of being risk averse, developers prefer to provide nothing inside the units.

The Probit Regression results also confirm our hypothesis that there is a significant positive relationship between repeated sales activities and furnished units. Nonetheless, the results of the interaction term BUILDONCE*LOG (CCMV) shows that not all "new" developers are equally keen to construct bare housing. The higher the construction cost as a proportion of property price, the higher would be the risk that the monetary return cannot cover the cost.

The literature indicated that shortages can lead to lower consumer requirement, and this is supported in this Probit model — there is an increase in likelihood of bare units construction in times of housing shortage. However, the insignificant relationship between GDP per capita and bare units, implies that people with higher information costs may not necessarily prefer well-furnished units.

Dependent Variable: BARE				
Method: ML - Binary Probit (Quadratic hill climbing)				
Included observations: 1704				
Convergence achieved after 5 iterations				
Covariance matrix computed using second derivatives				
Variable	Coefficient	Std. Error	z-Statistic	Prob.
C	2.320284	0.698542	3.321609	0.0009
LOG(LDMV)	-0.645794	0.233331	-2.767715	0.0056
BUILDONCE*LOG(CCMV)	0.614759	0.129308	4.754215	0.0000
BUILDONCE	0.919487	0.193384	4.754730	0.0000
LOG(A_CHANGEHSE)	-0.259066	0.100172	-2.586210	0.0097
LOG(GDP)	-0.007551	0.113477	-0.066540	0.9469
McFadden R-squared	0.106837	Mean dependent var		0.861502
S.D. dependent var	0.345523	S.E. of regression		0.325198
Akaike info criterion	0.725551	Sum squared resid		179.5704
Schwarz criterion	0.744708	Log likelihood		-612.1692
Hannan-Quinn criter.	0.732642	Restr. log likelihood		-685.3946
LR statistic	146.4507	Avg. log likelihood		-0.359254
Prob(LR statistic)	0.000000			
Obs with Dep=0	236	Total obs		1704
Obs with Dep=1	1468			

Table 6 Results of Probit regression

Summary

An entrepreneur is the major driving force behind production of goods as he/she combines the resources of land, capital and labour. All entrepreneurs share one characteristic: they bear risk and earn profit or suffer losses in return. It is assumed here that housing entrepreneurs are risk averse. They attempt to find ways to reduce their risk. Housing entrepreneurs in mainland China who provide bare units provide a vivid example of risk aversion. As there is no guarantee of making a profit and there is imperfect information in our business world, the entrepreneur must assess the risk and, like other entrepreneurs, he/she cannot accept a job that has too high a risk of loss (Brue et al., 2009).

Running a development business is risky; it requires substantial capital investment in land, professional training and recruitment. Nevertheless, this huge expenditure does not guarantee a great return, and it is possible to suffer a substantial loss if the entrepreneur makes a wrong decision. Provision of home fittings is one of the risks that the housing entrepreneur faces. Some home buyers may dislike the fittings and thus decide not to purchase the home. In some places such as Hong Kong, land supply is scarce, but demand is huge (7 million people live in a small city with a hilly landscape). Land price occupies a relatively large proportion of the costs of dwelling production. Provision of fittings amounts to only a small proportion of the total costs of construction. Entrepreneurs thus focus on land purchases more than provision of fittings. The risk of installing fittings is not very high. Furthermore, many developers have run their business for many years and have accumulated sufficient knowledge on buyers' taste. Provision of fittings to customers is common. Yet, in mainland China, especially in areas of low land costs, costs of fittings have become relatively high. The risk of supplying fittings that prospective owners dislike is higher and this leads to the phenomenon of the bare flats' sale in China's real estate market.

A housing shortage reduces the pressure on developers to provide more than the most basic dwellings hence increasing the probability that developments will be sold without fittings. Changes in per capita income do not change this tendency in any significant way.

Conclusion

In conclusion, developers exhibit a higher tendency to build bare units when:

1. There is shortage of housing so that developers do not need to face intense competition and supply extra stuff, i.e. fittings in addition to the building structure itself;
2. Land costs are low; the costs of providing fittings become relatively high and developers tend not to take the risk of speculating in consumers' taste of fittings;
3. There were no repeated sales during the observation period, hence consumers have little guide to predict the quality of fittings.
4. And Finally, there is an insignificant relationship between proportion of bare units construction and income.

This paper has also demonstrated that Probit is a powerful tool for this type of analysis.

References

Baltas, G. and Saridakis, C. (2009) 'Brand-name effects, segment differences, and product characteristics: an integrated model of the car market', *Journal of Product & Brand Management,* **18**, 143-151

Bechern, S. I. (2008) 'Asymmetric information in consumer contracts: The Challenge that is yet to be met', *American Business Law Journal,* **45**, 723-774

Berkelaar, A. B., Kouwenberg, R. and Post, T. (2004) 'Optimal portfolio choice under loss aversion', *The Review of Economics and Statistics,* **86**, 973-987

Blanchflower, D. G. and Oswald, A. J. (1998) 'What makes an entrepreneur?', *Journal of*

Labor Economics, **16**, 26-60

Brue, S. L., Mcconnell, C., R. and Flynn, S. M. (2009) *Essentials of Economics,* New York, McGraw Hill

Burnett, J. and Bruce, H., R. (2007) 'New consumers need new brands', *Journal of Product & Brand Management,* **16,** 342-347

Carland, J. W., Hoy, F., Boulton, W. R. and Carland, J. A. C. (1984) 'Differentiating entrepreneurs from small business owners: A conceptualization', *The Academy of Management Review,* **9**, 354-359

Chen, H. L. (2007) 'Gray marketing and its impacts on brand equity', *Journal of Product and Brand Management,* **16**, 247-256

Chongqing, S. I. O. (2008) *Chongqing Statistical Yearbook 2008* [Online] Available: http://www.cqtj.gov.cn/tjnj/2008/yearbook/indexe.htm [Accessed 30 April 2010].

Das, T. K. and Teng, B. S. (2001) 'Trust, control, and risk in strategic alliances: An integrated framework', *Organization Studies,* **22**, 251-283

Fellner, G. and Maciejovsky, B. (2007) 'Risk attitude and market behavior: Evidence from experimental asset markets', *Journal of Economic Psychology,* **28**, 338-350

Granlund, D. and Rudholm, N. (2011) 'Consumer information and pharmaceutical prices: Theory and evidence', *Oxford Bulletin of Economics and Statistics,* **73**, 230-254

Gujarati, D. N. (2006) *Essentials of Econometrics,* Boston, McGraw-Hill/Irwin

Gwin, C. R. and Ong, S. E. (2000) 'Homeowner warranties and building codes', *Journal of Property Investment & Finance,* **18**, 456-472

Hangzhou Soufun (2009) *Hangzhou Soufun* [Online] Available: http://hz.soufun.com [Accessed 1 May 2010]

Hangzhou Statistics Bureau (2008) *Hangzhou Statistical Yearbook* [Online] Available: www.hzstats.gov.cn/web/tjnj/nj2008 [Accessed July 2009]

Hawthorne, L. and Birrell, B. (2002) 'Doctor shortages and their impact on the quality of medical care in Australia', *People and Place,* **10**, 55-67

Horppu, M., Kuivalainen, O., Tarkiainen, A. and Ellonen, H. K. (2008) 'Online satisfaction, trust and loyalty, and the impact of the offline parent brand', *Journal of Product & Brand Management,* **17**, 403-413

Ingersoll, R. M. and Smith, T. M. (2003) 'The wrong solution to teacher shortage', *Education Leadership,* **60**, 30-33

Jensen, M. and Drozdenko, R. (2008) 'The changing price of brand loyalty under perceived time pressure', *Journal of Product & Brand Management,* **17**, 115-120

Jin, X. H. and Doloi, H. (2008) 'Interpreting risk allocation mechanism in public private-partnership projects: an empirical study in transaction cost economics perspective', *Construction Management and Economics,* **26**, 707-721

Laforet, S. (2011) 'Brand names on packaging and their impact on purchase preference', *Journal of Consumer Behaviour,* **10**, 18-30

Li, H., Tang, F. F., Huang, L. and Song, F. (2009) 'A longitudinal study on Australian online DVD pricing', *Journal of Product & Brand Management,* **18**, 60-67

Li, R. Y. M. (2009) 'Legal regulations, people's perceptions on law and scope of services provided by firms: a study on dwelling developers in Beijing and Hong Kong', *The Business Review, Cambridge,* **12**, 221-226

Li, R. Y. M. (2010) *Factors Influencing Developers' Decision to Sell Housing Units With Fittings - Empirical Evidence from China,* The University of Hong Kong, Unpublished Ph.D.

Thesis

Matzler, K., Kräuter, S. G. and Bidmon, S. (2008) 'Risk aversion and brand loyalty: the mediating role of brand trust and brand affect', *Journal of Product & Brand Management,* **17**, 154-162

Olson, E. L. (2008) 'The implications of platform sharing on brand value', *Journal of Product & Brand Management,* **17**, 244-253

Omar, M., Williams, R. L. and Lingelbach, D. (2009) 'Global brand market-entry strategy to manage corporate reputation', *Journal of Product & Brand Management,* **18**, 177-187

Pitta, D. A. and Franzak, F. J. (2008) 'Foundations for building share of heart in global brands', *Journal of Product & Brand Management,* **17**, 64-72

Powell, M. and Ansic, D. (1997) 'Gender differences in risk behaviour in financial decision-making: an experimental analysis', *Journal of Economic Psychology,* **18**, 605-628

Richard, S. F. (1975) 'Multivariate risk aversion, utility independence and separable utility functions', *Management Science,* **22**, 12-21

Shanghai Soufun (2009) *Shanghai Soufun* [Online] Available: http://sh.soufun.com/ [Accessed 1 July 2009]

Shanghai Statistics Bureau (2008) *Shanghai Statistics Yearbook 2008* [Online] Available: http://www.stats-sh.gov.cn/2004shtj/tjnj/tjnj2008.htm# [Accessed 2 October 2009]

Shannon, R. and Mandhachitara, R. (2008) 'Causal path modeling of grocery shopping in hypermarkets', *Journal of Product & Brand Management,* **17**, 327-340

Soufun (2009) *104 Cities* [Online] Available: http://www.soufun.com/SoufunFamily.htm [Accessed 10 August 2009]

Souiden, N. and Pons, F. (2009) 'Product recall crisis management: the impact on manufacturer's image, consumer loyalty and purchase intention', *Journal of Product & Brand Management,* **18**, 106-114

Tovar, P. (2009) 'The effects of loss aversion on trade policy: Theory and evidence', *Journal of International Economics,* **78**, 154-167

Wang, C. X., Webster, S. and Suresh, N. C. (2009) 'Would a risk-averse newsvendor order less at a higher selling price?', *European Journal of Operational Research,* **196**, 544–553

Wang, Y. P. and Murie, A. (1996) 'The process of commercialisation of urban housing in China', *Urban Studies,* **33**, 971-989

Wooldridge, J. M. (2003) *Introductory Econometrics: A Modern Approach,* Australia, South-western College Publications

Wu, W. (1999) 'Reforming China's institutional environment for urban infrastructure provision', *Urban Studies,* **36**, 2263-2282

Xiao, T. and Yang, D. (2008) 'Price and service competition of supply chains with risk-averse retailers under demand uncertainty', *International Journal of Production Economics,* **114**, 187–200

Zilg, A. (2011) 'THAT'S AMORE: Brand names in the Italian food market', *International Journal of Applied Linguistics,* **21**, 1-25

Zou, P. X. W., Zhang, G. and Wang, J. (2007) 'Understanding the key risks in construction projects in China', *International Journal of Project Management,* **25**, 601-614

Entry Location and Entry Timing (ELET) Decision Model for International Construction Firms

13

Che Maznah Mat Isa, Hamidah Mohd Saman (Universiti Teknologi MARA, Malaysia)

Christopher Nigel Preece (Universiti Teknologi Malaysia, Malaysia)

Abstract

This paper proposes a model for entry location (EL) and entry timing (ET) decisions to guide construction firms in accessing targeted international markets. Neglecting to properly choose the right combination of the entry location and entry timing (ELET) decisions can lead to poor performance of the firms' international ventures. The sampling frame was from the Malaysian construction firms that have undertaken and completed projects abroad. Survey questionnaires sent to 115 firms registered with Construction Industry Development Board (CIDB) Malaysia, operating in more than 50 countries, achieved a 39.1 per cent response rate. Based on a comprehensive statistical analysis of survey data it was found that the mutually inclusive significant factors that influenced the firms' ELET decisions were: the firm's ability to assess market signals and opportunities, international experience, financial capacity, competencies and capabilities (project management, specialist expertise and technology), resources (level of knowledge based on research and development), experience in similar works, financial support from the home country banks, technical complexities of projects and availability of funds for projects. Hence, the present research builds on and extends the literature on the ELET decisions in a more integrated way.

Keywords: Entry location, entry timing, resource-based view, international markets, Malaysian construction firms.

Introduction

Market expansion is one of the most critical business strategies made by firms to exploit opportunities in international markets. These firms have gained access to foreign countries using combinations of market entry strategies and have been gradually extending their operations including Malaysian firms. However, the Engineering News Record (2013) revealed none of Malaysian construction firms was listed in the top 250 international contractors. Despite government encouragement through various plans such as the 10th Malaysian Plan (10MP), 3rd Industrial Malaysia Plan (IMP) and Construction Industry Malaysian Plan (CIMP), only 115 Malaysian construction firms have worked abroad undertaking various construction projects, ranging from infrastructure, building and other construction related projects (CIDB Malaysia 2013) since 1986. CIDB Malaysia records about 35% of these firms have been operating within the ASEAN (Association of Southeast Asian Nations), while more than 65% of the firms have penetrated the non-ASEAN. The ASEAN include Malaysia, Singapore, Thailand, Vietnam, Laos, Myanmar, Cambodia, Indonesia, Philippines and Brunei.

Within the international market domain, some of the main obstacles or barriers to entry identified by previous researchers on Malaysian construction firms were trade and investment barriers (Kaur & Sandhu 2014), insufficient information to access markets, lack of financial, advanced technology and technical resources, lack of economies of scale and scope, high market structure and competition against other foreign firms (Che Ibrahim et al. 2009). In addition, there were studies on other aspects of international Malaysian construction firms such as the firms' business locations (Abdul-Aziz & Sing-sing 2008; Che

Ibrahim et al. 2009), risks and challenges, competitive assets (Abdul-Aziz & Wong 2010) and strength, weakness, opportunity and threat attributes (Mat Isa et al. 2012).

Research has shown that adopting suitable market entry strategies is crucial in a firm's decision to enter and subsequently to perform in international markets. Hence, many researchers have proposed different plans for crafting the right combination of international market entry strategies (Luo 1998; Luo & Peng1998; Yean et al. 2008; Chen & Orr 2009; Polat & Donmez 2010; Lee et al. 2011). During market expansion, firms addressed three major entry strategic questions specifically; which location to enter (EL), when to enter (ET) and how to enter (EM) (Gaba et al. 2002). Similarly, Ekeledo (2007) contended that the choice of which country to enter, when to enter and how to enter commits a firm to operating on a given terrain and lays the foundation for its future international expansion. Even though Huang and Sternquist (2007) have stressed that these interlinking decisions were among the dilemmas that challenge firms during their international expansion, previous studies have primarily addressed each entry decision in an isolated way as shown by Dacko (2002) on ET, Somlev and Hoshino (2005) on EL and Chen and Messner (2009) on EM decisions.

Generally, solving these problems was attempted by the researchers based on the three streams of research in international business and strategic management related to the EL, ET and EM decisions respectively. Out of these three entry decision dimensions, most of the classic and current literature regarding internationalization process has focused mainly on the EM decision (Agarwal & Ramaswami 1992; Ekeledo & Sivakumar 1998; Tawanda 2006; Chen et al. 2007; Che Ibrahim et al. 2009; Chen & Chang 2011). In comparison, the EL decision has received less attention from researchers in international business (Koch 2001a; Koch 2001b; Gallego et al. 2009; Gaston-Breton & Martín 2011). Even though the importance of EL decision that emerged from the relationship between EL and EM decision (Koch 2001a; Koch 2001b; Boeh & Beamish 2012) and segmentation in the process leading to the identification of promising European target markets (Gaston-Breton & Martín 2011) has been somewhat acknowledged, the ET decision dimension has been the most neglected in international research areas (Luo & Peng 1998; Gao & Pan 2010). It is important to understand the firms' ET decision as claimed by Dacko (2002), where firms normally face a particularly difficult decision when planning the right time to enter a foreign market. Thus, the ET decision of foreign direct investment (FDI) also plays a critical role in multinational corporations' (MNCs) market entry strategy (Luo & Peng 1998). Green, Sedef and Bjorn (2004) asserted that ET decision may affect the firm's competitive position, especially on the ability and competency of a firm to fulfill its objectives in order to attain or even sustain its competitive edge.

Gallego et al (2009) argued that very few studies have attempted to establish relationships between these three interlinking decisions in an integrative manner. Hence, the issues were further addressed by establishing a model that illustrates the influence of ET on EM, influence of ET on EL and influence of EL on EM into international markets (Gallego et al. 2009). However, this model only focused on the influence of dependent variables on each other and only considered five dimensions of independent variables, that is; knowledge, resources, product and process innovation, mimicry and situational uncertainty that have a bearing on these three entry decisions and has ignored the influence of significant factors on them. Thus there still remain a number of questions that need to be addressed, especially in choosing a different combination of market entry decisions such as the EL and ET decisions and determining the significant influential factors on these entry decisions to enter international markets. Hence, the research on which this paper is based, aims to empirically determine the factors influencing EL and ET decisions of construction firms in international markets.

Literature Review

Factors Influencing Entry Location (EL) Decisions

Based on the reviews of factors related to EL decision, the following discussions are grouped into four main factors: country, market/industry, firm and project. It was found that most of the researchers have considered firm factors in their studies.

Country factors were amongst the important factors considered in the EL decision. Pioneering research by El-higzi (1999) on Australian construction firms, focused on the risk factors that influenced the firms' decision to internationalize their operations, how the interplay to influence the EL decision of a foreign country is related to country factors such as the host country's political stability, foreign exchange control, trade barriers, tax discrimination, home government policy towards foreign investment, tax incentives with host country, and home country market demand fluctuation. Further, Abdul-Aziz and Wong (2010) evaluated a wider range of country factors influencing the Malaysian contractors in making the go/no go decision to enter foreign markets. The factors include political stability, taxation and incentive, law and order, host government delivery system/bureaucratic efficiency, host government's integrity and transparency, host government's encouragement, language and culture similarity and visa requirements. More recent literature focuses on more in-depth research related to factors influencing the EL decision, however still dwelling on the similar country factors namely; home country government support, well-established host country institutions (Lu et al. 2014), attitude and intervention of host government, host government control on licensing, restrictions and other FDI requirements (Mat Isa et al. 2013), travel time and liability of distance (Boeh & Beamish 2012), host–home country linkages, host–home country specific advantages (Buckley, Forsans & Munjal 2012), local density of home-country affiliates (Zhu et al. 2012), and distance factors (Malhotra, Sivakumar & Zhu 2009).

The next important influential factor is related to the market factors, for example the market size, high economic performance, competition intensity in host country, availability of finance, intensity of competition in home country, and reliable and timely information (El-higzi 1999); market growth, rapid economic development, market size, business cost, financial freedom, market openness, exchange rate, foreign competition, connection and degree of business interaction (Abdul-Aziz & Wong 2010). Sakarya et al. (2007) claimed that traditional market EL analysis relies on purely macroeconomic and political factors and fails to account for the market's dynamism such as growth and opportunities for future works. The findings indicated that strong future market potential, manageable level of cultural distance, supportive and developing local industry and positive customer receptiveness for foreign products and business were very important in the firms' market EL.

The firm factors being studied were the profit repatriation, desire to expand strategically, firm's strength in know-how (El-higzi 1999), capital requirement, local competition, technological capability, geographical distance, trained workforce, ease of obtaining financial funding (Abdul-Aziz & Wong 2010), enhanced organizational capabilities, accumulation of experiential knowledge and capabilities based on prior entry experience (Lu et al. 2014), strong financing capacity, experience in similar works (Mat Isa et al. 2013), prior experience in the local market (Zhu et al. 2012), resources of firms, firm internationalization market-seeking and labor-seeking strategies (Jain 2010), financial strength, project management skill, and international network (Che Ibrahim et al. 2009).

Relative to the resource based view, firm factors such as resources and capabilities are the main elements being investigated by the majority of the researchers. These firm resource-based capabilities are the internal factors that shape the firm's competitive advantages commonly used during their international operations. It started with the firms' vision and mission with the desire to expand using market and labor seeking strategies. Targeting profit repatriation and using the firm's know-how and project management skill, international network, accumulated experiential knowledge and enhanced organizational capabilities the firms were encouraged to penetrate the highly risky international markets. However, based on the project capital requirements, they must have strong financing capacity or be able to get financial funding in order to compete and sustain their place in the international markets.

In addition, some of the project factors based on previous research were project funding, project nature and future potential (El-higzi 1999), infrastructure and other related and supporting industries (Abdul-Aziz and Sing-sing 2008), and the availability of project funds (Mat Isa et al. 2013a). Hence, to address the gap, the research question in this paper was whether and how significantly these factors influenced the firms' EL decision.

Factors Influencing Entry Timing (ET) Decisions

This section further analyzes the influential factors related to ET decisions based on previous studies. The application of the resource-based view is also incorporated in the firm factors discussion, since the majority of the researchers have also considered these factors in their studies.

Previously, a number of researchers have attempted to determine influential factors for ET decisions for international firms under various industries. Under country factors, in their study, Lilien and Yoon (1990) have included the competition elements such as the entry competition and demand potential. The reviews revealed that the researchers also focused on the market factors such as the demand potential, market evolution and marketing rivalry (Lilien & Yoon 1990), foreign market stability, and degree of globalization in the industry (Petersen & Pedersen 1999), environmental conditions, namely market dynamism and market rivalry (Villaverde & Ortega 2007) and type of industry (Tsou, Yu & Lin 2009). In addition to the recent literature, Stevens and Dykes (2013) studied the influence of the country factors namely; home country cultural attributes, host country's political environment.

Next, under the firm factors Stevens and Dykes (2013) focused on the firms' high performance orientation, high power distance and high uncertainty avoidance cultures. Other firm factors studied by other researchers are international experience, level of knowledge, research & development, project management, specialist expertise and technology, financing capacity (Mat Isa et al. 2013b), managerial, marketing, technological capabilities (Villaverde & Ortega, 2007), firms' resources and capabilities (Lieberman & Montgomery 1998), whether the company is producing manufactured goods or services, foreign market entry motives, company size, similar experience with foreign markets entered (Petersen & Pedersen 1999), firms' product technology strategies for example in offering products based on the technology standard and products incorporating the latest technology, pre-entry experience (Bayus & Agarwal 2007), firm size, research and development intensity (Tsou, Yu & Lin 2009; Lilien & Yoon 1990); firm factor (research and development competition) and also the project factor (product competition).

The earlier study by Lieberman and Montgomery (1998) explained that the firms' ET decisions were subject to the firms' internal factors, whether they are strong and confident enough to be the early movers or whether they have to wait-and-see to be the late movers. Hence, larger firms with strong tangible assets, having greater access to financing, were found to enter the foreign market early (Petersen and Pedersen 1999). In his review, Peng (2001) claimed that the resource-based view (RBV) of the firm has influenced the theoretical perspective in international business research. The RBV indicates that firms' resource capabilities allow the firms to achieve sustainable competitive advantages and successful firms usually possess proprietary assets to sustain better performance. Hence, the previous studies show that firms with strong resource capabilities and competencies have significantly influenced that firm's ET decision into an international market. Thus, these firms are likely to be among the early movers entering into a particular market, since their competitive advantages will offset the uncertainty and information disadvantages that are most profound for international entrants (Delios, Gaur & Makino 2007). In addition, Bayus and Agarwal (2007) indicated that ET decision plays an important role related to the pre-entry experience and firm survival, while Tsou, Yu and Lin (2009) later found that in addition to the industry type and host country environment, the firm size, and research and development intensity were the factors influencing the ET decision for Taiwanese firms in China. Hence, firms with competitive advantages entered early and performed better as compared to the late entrants. However, low growth market was also found to influence the firms to enter late into foreign markets.

In summary, the internal factors that commonly influence the ET decision are related to the firm's resource capabilities in terms of size, financial, physical and intangible assets. These are the manifestation of the firm's strengths in terms of skills, competencies and competitive advantages in order to be sustainable and perform in the international markets. On the other hand, external factors found to relate to the international market environment were identified as competition, market growth, uncertainties and information. Hence, to address the gap, the research question in this paper was whether and how significantly these factors influenced the firms' ET decision.

A Proposed Model of Factors influencing ELET Decisions

The evaluation on the literature reviewed has identified forty-four (44) factors influencing the EL and ET decisions related to both construction and non-construction businesses in international markets. These are the independent variables representing the following: country factors such as attitude and intervention of host government, similarity of host country/market (social/cultural/religious), proximity to host country, anticipated non-economic risks (political, technology) and economic risks (currency fluctuations, interest rates), other foreign competitors in the host country, promotion of export efforts of home government, financial support from home country banks, trade relationship between two countries, diplomatic relationship between two countries, host government control: licensing, restrictions and other foreign direct investment requirements, market/industry factors such as market profit potential/attractiveness, market intensity on competition, product/service market growth, market entry barriers, availability of innovative and entrepreneurial opportunities, and construction demands related to finance, labor, material, transport, utilities, firm factors such as, size, ability to assess market signals and opportunities, level of international experience, long-term and strong management strategic orientations/objectives, superior management & organizational dynamic capabilities, financing capacity, competencies in project management, specialist expertise and technology, resources based on level of knowledge and research & development, risk management attitude, quality management of product, service, human resource, profit targets (return on investment/sales/assets), level of knowledge and international experience, uncertainty avoidance, international business network for example relationship with foreign partners, product differentiation with strong brand name, reputation and good track record /competitive advantage, and project factors such as, project size, project types (building, manufacturing), technical complexity of projects, type of clients (public, private), availability of funds for projects, contract types /procurement methods: lump sum, cost-plus, D&B, experience in similar works, existence of strict time and quality requirements, and availability of partner/alliance.

The proposed ELET decision model will be developed by identifying the common shared factors significantly influencing both EL and ET factors using mutually inclusive principle as shown in Figure 1.

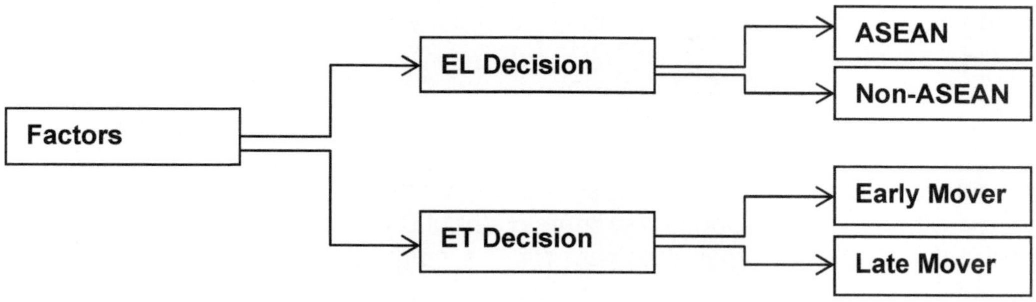

Figure 1 Conceptual framework for EL and ET decision model (Developed for this study)

The proposed ELET decision model consists of two dependent variables which are the EL and ET decisions adopted by the firms. These are the variables of primary interest, the variances in which are attempted to be explained by the forty four factors as independent variables identified from the literature review.

Research Method

The study adopts an exploratory approach utilizing a quantitative method. This approach is a common strategy in business and management research and particularly suitable when the aim is to understand the "what" and "how" factors influencing the firms' EL and timing decisions (Saunders, Lewis & Thornhill 2009). The target population was chosen from the cross-section of Malaysian construction firms that had undertaken and completed projects in international markets. However, an official registry that contains the total number of construction firms in international markets is not available. Hence, the sampling frame of the population was based on CIDB (2103) record with 115 firms registered as international players operating in more than 49 countries. Their involvement in international projects includes various sectors such as building, infrastructure, branches of engineering, mechanical and electrical, power transmission and plant, and oil and gas. The unit of analysis is an individual construction firm. Thus, the target respondents from these firms were the general managers, senior managers, project managers, assistant project managers, project engineer, project planners, contract managers and project coordinators; those directly involved who have acquired international experience in handling construction projects in international markets.

This paper is part of on-going research based on a section of the survey questionnaire to seek the experts' opinion on their firms' EL and ET decisions. The purposes of the questionnaire are: (1) to determine individually the significant factors influencing EL and ET decisions and, (2) to determine the commonly shared or mutually inclusive significant factors that the respondents considered in their EL and ET decisions using factor analysis from each model. Hence, the respondents were asked to select their firms' (a) preferred ET (early mover/late mover) and (b) international business locations (a list of countries was given as recorded by CIDB Malaysia and respondents were encouraged to state other non-listed locations). The preceding question seeks the experts' opinions on the significant level of 44 factors on their EL and ET decisions. The level of significance of influence for each opinion was measured using a 5-point Likert scale (1: Not critical; 2: A little critical; 3: Critical; 4: Very critical; and 5: Extremely critical).

The purpose of each analysis, results and discussion on factors influencing the EL and ET decisions are outlined in the next section. The binary logistic regression analysis dependent (EL and ET decisions) was used while factor analysis was carried out to measure the independent variables (factors influencing the EL and ET decisions). Other statistical analysis techniques such as descriptive analysis validity test, reliability test, normality test, correlation were adopted.

Results and Discussions

Based on a total of 115 firms, 45 firms responded giving a 39.1 percent response rate. In order to increase the rate of response, personal distribution, follow-up letters and phone calls were carried out. Hence, the response rate for this study is acceptable since most of the surveys done in Malaysia generated a rate between 10 to 20 per cent for example, 10.8% from previous studies by Ahmed et al. (2002), 19.8% from Abdul Aziz, Wong and Awil (2008) and 12.1% by Abdul-Aziz and Awil (2010).

The respondents' designation indicates a diverse background is required during international operations. In general, the respondent's designation varied from being managing director (2), technical director (2), vice president (2), general manager (1), senior project manager (3), project manager (5), project/architecture coordinator (2), senior project engineer (1), project engineer (3), design/civilvengineer (9), contract manager (3), quality/financial/human resource manager (3), quantity surveyor (2), project planner(1) and other managerial positions (6). Results demonstrate that 25% of the respondents have more than 10 years of international experience, 29% having experience between 5 to 10 year and the rest (47%) have between 1 to 4 years of experience. Hence, the respondents have the required international related construction background to participate and give reliable opinions.

They were asked to choose their international business locations based on a list of 43 countries adopted from the CIDB record (2013) which are: Algeria, Australia, Bangladesh, Cambodia, China, Egypt, Hong Kong, India, Indonesia, Iran, Iraq, Ireland, Japan, Kuwait, Libya, Maldives, Mauritius, Myanmar, Mongolia, Morrocco, Nepal, Nigeria, Oman, Pakistan, Philiphines, Qatar, Saudi Arabia, Seychelles, Singapore, Spain, South Africa, South Korea, Sudan, Syria, Sri Lanka, Taiwan, Thailand, Turkmenistan, United Arab Emirates, United Kingdom, Vietnam and Yemen and were also encouraged to state other non listed locations.

The results indicate about 88% similarity in terms of countries recorded by CIDB and the countries penetrated by the firms from this study. However, eight (8) additional countries were found in this study namely; Austria, Botswana, Brunei, France, Germany, Tobago, United States of America and United Kingdom which were not in the CIDB list.

To ensure consistency and a proper comparison with the CIDB Malaysia (2013) report on the international business locations of Malaysian construction firms, this study also grouped the countries under ASEAN and non-ASEAN. This classification alllows the measure of the EL and ET decisions as binary variables as required by the logistic regression analysis. Using descriptive analysis, it was found that about 73% of firms (33 firms) have chosen the non-ASEAN while the other 27% (12 firms) have chosen the ASEAN. The findings in this study are supported by the CIDB Malaysia (2013) statistics where more than 65% of the construction firms have penetrated non-ASEAN.

Measurement of Independent Variables

Validity Tests

In order to determine the suitability of data using factor analysis, two main issues were validated, accordingly; (a) a sampling adequacy issue and (b) the strength of the correlations among the independent variables (factors) using Kaiser-Mayer-Olkin (KMO) measure of sampling adequacy (MSA) and Bartlett's tests of Sphericity (Pallant 2011). William and Brown (2012) stated that KMO static varies between 0 and 1 and recommend accepting values greater than 0.5, which indicates that the sample meets the fundamental requirements for factor analysis. In addition, the Bartlett's test of Sphericity should be significant ($p < 0.05$) for factor analysis to be considered appropriate. Table 1 shows the values for KMO MSA and Bartlett's test of Sphericity for factors influencing the EL and ET decisions.

Table 1: KMO Measure of Sampling Adequacy and Bartlett's Test of Sphericity for EL and ET Decisions

Entry Decisions		**EL Decision**	**ET Decision**
Kaiser-Meyer-Olkin Measure of Sampling Adequacy		0.727	0.779
Bartlett's Test of Sphericity	Approx. Chi-Square	553.194	725.122
	Df	136	190
	Sig.	.000	.000

The results show that the KMO MSA values are 0.727 and 0.779 for EL and ET decisions, respectively; both are greater than 0.5 (Williams & Brown 2010). Further, the results for Bartlett's test of Sphericity have given χ^2 (136) = 553.194, $p < 0.001$, and χ^2 (190) = 725.122, $p < 0.001$, for EL and ET decisions, respectively which indicate that correlations between items were sufficiently large and strong for PAF.

Total Variance Explained

In this study, the data reduction process follows three criteria. First, Kaiser's criteria that only factors with eigenvalue greater than one are retained. Second, factors with just one item were excluded from the analysis and thirdly, the cumulative percent of variance extracted are presented. Table 2 and Table 3 show the results for the EL and ET decisions respectively, extracted from the PAF analysis.

The results from Table 2 for EL decision reveal the presence of five (5) components with eigenvalue exceeding 1. These five factor components explain a total of 71.032% of the variance contributed by component 1 (42.763%), component 2 (9.182%), component 3 (7.865%), component 4 (6.527%) and component 5 (4.695%).

Table 3 also reveals the presence of five components with eigenvalue exceeding 1 for ET decision. These five components explained a total of 72.186% of the variance contributed by component 1 (48.247%), component 2 (7.984%), component 3 (6.339%), component 4 (5.212%) and component 5 (4.404%).

Thus, results for the EL and ET decisions demonstrate a good cumulative percentage of variance of 71.032% and 72.186%, respectively; both are well above the common percentage of the explained variance for humanities research which is commonly in the range of 50% up to 60% (Hair et al. 1995).

Table 2: Total Variance Explained for EL Decision

Factor	Initial Eigenvalues			Extraction Sums of Squared Loadings			Rotation Sums of Squared Loadings [a]
	Total	% of Variance	Cumulative %	Total	% of Variance	Cumulative %	Total
1	7.556	44.449	44.449	7.270	42.763	42.763	5.561
2	1.838	10.810	55.260	1.561	9.182	51.945	3.152
3	1.609	9.464	64.724	1.337	7.865	59.810	3.504
4	1.357	7.981	72.705	1.110	6.527	66.337	3.129
5	1.090	6.414	79.119	0.798	4.695	71.032	2.728
6	0.773	4.549	83.668				
7	0.566	3.327	86.994				
8	0.496	2.916	89.910				
9	0.426	2.507	92.417				
10	0.318	1.871	94.288				
11	0.241	1.418	95.706				
12	0.201	1.183	96.889				
13	0.168	0.988	97.877				
14	0.133	0.785	98.662				
15	0.109	0.640	99.302				
16	0.070	0.410	99.712				
17	0.049	0.288	100.000				

Extraction Method: Principal Axis Factoring.
a. When factors are correlated, sums of squared loadings cannot be added to obtain a total variance.

Table 3: Total Variance Explained for ET Decision

Factor	Initial Eigenvalues			Extraction Sums of Squared Loadings			Rotation Sums of Squared Loadings [a]
	Total	% of Variance	Cumulative %	Total	% of Variance	Cumulative %	Total
1	9.920	49.599	49.599	9.649	48.247	48.247	4.744
2	1.842	9.212	58.811	1.597	7.984	56.231	4.506
3	1.523	7.615	66.427	1.268	6.339	62.571	4.830
4	1.327	6.634	73.061	1.042	5.212	67.782	5.733
5	1.166	5.828	78.889	0.881	4.404	72.186	5.883
6	0.738	3.690	82.578				
7	0.604	3.019	85.597				
8	0.529	2.645	88.243				
9	0.415	2.074	90.316				
10	0.340	1.699	92.016				
11	0.291	1.453	93.469				
12	0.268	1.340	94.809				
13	0.236	1.178	95.987				
14	0.193	0.965	96.952				
15	0.170	0.848	97.800				
16	0.141	0.707	98.508				
17	0.121	0.606	99.113				
18	0.088	0.438	99.552				
19	0.058	0.292	99.844				
20	0.031	0.156	100.000				

Extraction Method: Principal Axis Factoring.
a. When factors are correlated, sums of squared loadings cannot be added to obtain a total variance.

Factor Analysis: Principal Axis Factoring Analysis (PAF)

Factor analysis involves factor extraction and rotation methods. The extraction methods that are commonly used in the published literature for both exploratory factor analysis (EFA) and confirmatory factor analysis (CFA) are the principal component analysis (PCA) and principal axis factoring (PAF) also known as factor analysis (FA)(Williams & Brown 2010).

Firstly, following the PCA analysis, PAF was conducted for comparison and assessment for best fit as suggested by William and Brown (2010). According to Williams and Brown (2010), PAF is more correct theoretically but more complicated than PCA. In other words, whichever rotated solution produces the best fit and factorial suitability, both intuitively and conceptually, should be used. The difference between these analyses lies on the communalities used where in PCA, the communalities are assumed to be 1, assuming that the total variance of the variables can be accounted for by means of its components and hence, there is no variance. However, in the PAF method, the initial communalities are not assumed to be 1, hence the variables do not account for the 100 percent of the variance. Since, the sample size in this study (45 respondents) is less than 100; communalities above 0.60 were sought.

The second step is factor rotation method, in which the interpretation and naming of factors or component are carried out by altering the pattern of the factor loadings. Instead of orthogonal, the oblique rotation with direct-oblimin was used in this study. Oblique rotations assume that the factors are correlated. Oblique rotation is the best method to use in factor rotation as it is also looks at the correlations among the factors. The results are presented in pattern matrix and used later in the interpretation of factors.

In this study, the PAF was empirically found to offer best fit and used in this study to analyze the responses to the forty four (44) factors used in the questionnaires. Table 4 and Table 5 show the results from the PAF used as the data reduction technique using direct-oblimin with Kaiser Normalization for EL and ET decisions respectively.

Table 4 reveals a five-factor solution which has resulted in 17 items with factor loadings above 0.50 with eigenvalues above Kaiser's criterion of 1, thus in combination explained 71.032% of variance. According to Williams and Brown (2012), for study that has sample size less than 100 the communalities values must be above 0.6. The results indicate that all seventeen (17) factors have values above 0.6 hence there is no violation of the assumption of communalities. The good communalities shown in the last column indicate that the factor analysis is suitable even though the sample size for this study is only 45 respondents (Williams & Brown 2010). The five components extracted were grouped as: (1) firm factor; (2) country factor; (3) market factor; (4) project factor; and (5) management factor.

Table 5 also reveals a five-factor solution which has resulted in 20 items with factor loadings above 0.50 with eigenvalues above Kaiser's criterion of 1, thus in combination explained 72.186 percent of variance. The results indicate that all 20 factors have values above 0.6 hence there is no violation of the assumption of communalities. The five components extracted were grouped as: (1) firm factors; (2) project factors; (3) performance factors; (4) management; and (5) market factors.

Reliability Test

Once the factors were obtained and interpreted, the tests for consistency and stability took place by conducting reliability test using Cronbach's alpha. It is a reliability coefficient to indicate the extent to which all item in a set are positively correlated to one another and is obtained by computing the average inter-correlation among all items. The closer Cronbach's alpha is to 1, the higher the internal consistency reliability. In general, reliability less than

0.60 is considered poor, a result of 0.70 is acceptable and above 0.80 is considered as good (Sekaran & Bougie 2013).

Results of the reliability test for EL decision in Table 4 demonstrates that the components extracted from this analysis show high reliability of internal consistency where the value for each component exceeds 0.70; Firm factors (α= 0.912); country factors (α= 0.792); market factors (α= 0.812); project factors (α= 0.819) and management factors (α= 0.791). Moreover, all items when combined indicate a very good overall internal consistency (α= 0.917). Results for ET decision in Table 5 show that the components extracted from factor analysis indicate high reliability of internal consistency where the value for each component exceeds 0.70; Firm factors (α= 0.873); Project factors (α= 0.823); Performance (α= 0.905); Management factors (α= 0.890) and Market factors (α= 0.862). In addition, all items when combined indicate a very good overall internal consistency (α= 0.945). Hence, the factors for EL and ET decisions have high reliability of internal consistency.

Table 4: Factor loadings using Principal Axis Factoring (PAF) for EL Decision

Component	***Factor Loading***					***Commu-nalities***
	1	**2**	**3**	**4**	**5**	
Firm Factors						
Strong competencies (project management, specialist expertise and technology)	0.823					0.812
Strong financing capacity	0.760					0.620
Strong resources : (Level of knowledge and Research & Development)	0.750					0.712
Experience of firm in similar works	0.667					0.715
Management quality (product, service, human resource)	0.611					0.621
Availability of funds for projects	0.595					0.777
International business network : Strong relationship with foreign partners in host countries	0.580					0.647
Country Factor						
Trade relationship between two countries		0.933				0.932
Diplomatic relationship between two countries		0.761				0.699
Financing support of home country banks		0.568				0.671
Market Factor						
Product/Service market growth			0.863			0.801
Market entry barriers			0.750			0.685
Project Factor						
Firm ability to assess market signals & opportunities				-0.786		0.687
Firm level of international experience				-0.686		0.755
Technical complexity of projects				-0.569		0.699
Management Factor						
Uncertainty avoidance					0.665	0.686
Long-term and strong management strategic orientation/objectives					0.620	0.756
Eigenvalues	7.556	1.838	1.609	1.357	1.090	
% of Variance	42.763	9.182	7.865	6.527	4.695	
Cumulative of Variance%	42.763	51.945	59.810	66.337	71.032	
Cronbach's Alpha (n)	0.912 (7)	0.792 (3)	0.812 (2)	0.819 (3)	0.791 (2)	
Overall items Cronbach's Alpha (n)	0.917 (17)					
Extraction Method: Principal Axis Factoring Rotation Method: Direct-Oblimin with Kaiser Normalization						

Table 5 Factor loadings using Principal Axis Factoring (PAF) for ET Decision

Component	***Factor Loadings***					***Commu-nalities***
	1	2	3	4	5	
Firm Factors						
Firm's international experience	0.795					0.686
Firm's resources: Level of knowledge and Research & Development	0.622					0.768
Firm's competencies: Project management, specialist expertise &technology	0.592					0.703
Firm's financing capacity	0.515					0.749
Project Factors						
Performance: Increase level of knowledge and international experience		0.783				0.775
Availability of funds for projects		0.588				0.731
Technical complexity of projects		0.505				0.742
Performance factors						
Project size			-0.859			0.855
Good track record and competitive advantage			-0.806			0.844
Type of clients: public, private			-0.544			0.845
Firm's reputation			-0.505			0.682
Management factors						
Financing support of home country banks				-0.796		0.704
Experience of firm in similar works				-0.746		0.840
Existence of strict time limitations				-0.706		0.821
Superior management and organizational capabilities				-0.548		0.715
Market Factors						
Construction demand: Finance, labor, material, transport and other utilities					0.805	0.748
Availability of partner/alliance					0.625	0.677
Attitude and intervention of host governments					0.598	0.666
Similarity of host country/market: social, cultural, religions					0.575	0.513
Firm's ability to assess market signals and opportunities					0.558	0.624
Eigenvalues	9.920	1.842	1.523	1.327	1.166	
% of Variance	48.247	7.984	6.339	5.212	4.404	
Cumulative of Variance %	48.247	56.231	62.571	67.782	72.186	
Cronbach's Alpha (n)	0.873 (4)	0.823 (3)	0.905 (4)	0.890 (4)	0.862 (5)	
Overall items Cronbach's Alpha (n)			0.945 (20)			
Extraction Method: Principal Axis Factoring Rotation Method: Direct Oblimin with Kaiser Normalization						

Normality Test

A univariate analysis known as Skewness and Kurtosis was used to determine whether the data is normally distributed or not. The distribution is considered normal if the value lies between -1 and +1. In addition, the normal quantile-quantile (Q-Q) plot method was also used. If the majority of values (smaller dots) lie on the straight line in the plot, the data are approximately normally distributed. In order to meet the assumption of normality, the Skewness and Kurtosis statistics was performed on the factors for the EL and ET decisions.

For the EL decision factors, the results revealed the followings; Firm factors (Skewness = -0.809; Kurtosis = 0.298), country factors (Skewness = -0.192; Kurtosis = -0.745), market factors (Skewness = -0.772; Kurtosis = 0.4770, project factors (Skewness = -0.698; Kurtosis = 0.030), and for management factors (Skewness = -0.046; Kurtosis = -0.680). While for the ET decision factors, the results are as follows; the firm factors (Skewness = -0.360; Kurtosis = -0.461), project factors (Skewness = -0.001; Kurtosis = -1.092), performance factors (Skewness=-0.524; Kurtosis= -0.659), management factors (Skewness = -0.115; Kurtosis = -0.431) and market factors (Skewness = -0.365; Kurtosis = -0.223).

The descriptive statistics shows that all variables are normally distributed since the values of Skewness and Kurtosis coefficients are in the range of ±1.0 with majority of values lying on the straight line in the Q-Q plots. Hence, all factors influencing the EL and ET decisions are considered approximately normally distributed.

Omnibus Tests of Model Coefficient

The omnibus test indicates an overall significant of the influential factors on the EL and ET decisions. For EL decision, the result shows that the Omnibus test of Model Coefficients is significant; [χ^2 (5) = 22.207, $p < 0.001$]. While for the ET decision, the result reveals that the Omnibus test of Model Coefficients is also significant; [χ^2 (5) = 21.792, $p < 0.05$]. Therefore, the models have a good set of independent variables for both EL and ET decision factors.

Model Summary Using Cox and Snell R ^{2} and Nagelkerke R^2

The Cox and Snell R^2 and Nagelkerke R^2 are statistics that provide an indication and quantify the 2R proportion of explained "variation" in the logistic regression model. Precisely, for the EL decision, the value of Cox and Snell R^2 is 0.418 which reveals about 41.8% of the variation in the outcome variable is explained by the model. Likewise, the Nagelkerke R^2 is 0.642 which indicates that about 64.2% of the variation in the outcome variable is explained by the logistic regression model. Similarly, for the ET decision, the Cox and Snell R^2, and Nagelkerke R^2 suggest that between 43.6% and 59.6% of the variability is explained by the logistic regression model.

Goodness-of-Fit of the Model using Hosmer and Lemeshow Test

Using Hosmer-Lemeshow Goodness of Fit Test, a good fit is indicated by a significant value more than 0.05. For the EL decision, results indicate that the data supports the model with $\chi2$ (8) = 5.734 with a significance level of 0.263 ($p > 0.05$). While for the ET decision, the results also indicate that the data supports the model with $\chi2$ (7) = 8.857 with a significance level of 0.263 ($p > 0.05$). Thus, both results for the EL and ET decisions indicate sufficient evidence to claim that the model is worthwhile and fits the data adequately.

Assumptions of Logistic Regression Model (LRM)

Prior to performing the logistic regression analysis, a correlation statistics and outlier diagnosis were prepared to investigate possible signs of multi-collinearity and the presence of outliers. The following sections discuss on the multi-collinearity and outliers assumptions that have to be met in order to have a valid model.

Multi-collinearity Diagnosis for Factors influencing EL and ET Decisions

Multi-collinearity problems exist when there are strong relationships among independent variables. The variance inflation factor (VIF) is calculated for all variables with the aim of verifying the possible existence of multi-collinearity. This test measures the extent to which the variances of the coefficients estimated in a regression are inflated when compared to the cases in which the independent variables are not linearly related. When the VIF values are greater than 10, the cut-off point can become indicators of the existence of multi-collinearity (Pallant 2011).

Table 6: Coefficients [a] and collinearity statistics for EL Decision

Model		Unstandardized Coefficients		Standardized Coefficients	t	Sig.	Collinearity Statistics	
		B	Std. Error	Beta			Tolerance	VIF
Step 1	(Constant)	11.643	12.249		.951	.348		
	Firm	.050	3.500	.003	.014	.989	.507	1.971
	Country	-1.555	2.699	-.103	-.576	.568	.754	1.326
	Market	-3.798	2.668	-.257	-1.424	.163	.737	1.357
	Project	3.946	3.141	.247	1.256	.217	.617	1.620
	Management	4.274	3.465	.254	1.233	.226	.565	1.769
a. Dependent Variable: series								

Table 6 and Table 7 depict the coefficients and linear statistics which explain whether there is existence of multi-collinearity for EL and ET decisions respectively. The results show the highest VIF values of 1.971 and 2.427 respectively, which are well below 10, the cut-off point recommended by Pallant (2011). Furthermore, all tolerance values for both tables are greater than 0.1 which rules out the presence of multi-collinearity in the data. Hence, multi-collinearity problems do not exist for either EL or ET decision factor models.

Table 7: Coefficientsa and collinearity statistics for ET Decision

Model		Unstandardized Coefficients		Standardized Coefficients	t	Sig.	Collinearity Statistics	
		B	Std. Error	Beta			Tolerance	VIF
Step 1	(Constant)	12.019	15.687		.766	0.449		
	Firm	-7.754	3.818	-.428	-2.031	0.051	0.583	1.716
	Project	3.811	5.045	.190	.755	0.455	0.412	2.427
	Performance	2.536	3.727	.154	.680	0.501	0.508	1.967
	Management	7.087	5.465	.319	1.297	0.204	0.428	2.335
	Market	-2.954	4.517	-.142	-.654	0.518	0.552	1.813
a. Dependent Variable: series								

The Presence of Outliers

Preliminary analyses were performed to ensure no violation on the assumptions of normality and outlier cases. For the EL decision, the analysis detected 1 outlier which can harm the logistic regression analysis and correlation analysis results for a small sample size. The outlier was removed leaving only 44 out of 45 items. For the ET decision, the analysis detected 7 outliers. Similarly, the outliers were removed leaving only 38 out of 45 items. Consequently, the results depicted by the Box-plots indicate no outlier for the entire variables which shows the data set is clean from any outlier cases. The output in the logistic regression tables also shows that the case wise plot was not produced due to the absence of outliers. Since these two assumptions were met, both logistic regression models for EL and ET decisions together with their results are valid. In summary, the findings have proven that the multi-collinearity problems did not exist in either EL or ET decisions models and are supported by all VIF values (less than 10.00) with tolerance values, all above 0.10.

Measurement of Dependent Variable

The objective of this study is to examine the influence of factors on firms' EL decisions in choosing either ASEAN or non-ASEAN countries and on the firms' ET decisions to be either an early or late mover. Descriptive analysis shows that more than 70% of the firms have penetrated the non-ASEAN with more than 50% of the firms choosing to be late movers. Thus, in order to have a better understanding of the factors influencing these decisions, a binary logistic regression analysis was carried out. As a limitation to this study, the actual distance from Malaysia to the foreign locations was not considered. However, the actual distance from host to home country is proposed as an observable measurement for future research on EL decision to develop a decision tool for construction firms.

A binary logistic regression is similar to a linear regression except that it is used when the dependent variable is dichotomous/binary, while multinomial logistic regression is used when the dependent or outcome variable has more than two categories (Leech, Barrett, & Morgan 2005). It is used as an appropriate multivariate procedure for describing and testing relationships between a dichotomous/binary (0/1) outcome variable and a number of categorical and/or continuous variables. Hence, a binary logistic regression model was used in this study to determine the effect an increment of each independent variable (factors) on how likely the binary variable (EL) decision is to take value 1 (non-ASEAN countries) as opposed to value 0 (ASEAN countries). Similarly, a binary logistic regression model was also used to determine the effect an increment of each independent variable (factors) on how likely the binary variable (ET) decision is to take value 1 (late movers) as opposed to value 0 (early mover). An assessment of the goodness-of-fit of the model using Hosmer and Lemeshow test carried out earlier has determined the appropriateness of the model.

Table 8 and Table 9 known as the classification tables summarize the results with the predictor variables in the EL and ET decision models respectively. Table 8 shows that the EL decision model has correctly classified 90.6% of construction firms have chosen the non-ASEAN countries, while 55.6% have chosen the ASEAN countries. As a result, in the overall model, 82.9% of the sample population has been correctly classified.

Table 8: Classification Table with predictor variables for EL Decision[a]

Observed			Predicted		
			EL		Percentage
			Non-ASEAN	ASEAN	Correct
Step 1	EL	Non-ASEAN	29	3	90.6
		ASEAN	4	5	55.6
	Overall Percentage				82.9

a. The cut value is .500

Similarly, as shown in Table 9, 95.8% of the construction firms were correctly classified in the late mover group, while 78.65% in the early mover group. As a result, in the overall model, 89.5% of the sample population has been correctly classified. Hence, both classification tables have indicated that each model has predicted the correct category for each case well.

Table 9: Classification table with predictor variables for ET Decision[a]

Observed			Predicted		
			ET		Percentage Correct
			Early Mover	Late Mover	
Step 1	ET	Early Mover	11	3	78.6
		Late Mover	1	23	95.8
	Overall Percentage				89.5
a. The cut value is .500					

Variables in the Equation for EL Decision

The binary logistic regression was also performed to assess the impact of a number of factors on the likelihood of the firms' EL decision in international market. Table 10 known as "variables in the equation" provides the information on the contribution or importance of each independent variable (factors for EL decision) on the model. The Wald statistics (fourth column) was also used to identify the independent variables that are good predictors. This method of assessing the successive accuracy of a model is to evaluate its ability to correctly predict the category for cases for which the outcome is known.

Table 10 shows a model contains five independent variables named as the firm, country, market, project and management factors. The factors corresponding to the values under the sixth column labeled Sig. which are less than 0.05 are the variables that contribute significantly to the predictive ability of the model. This suggests that the model was able to distinguish between firms that chose to enter the ASEAN or Non-ASEAN countries. However, the model shows only three (3) predictors namely; firm, country and market were statistically significant, as shown earlier, where $\chi 2(5) = 22.207$, $p < 0.001$, with B values of -3.177, -4.780 and 4.150 respectively. Hence, for the firms to choose either to enter the ASEAN or non-ASEAN countries depends, among other factors, on the firm, country and market factors that are needed to be established before stepping into the chosen foreign market. While, the other two factors namely project and management with Sig. values of 0.073 and 0.224 did not influence the EL decision of the firms in international market expansion.

Table 10: Variables in the Equation for EL Decision

	Factors	B	S.E.	Wald	df	Sig.	Exp (B)	95% C.I.for EXP(B)	
								Lower	Upper
Step 1[a]	Firm	-3.177	1.554	4.181	1	.041	0.042	.002	.877
	Country	-4.780	2.136	5.007	1	.025	0.008	.000	.552
	Market	4.150	1.920	4.672	1	.031	63.404	1.472	2730.816
	Project	-2.550	1.422	3.214	1	.073	0.078	.005	1.269
	Management	1.725	1.418	1.479	1	.224	5.611	.348	90.368
	Constant	14.238	6.981	4.159	1	.041	1525167.612		
a. Variable(s) entered on step 1: firm, country, market, project, management.									

The other useful information in Table 10 is provided in the Exp (B) (seventh column). The B values provided in the second column are the values that are used in an equation to calculate the probability of a case falling into a specific category (independent variables influencing EL decision). These values are the odds ratios (OR) for each of the independent variables. The odds ratio represents 'the change in odds of being in one of the categories of outcome when the value of a predictor increases by one unit. It is shown as the strongest predictor. Since each predictor is a continuous variable, "increase" is reported for value more than 1 (decrease if less than 1) of the odds for each unit increase in the predictor variable. Hence, based on the overall results, the strongest predictor is the country factor (B = -4.780, $p < 0.05$), recording an odds ratio of 0.008. The odds ratio for this variable, however, as .008, is a value less than 1. This indicates that a firm with less knowledge on the country factors is 1 times less likely to choose the ASEAN countries as compared to those who have more knowledge on the country factors, all other factors being equal.

Variables in the Equation for ET Decision

Similarly, Table 11 provides information about the contribution or importance of each of the independent variables (factors) on the model for the ET decision. The factors corresponding to the values under column labeled Sig. which are less than 0.05 are the variables that contribute significantly to the predictive ability of the model.

Table 11: Variables in the Equation for ET Decision

Factors		B	S.E.	Wald	df	Sig.	Exp (B)	95.0% C.I.for EXP(B)	
								Lower	Upper
Step 1[a]	Firm	-4.660	2.024	5.300	1	0.021	0.009	0.000	0.500
	Project	3.501	1.760	3.959	1	0.047	33.155	1.054	1.043E3
	Performance	1.669	1.342	1.546	1	0.214	5.304	0.382	73.624
	Management	-4.035	1.995	4.090	1	0.043	0.018	0.000	0.883
	Market	3.786	1.769	4.578	1	0.032	44.064	1.374	1.413E3
	Constant	1.014	3.436	0.087	1	0.768	2.756		

a. Variable(s) entered on step 1: firm, project, performance, management and market

Choosing either to enter early or late depends, among other factors, on the firms' background and resources to be established before stepping into the chosen foreign market. Hence, the logistic regression model fitted well with the predictor variables (ET decision) under components namely; firm, project, performance, management and market factors. In this case, the model reveals that four out of five independent variables which are the firm, project, management and market factors with Sig. values of 0.021, 0.047, 0.043 and 0.032 respectively, have made a unique statistically significant contribution to the model. However, another factor, performance (Sig. = 0.214) did not influence the ET decision. While the B value provided in the second column are the values that are used in an equation to calculate the probability of a case falling into a specific category (an independent variable that influences the ET decision).

Similarly for the ET decision based on Table 11 (Exp (B) column) the strongest predictor is the firm factor (B = -4.660, $p < 0.05$), recording an odds ratio of 0.009. The odds ratio for this variable, however, as .009, is a value less than 1. This indicates that the more knowledge a firm has on the firm factors, the less likely the firm chooses to be the late mover. Hence, the odds of a firm choosing to be a late mover decrease by a factor of 0.009, all other factors being equal.

Proposed ELET decision model

Common factors influencing both EL and ET decisions are consolidated towards developing the ELET decision model based on extracted loadings from factor analysis. Due to the limitation of using logistic regression analysis in the SPSS software, the analysis has to be carried out individually for both EL and ET decision models. Hence, the ELET decision model is developed by consolidating the common factors shared by both EL and ET decisions. The integration of decision using structural equation modelling that can model the constructs/factors influencing the two dependent variables simultaneously, is in progress.

The logistic regression models have successfully determined the effect of factors on EL and ET decisions in which 90.6% of the respondents have been correctly classified under non-ASEAN while 78.5% of the respondents have been correctly classified as late movers. Factor analysis has resulted in 17 significant factors for the EL decision and 20 significant factors for the ET decision. Table 12 shows the significant factor loadings for EL and ET decisions extracted from Tables 4 and 5.

Table 12: Summary of significant factors influencing EL and ET decisions based on factor loadings

	Factors influencing EL and ET decisions	**Loadings for EL**	**Loadings for ET**
C1	Host government attitude and intervention		0.595
C2	Similarity with host country/market (social/cultural/religious) environment		0.575
C8	Financial support from home country banks	0.568	0.796
C9	Trade relationship between two countries	0.933	
C10	Diplomatic relationship between two countries	0.761	
M3	Product/service market growth	0.863	
M4	Market entry barriers	0.750	
M6	Construction demand (e.g. finance, labor, material, transport and other utilities)		0.805
F2	Firm's ability to assess market signals and opportunities	-0.786	0.558
F3	Firm's level of international experience	-0.686	0.795
F4	Firm's long-term and strong management strategic orientation/objectives	0.620	
F5	Firm's superior management & organizational dynamic capabilities		0.548
F6	Firm's financing capacity	0.760	0.515
F7	Firm's competencies (project management, specialist expertise and technology)	0.823	0.592
F8	Firm's resources (level of knowledge and Research & Development)	0.750	0.622
F10	Firm's management of quality (product, service, human resource)	0.611	
F12	Firm's perfomance in terms of increase level of knowledge and international experience		0.783
F13	Uncertainty avoidance	0.665	
F14	International business network (strong relations with foreign partners)	0.580	
F16	Firm's reputation		0.505
F17	Firm's good track record /competitive advantage		0.806
P1	Project size		0.859
P3	Technical complexity of projects	-0.569	0.505
P4	Type of client (public vs. private)		0.544
P5	Availability of funds for projects	0.595	0.588
P7	Experience of company in similar works	0.667	0.746
P8	Existence of strict time limitations		0.706
P10	Availability of partner/alliance		0.625
	Total number of factors	17	20

Further consolidation has resulted in nine (9) factors that are commonly shared or mutually inclusive for both EL and ET decisions as shown in Table 13. These factors were found to significantly influence the majority of firms' ET decision as late movers by entering the selected market locations mostly in the non-ASEAN countries far from their home country as shown in the ELET model in Figure 2.

Table 13: Mutually inclusive significant factors influencing ELET decisions of firms in international market expansion

No.	Factors influencing ELET decisions
1	Financial support from home country banks
2	Firm's ability to assess market signals and opportunities
3	Firm's level of international experience
4	Firm's financing capacity
5	Firm's competencies (project management, specialist expertise & technology)
6	Firm's resources (level of knowledge and Research & Development)
7	Technical complexity of projects
8	Availability of funds for projects
9	Firm's experience in similar works/projects

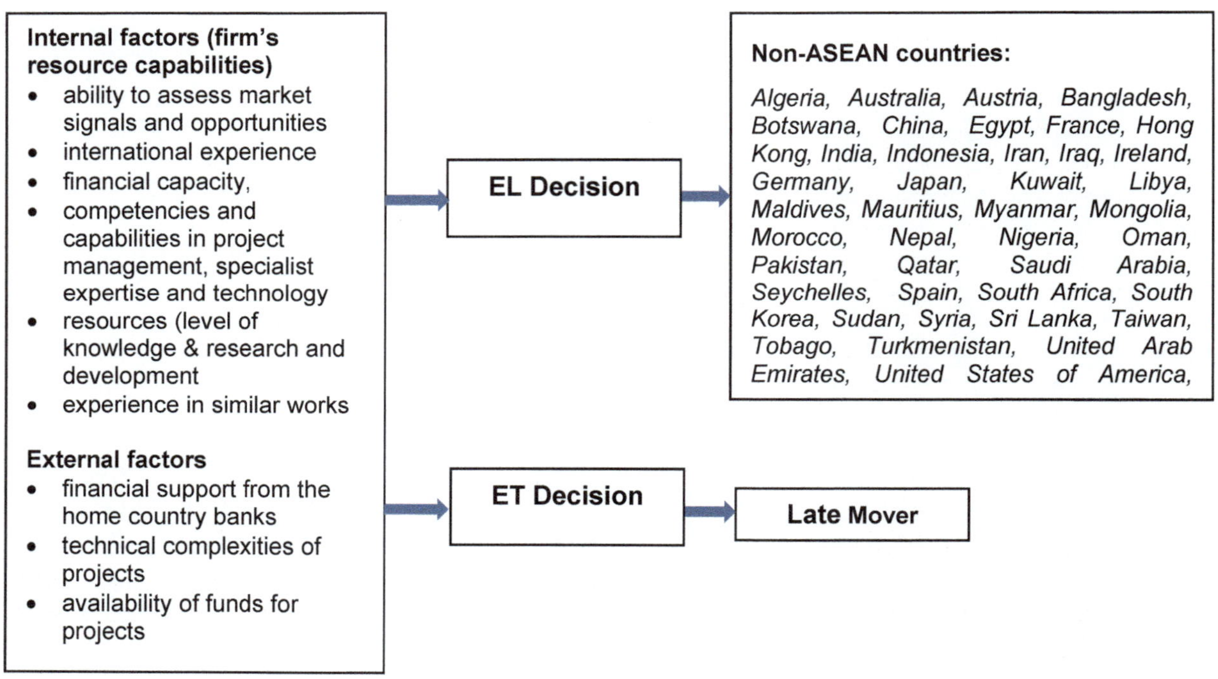

Figure 2: ELET decision model for Malaysian construction firms in international market (Developed from this study)

These nine (9) factors are further grouped into six (6) internal factors related to the firms' resource capabilities and three (3) external factors which contributed to their entry decisions. The factors related to the firms' resource capabilities were the firms' ability to assess market signals and opportunities, international experience, financial capacity, competencies and capabilities in project management, specialist expertise and technology, resources (level of knowledge & research and development) and experience in similar works, while the external factors are the financial support from the home country banks, technical complexities of projects and availability of funds for projects.

The ability of a firm to assess the emerging market growth and sourcing opportunities was crucial and should not be overlooked, as agreed by Sakarya, Eckman and Hyllegard (2007). Firms should not rely purely on macroeconomic and political factors. Gaston-Breton and Martín (2011) have recommended that firms should look for market signals such as the market size and its potentials in order to select the right location for their businesses. Hence, the findings have fulfilled the suggestion that the significant firm's resource based capabilities which have influenced the firms' ELET decisions to enter international markets, include the ability to assess market signal, financial capacity, competencies and capabilities in project management, specialist expertise and experience in similar works with no significant factor related to the political factor. According to previous studies, many business opportunities exist for construction firms to operate in the international market due to its high volume of construction demand and its growing economy either within the ASEAN or non-ASEAN regions. Hence, in a firm's strategic planning it is very important to assess the market signals and explore opportunities by gathering the required information on ELET decisions in potential international markets.

As indicated by Gunhan and Arditi (2005), financial strength is one of the essential firm resources in international construction. Capital requirement during market entry is very high and firms with large resources are able to cover the capital requirement and enter the foreign market earlier. However, these findings from this study are in line with Söderblom (2011) who pointed out that firms with strong financing capability or that have easier access to financing were able to enter the foreign market early.

International construction projects are known to be more complex with many known and unknown risk factors as compared to domestic projects. Hence, project management skills including specialist expertise and technological capabilities are very much needed to handle the complex nature of international projects (Gunhan & Arditi 2005). Thus, firms with a high level of competencies were among the early movers entering into a foreign market, since its competitive advantages will offset uncertainty and information disadvantages which are most profound for international entrants (Villaverde & Ortega 2007). Firms' capabilities, measured by the firm size, project management skills, specialist expertise level, firm reputation, technology knowledge and firm network, are some of the factors that influence the ET decision. Hence, when firms have low levels of competencies and capability, the implication is late entry (Dacko 2002).

The findings show that the firms' resources related to their level of knowledge based on Research and Development has significantly influenced their ELET decisions, which may indicate that low level of knowledge has resulted in lack of power to gain access to suppliers, markets, customers and other assets as accentuated by Soderblom (2011). However, Soderblom emphasized that learning creates substantial entry barriers for the late movers as compared to the early movers especially in unstable situations related to customer needs, where the early movers grasp opportunities that exist when entering the market that later will limit market opportunities for late movers. Guler and Guillén (2009) deliberated that the level of knowledge and technology increases as the firm's international experience increases. Hence, firms should plan properly to increase their knowledge and overcome the entry barriers in order to understand the needs of customer and predict the market trends (Kerin, Varadarajan & Robert 1992). Hence, late movers must acquire greater knowledge and other intangible assets to help reduce the risks and competition during the exploitation of opportunities in international markets. The factors affecting the firm's resources availability include satisfying capital requirement, lowering risk, more flexibility in decision making and increasing market power.

The findings indicate that firms' international experience significantly influenced the firms' ELET decisions to enter international markets. This finding is consistently supported by another study where firms with a high level of experience in similar projects have entered early compared to those having less experience (Schwens & Kabst 2009). As found by Guler and Guillén (2009), as firms gained more international experience they were more likely to overcome the entry barriers by entering as an early mover and to prepare for cross-border by improving their knowledge and experience (Liu, Low & Niu 2011). The findings suggest that the Malaysian firms' resource capabilities such as experience in similar projects plays important roles related to the ELET decisions. Hence, the construction firms contended that based on their firm's international experience in similar projects, they have chosen to be late entrants to the non-ASEAN countries.

Conclusions

In light of the work presented in this paper a number of conclusions can be made. As the Malaysian construction firms go international, the EL and timing decisions are perceived as very important strategies to fulfill the firms' missions and long-term objectives for their global operations. Before expanding internationally, the firms need to decide on suitable market EL and ET business' strategies. Both entry decisions were found to be contingent upon qualities of the firms' resource based capabilities. The findings revealed resource-based factors such as a firm's financial capacities, human capabilities, competencies and specialist expertise, knowledge and experience complemented by some external factors such as the financial supports from banks, are crucial in order to enter international markets.

The decision to enter a foreign market and concurrently choose the right time is a complex decision making process for construction firms. Malaysian construction firms contended that they have chosen to be the late entrants to enter the non-ASEAN countries based on the level of a firm's resource capabilities. This present study builds on and extends the literature on EL and ET decisions in a more integrated way and brings forward the significant mutually inclusive factors influencing both the EL and ET dimensions toward developing an integrated ELET decision model.

If firms require new resources in order to enter foreign markets, entry modes via acquisition and joint venture may enable them to be the early movers. Therefore, a further research to add entry mode (EM) decision in the ELET model would seem to be needed in order to develop a complete and comprehensive decision model, namely ELETEM model for international markets entry. It is recommended that a complex analysis such as structural equation modeling that can model constructs (factors) influencing two or more dependent variables simultaneously should be adopted to integrate these ELETEM decisions.

References

Abdul Aziz, A.R, Wong, S, & Awil, A. 2008, 'Formulating a holistic model to describe international diversification of construction companies', In Conference on Building Abroad: Procurement of Construction and Reconstruction Projects in the International Context, IF Research Group and University of Montreal, 23-25 Oct. Montreal Canada, (9), 27-36.

Abdul-Aziz, A, & Wong, S.S. 2010, 'Exploring the internationalization of Malaysian contractors: the international entrepreneurship dimension', Construction Management and Economics, 28 (1), 51-61.

Abdul-Aziz, A.R. & Awil, A.U. 2010, 'Locational considerations and international Malaysian housing developers', Journal of Financial Management of Property and Construction, 15 (1), 7-20.

Abdul-Aziz, A.R. & Sing-sing, W. 2008, 'Locational Factors and Foreign Market-entry Considerations for Malaysian Contractors', The Australasian Journal of Construction Economics and Building, 9 (1), 27-36.

Agarwal, S. & Ramaswami, S.N. 1992, 'Choice of Foreign Market Entry Mode: Impact of Ownership, Location and Internalization Factors', Journal of International Business Studies, First Quar (8), 1-27.

Ahmed, Z.U, Mohamad, O, Tan, B. & Johnson, J.P. 2002, 'International risk perceptions and mode of entry: a case study of Malaysian multinational firms', Journal of Business Research, 55 (10), 805-813.

Bayus, B.L. & Agarwal, R. 2007, 'The Role of Pre-Entry Experience, Entry Timing, and Product Technology Strategies in Explaining Firm Survival', Management Science, 53 (12), 1887-1902.

Boeh, K.K. & Beamish, P.W. 2012, 'Travel time and the liability of distance in foreign direct investment: Location choice and entry mode', Journal of International Business Studies, 43(5), 525-535.

Buckley, P.J, Forsans, N. & Munjal, S. 2012, 'Host-home country linkages and host-home country specific advantages as determinants of foreign acquisitions by Indian firms', International Business Review, 21(5), 878-890.

Che Ibrahim, C.K.I, Ayub, A.R, Ahzahar, N. & Hassan, S.H. 2009, 'Exploring International Construction Projects : Views from Malaysian Construction Firms', Esteem Academic Journal, 5 (2), 1-23.

Che Ibrahim, C.K.I, Mohd Baki, A, Mat Isa, C.M, Endut, I.R. & Ghani, M.K. 2009, 'Mode of Entry into International Construction Contracts: Views from Malaysian Construction Firms', In Second Construction Industry Research Achievement International Conference CIRIAC2009, 1-11.

Che Ibrahim, C.K.I, Othman, M, Baki, A.M. & Ghani, M.K. 2009, 'Location Consideration in International Construction Market by Malaysian Firms: Specific Locale Perspective', In Conference on Scientific and Social Science Research CSSR08-09, 1-10.

Chen, C. & Messner, J.I. 2009, 'Entry Mode Taxonomy for International Construction Markets', Journal of Management in Engineering, 25 January, 3-11.

Chen, C. & Orr, R.J. (2009), 'Chinese Contractors in Africa : Home Government Support, Coordination Mechanisms and Market Entry Strategies', Journal of Construction Engineering and Management, 135 no 11, 1 November 2009, 1201-1210.

Chen, C, Pi-Chu, C, Orr, R.J. & Goldstein, A. 2007, 'An Empirical Analysis of Chinese Construction Firms' Entry into Africa', In The CROCM International Symposium on Advancement of Construction Management and Real Estate 451-463.

Chen, M.Y. & Chang, J.Y. 2011, 'The choice of foreign market entry mode: An analysis of the dynamic probit model', Economic Modelling, 28 (1-2), 439-450.

CIDB Malaysia 2013, Construction Industry Development Board Malaysia Record, Retrieved from www.cidb.gov.my

Dacko, S.G. 2002, 'Understanding market entry timing decisions: the practitioner-academic gap', Marketing Intelligence & Planning, 20 (2), 70-81.

Delios, A, Gaur, A.S. & Makino, S. 2007, 'The Timing of International Expansion: Information, Rivalry and Imitation Among Japanese Firms, 1980–2002', Journal of Management Studies, 117592 January.

Ekeledo, I. & Sivakumar, K. 1998, 'Foreign Market Entry Mode Choice of Service Firms: A Contingency Perspective', Journal of the Academy of Marketing Science, 26 (4), 274-292.

El-higzi, F.A. 1999, 'Foreign Market Selection Factors in the Australian Construction Service Sector', Australian Journal of Construction Economics and Buildings, 2 (1), 107-120.

Ekeledo Ellis, P.D. 2007, 'Paths to foreign markets: Does distance to market affect firm internationalisation?' International Business Review, 16 (5), 573-593.

Engineering News Record 2013, The top 225 international contractors, McGraw- Hill, New York, Vol. 104, New York doi:10.1093/jnci/djs068

Gaba, V, Pan, Y. & Gerardo R. Ungson 2002, 'Timing of Entry in International Market: An Empirical Study of US Fortune 500 Firms in China', Journal of International Business Studies, 33 (1), 39-55.

Gallego, M.A, Hidalgo, E.R, Acedo, F.J, Casillas, J.C. & Moreno, A.M. 2009, 'The relationship between timing of entry into a foreign market, entry mode decision and market selection', Time & Society, 18 (2-3), 306-331.

Gao, G.Y. & Pan, Y. 2010, 'The pace of MNEs' sequential entries: Cumulative entry experience and the dynamic process', Journal of International Business Studies, 41 (9), 1572-1580.

Gaston-Breton, C. & Martín, O.M. 2011, 'International market selection and segmentation: a two-stage model', International Marketing Review, 28 (3), 267-290.

Green, A, Sedef, K. & Bjorn, R. 2004, Factors indicating first-mover advantages and second-mover advantages, Kristianstad University College.

Guler, I. & Guillén, M.F. 2009, 'Institutions and the internationalization of US venture capital firms', Journal of International Business Studies, 41 (2), 185-205.

Gunhan, S. & Arditi, D. 2005, 'Factors Affecting International Construction', Journal of Construction Engineering and Management, 131 (3), 273-282.

Huang, Y. & Sternquist, B. 2007, 'Retailers' foreign market entry decisions: An institutional perspective', International Business Review, 16 (5), 613-629.

Jain, N.K. 2010, Resource, Strategies, Location Determinants, And Host Country Location Choice by Emerging Market Firms, Florida International University, Retrieved from http://digitalcommons.fiu.edu/etd/147

Kaur, S. & Sandhu, M.S. 2014, 'Internationalisation of born global firms: Evidence from Malaysia', Journal of the Asia Pacific Economy, 19 (1), 101-136.

Kerin Roger A, Varadarajan, R. & Robert, P. 1992, 'First-mover Advantage: A Synthesis, Conceptual Framework and Research Propositions', Journal of Marketing, 56, 33-52.

Koch, A.J. 2001a, 'Factors influencing market and entry mode selection: developing the MEMS model', Marketing Intelligence & Planning, 19 (5), 351-361.

Koch, A.J. 2001b, 'Selecting overseas markets and entry modes: two decision processes or one?' Marketing Intelligence & Planning, 19 (1), 65-75.

Lee, S, Jeon, R, Kim, J. & Kim, J. 2011, 'Strategies for Developing Countries to Expand Their Shares in the Global Construction Market: Phase-Based SWOT and AAA Analyses of Korea', Journal of Construction Engineering and Management, 137 (6), 460-470.

Lieberman, M.B. & Montgomery, D.B. 1998, 'First Mover Disadvantages: Retrospective and link with the Resource-Based View', Strategic Management Journal, 19, 1111-1125.

Lieberman, M.B. & Montgomery, D.B. 2013, 'Conundra and Progress: Research on Entry Order and Performance', Long Range Planning, 46 (4-5), 312-324.

Lilien, G.L. & Yoon, E. 1990, 'The Timing of Competitive Market Entry: An Exploratory Study of New Industrial Products', Management Science, 36 (5), 568-585.

Liu, J.Y, Low, S.P. & Niu, M. 2011, 'Challenges and opportunities for cross-border acquisitions by Chinese construction enterprises', Journal of Technology Management in China, 6 (2), 140-152.

Lu, J, Liu, X, Wright, M. & Filatotchev, I. 2014, 'International experience and FDI location choices of Chinese firms: The moderating effects of home country government support and host country institutions', Journal of International Business Studies, 1 (1), 1-22.

Luo, Y. & Peng, M.W. 1998, 'First Mover Advantages in Investing in Transitional Economies', Thunderbird International Business Review, 40 (2), 141-162.

Malhotra, S, Sivakumar, K. & Zhu, P. 2009, 'Distance factors and target market selection: the moderating effect of market potential', International Marketing Review, 26 (6), 651-673.

Mat Isa, C.M, Idris, A, Mohd Saman, H. & Mohd Nasir, S.R. 2013a, 'Dynamic Factors for Entry Location Decision by Malaysian Construction Firms in International Market Expansion', In International Post Graduate Seminar 2013.

Mat Isa, C.M, Mohd Saman, H, Mohd Nasir, S, Nasir, & Jaapar, A. 2013b, 'Strength, Weakness, Opportunity and Threat (SWOT) Attributes of Malaysian Construction Firms in International Market', In International Conference in Infrastructure Engineering and Construction.

Mat Isa, C.M, Mohd Saman, H, Mohd Nasir, S.R. & Jaapar, A. 2013c, 'Factors for Entry Timing Choices Using Principal Axis Factorial Analysis and Logistic Regression Model', International Journal of Mechanical, Industrial Science and Technology Engineering, 7 (12), 1387-1397, Retrieved from www.waset.org.

Mat Isa C.M, Mustaffa, N.K, Saman, H.M, Rashidah, S, Nasir, 2012, 'Determining Malaysian Construction Firms' Internal and External Factors to Venture into International Market', ICIMTR 2012

Pallant, J. 2011, SPSS SURVIVAL MANUAL A step by step guide to data analysis using SPSS, Fourth Edition, Allen & Unwin, Crows Nest NSW.

Peng, M.W. 2001, 'The resource-based view and international business', Journal of Management, 27 (6), 803-829.

Polat, G. & Donmez, U. 2010, 'Marketing Management Functions of Construction Companies: Evidence from Turkish Contractors', Marketing Management, 16 (2), 267-277.

Sakarya, S, Eckman, M. & Hyllegard, K.H. 2007, 'Market selection for international expansion: Assessing opportunities in emerging markets', International Marketing Review, 24(2), 208-238.

Saunders, M, Lewis, P. & Thornhill, A. 2009, Research Methods for Business Students, Fifth Edition, Prentice Hall.

Schwens, C. & Kabst, R. 2009, 'How early opposed to late internationalizers learn: Experience of others and paradigms of interpretation', International Business Review, 18 (5), 509-522.

Sekaran, U. & Bougie, R. 2013, Research Methods for Business : A Skill Building Approach, Fifth Edition, John Wiley & Sons Inc.

Söderblom, A. 2011, Private Equity Fund Investing, Stockholm, Published Thesis.

Somlev, I.P. & Hoshino, Y. 2005, 'Influence of location factors on establishment and ownership of foreign investments: The case of the Japanese manufacturing firms in Europe', International Business Review, 14 (5), 577-598.

Stevens, C.E. & Dykes, B.J. 2013, 'The Home Country Cultural Determinants of Firms' Foreign Market Entry Timing Strategies', Long Range Planning, 46 (4-5), 387-410 doi:10.1016/j.lrp.2013.06.003.

Tawanda, M. 2006, Modes of market entry and strategies for South African Companies doing business in Tanzania, University of Pretoria.

Tsou, S.H, Yu, J.C. & Lin, Y. 2009, 'Entry Timing and Performance under Uncertainty: Taiwanese Firms Investing in China', Asia Pacific Management Review, 14 (3), 263-277.

Villaverde, P.M.G. & Ortega, M.J.R. 2007, 'Determinants of Entry Timing: Firm Capabilities and Environmental Conditions', Management Research, 5 (2), 101-112.

Williams, B. & Brown, T. 2010, Exploratory factor analysis : A five-step guide for novices, 8(3), 1-12.

Luo, Y. 1998, 'Timing of Investment and International Expansion Performance in China', Journal of International Business Studies, 29 (2), 391-408.

Yean, F, Ling, Y, Ibbs, C.W. & Chew, E.W. 2008, 'Strategies Adopted by International Architectural, Engineering, and Construction Firms in Southeast Asia', Issues in Engineering, (July), 248-256.

Zhu, H, Eden, L, Miller, S.R, Thomas, D.E. & Fields, P. 2012, 'Host-country location decisions of early movers and latecomers: The role of local density and experiential learning', International Business Review, 21 (2), 145-155.

Prioritising Project Scope Definition Elements in Public Building Projects

14

Mohammed K. Fageha, Ajibade A. Aibinu (University of Melbourne, Australia)

Abstract

A complete definition of the scope of a project upfront during early stages ensures smooth and successful implementation during the project execution. This research identifies and prioritises project scope definition elements for public buildings in Saudi Arabia. Elements that could significantly contribute to complete project scope definition package at pre-project planning stage are identified and their interrelationship determined and prioritised. Using the Project Definition Rating Index (PDRI) as a basis, the study uses analytical network process (ANP) technique based on data obtained from project managers who have been involved in public sector projects in Saudi Arabia. Data collection and analysis was conducted in three steps. The first step involved identification of scope definition elements while the second involved an investigation into interrelationships among the elements. In the third step, ANP was used to determine the weight of the elements' importance in terms of contribution to project scope definition completeness. Finally, Pareto analysis was used to prioritise and assess the distribution pattern of the elements. The outcome from this research is the prioritisation of project scope definition elements for public building projects in Saudi Arabia. The prioritised list developed indicates the importance of project scope definition elements. It should help project management teams identify elements to consider when evaluating project scope definition for completeness at the pre-project planning stage.

Keywords: Project scope definition, pre-project planning, prioritising, public building projects, Saudi Arabia, Analytical Network Process (ANP)

Introduction

Project scope definition is the process whereby the work that is needed to produce a building is identified and described in sufficient detail to facilitate project execution. It gives the project team an understanding of what needs to be done while at the same time helping the team in setting up management control systems that can be applied during project execution; and could impact on project outcomes. In the construction industry, having a better project outcome is significant because construction is one of the most important sectors in many economies and a significant contributor to the gross domestic product (GDP) of most countries. This is especially valid in developing countries such as Saudi Arabia, where the construction industry is one of its largest. While the construction industry in Saudi Arabia contributes approximately 4.8% to the GDP and 9.4% of non-oil sectors, almost 30% of the non-oil sectors' activities occur in the public sector (Central Department of Information and Statistics 2013). Due to the importance of the construction sector, it is necessary to ensure that construction projects are completed successfully. However, the Saudi construction sector has been experiencing problems in productivity, innovation, schedule slippage, rework, mistakes and disputes, which have all increased construction costs (Abdul-Hadi, Al-Sudairi & Alquahtani 2005). Project abandonments are very common and are often symptoms of failed processes in the early stages of a project. Purportedly, in 2011 there were around 2262 abandoned public building projects in the Makkah region, which is only one of the thirteen regions. Inadequate pre-project planning and poor definition of project elements have been identified as the major reasons for the problem (Al-Humaidan 2011). In fact, up to 70% of poor time performance in Saudi Arabian projects caused by changes in the project scope (Assaf & Al-Hejji 2006; AlKharashi & Skitmore 2009). Alsehaimi, Koskela and Tzortzopoulos (2013) investigated a number of studies on delay in construction projects in developing countries including Saudi Arabia. Their study cited poor project management as one of the main causes of delay. They reported that poor planning and control is, specifically, the factor that had been identified in most studies. The study concluded that action research is needed to generate practical managerial approaches to address delay issues and enhance project management practice in Saudi Arabia. This study contributes in the area of project scoping and project definition in the pre-project planning stage, which could remedy the problem if properly approached.

Most public sector projects in Saudi Arabia frequently rely on the traditional procurement method (also known as the design-bid-build method), where the client develops the business case, provides a brief, budget and tender document. Thereafter, the client appoints a design consultant in the design stage and a contractor through competitive bidding, in which the lowest bidder is awarded the contract (Hatush & Skitmore 1998). One of the critical problems faced by government authorities is the frequent and lengthy delays in their projects (Al-Khalil & Al-Ghafly 1999). Lengthy delays are often caused by a number of issues such as unqualified contractors, changes in the scope of work, rework and inappropriate parties involvement due to the procurement method (Assaf & Al-Hejji 2006). Traditional procurement has been criticised due to the sequential approach to delivering a project (Love 2002). Love referred to the time gap between design and construction as 'procurement gap'. Changes and reworks may occur due to the time gap. In fact, Arain, Pheng & Assaf (2006) identify that the inconsistencies between design and construction, which occur due to the procurement gap, have a significant impact on construction project performance in Saudi Arabia. In order to reduce these changes and reworks, the project scope should be well-defined at the pre-planning stage of the project.

Inadequate pre-project planning and poor scope definition continue to emerge as major causes of expensive changes, delays, rework, cost overruns and schedule overruns, and they often lead to project failure (Mirza, Pourzolfgaghar & Shahnazari 2013; Lordsleem Jr & Melhado 2014). Changes during project execution often reflect the uncertainties that occur during the early stages of the project (Assaf & Al-Hejji 2006). Changes are requested during the construction stage as a result of the differences in the perspectives that each stakeholder has on the project. The fundamental reason for such change in orders is either poor project definition, or poor idea of how the work has to be handled. Thus, defining the scope of a project at the early stage using input from all stakeholders is vital. The purpose of project scope definition is to generate adequate information that is needed to identify and describe the work to be performed, in order to avoid major changes that may negatively affect project outcome (Gibson et al. 2006). This information is needed before making the decision whether or not to proceed with the project execution (Kähkönen 1999). They also form the basis for project design and therefore project execution.

Lack of a clear project scope definition, as well as improper control of these, have been recognised by recent researchers as major barriers to project success (Mirza, Pourzolfgaghar & Shahnazari 2013). Therefore, having a well-defined project during the pre-project planning stage is crucial for success during project execution and for achieving a satisfactory project outcome. One of the first steps in the pre-project planning process is to understand what needs to be defined in order to ensure that project scope is clear upfront, thereby facilitating project success. Accordingly, the objectives of the study reported here are:

- To identify the elements that should be considered in defining public building projects in Saudi Arabia;
- To investigate the interrelationships among elements; and
- To determine the level of significance of each element in terms of their contribution to the overall project definition completeness.

Literature Review

Significance of Project Scope Definition

The purpose of project definition is to provide adequate information that is needed to identify the work to be performed without major changes, and it may affect performance of a project (Chritamara, Ogunlana & Bach 2001; Gibson et al. 2006). Success during detailed design and execution stage of a project is highly dependent on the level of effort expended during the scope definition stage. When a project proceeds to the implementation stages with inadequate definition of one or more project elements, it may be subjected to differing expectations and interpretation by different stakeholders (Atkinson et al. 2006). Thus, a poor definition of project scope can lead to dissatisfaction by project stakeholders, simply because their expectations have not been fulfilled (Cano & Lidón 2011). It can also lead to design errors, owner changes and rework, which are often sources of schedule slippage and cost overruns (Love, Irani & Edwards 2004; Hwang et al. 2009).

Project Scope Definition and Pre-project Planning

Construction projects procedures are conducted in the same manner, though each has its own characteristics. Different researchers classified the pre-project planning early phase into stages (Table 1). Despite the differences in classification, there is general agreement in the literature that project scope definition documentation should be developed prior to making the final decision on whether to proceed with the project or not. In order to make this decision, an evaluation of the completeness of the project scope definition document should be conducted. Therefore, knowing the significance of each element of the project scope definition document is important for the evaluation task, which is the purpose of this research.

Table 1: Stages within the pre-project planning phase

Source	Stages within the pre-project planning phase				
Gibson et al. (1995)	Organise for pre-project planning	Select project alternative(s)		Develop a project definition package	Decide whether to proceed with project
Kähkönen (1999)	Plan project definition	Generate and analyse alternatives	Select alternative(s)	Prepare documents	Decide on project
Woodhead (2000)	Initial idea	Capital proposal	Outline case	Full case	Decision approval
Haponava and Al-Jibouri (2009)	Initiative		Feasibility		Project definition

The Construction Industry Institute (CII) defined pre-project planning as "the process of developing sufficient strategic information with which owners can address risk and decide to commit resources to maximize the chance for a successful project" (CII 1994). Additionally, it is the process that combines all tasks between project initiation phase to the beginning of detailed design phase (Gibson et al. 2006). It begins with a project concept that reflects a business need and ends with a decision whether to proceed and start the execution of the project by developing the detailed design (Gibson, Kaczmarowski & Lore 1995). Many experts and industry practitioners believe that pre-project planning efforts in the project life cycle have significantly greater impact on the whole project life cycle, thus improving project final outcomes. Specifically, Cho and Gibson (2001) studied project performances of 53 capital facility projects. They investigated the total cost, scheduling and operational characteristics, and then compared them to the effort spent on the pre-project planning. The study concluded that up to 20% of cost savings and 39% of schedule savings could occur when a high level of pre-project planning effort is implemented. Therefore, better pre-project planning stage has a greater influence on the project life cycle when its expenditure is low compared to the other stages. Prior to deciding on a project's implementation, scope definition is the last activity undertaken in the pre-project planning stage. Because of its importance, many studies have been conducted to find better approaches that can be adopted for defining and scoping a project.

Approaches for Measuring the Project Scope Definition Completeness

In order to improve project scope definition completeness, the CII developed the Project Definition Rating Index (PDRI) (Wang 2002). PDRI is a scoring tool for measuring the adequacy of project scope definition and assessing pre-project planning effort. It is one of the most comprehensive and established tools for measuring project scope definition for completeness. When used, it allows the project team to take actions to improve the scope definition of those elements that have high scores, which indicate the areas of risk to the project (Gibson et al. 2006).

The PDRI started with a version that was designed specifically for industrial projects. After the success of PDRI for industrial projects, a similar tool was developed for buildings. The validation of PDRI on more than 190 projects with total estimated cost of more than $6.5 billion dollars (Cho & Gibson 2001; Wang & Gibson 2010), shows that there is a positive relationship between PDRI score for a project with cost and time performances. PDRI for building consists of three sections, eleven categories and sixty-four elements. During the planning phase, the project team would pay special attention to these elements to reduce the uncertainties in the subsequent phases. In order to calculate the overall score of scope definition elements for the project, the project management team would evaluate the completeness level of each element's definition on the list. The maximum score is 1000 points, and the lower score represents a more complete and well-defined project (Cho & Gibson 2001).

Construction organisations in Saudi Arabia lack a systematic approach for defining the project scope elements in the pre-project planning stage. Using a well-established list of project scope definition elements such as PDRI is a good approach. However, the PDRI was developed in the US and mainly for the same context. Even though the development process for construction projects may share the same procedure globally, it still requires a different set of information that respects the nature and the environment of the project. This study is focused on public building projects in Saudi Arabia. The differences between the two contexts could be from the tendering approach, project organisational structure or the allocation of responsibilities among a project's parties. Therefore, it is prudent to identify project scope definition elements and weights that are compatible with public building projects in the Saudi Arabian context. The current research uses the PDRI for building elements as a point of departure to develop a model for the Saudi Arabian public construction projects.

Analytical Approach

This research aims to develop an analytical decision-support model in the form of prioritised project scope definition elements, to assist public construction organisations in Saudi Arabia when assessing project scope defined for completeness. The study uses Analytical Network Process (ANP) approach. ANP is a developed version of what is known as AHP (Analytical Hierarchical Process). Saaty developed AHP as a decision-making technique used in the military for allocating resources and planning needs in the 1970s (Cheng & Li 2001). He stated that AHP is a general theory of measurement (Saaty 1994). It is a technique that helps break down a complex, unstructured situation into its component parts in hierarchical structures. After launching the ANP in 1996, and due to its flexibility to solve complex forms of decision-making problems, AHP became a special case of ANP because it contains neither feedback nor loops within the same cluster representing inner dependence. Saaty (1996) defined ANP as a general theory of relative measurement used to derive composite priority ratio scales from individual ratio scales that represent relative measurements of the influence of elements that interact with respect to control criteria. By replacing the hierarchies with networks, ANP has been widely used as a multi criteria decision-making tool, instead of AHP. AHP is a technique that constructs a problem into several levels in a form of hierarchy and each element is independent (Saaty 1994). However, ANP can be used as an effective tool in cases where there are interactions between elements of a system network structure (Saaty 1996).

Despite the fact that prioritising and weighting can be generated by other methods, ANP is found to be more appropriate for the current research because the relationships between elements appear better in a network structure. ANP is a multi-criteria decision-making tool that allows representation of any decision-making problem in a network of criteria, where interdependent relationships exist within and between all criteria. Experts' experiences are used to estimate relative magnitudes of tangible and intangible factors through paired comparisons, in order to make rational and consistent decision (Saaty 1996). ANP provides weights and priorities to these elements, taking into account the interdependent relationships among elements. However, instead of using ANP as a selection tool between alternatives, this study uses ANP as a decision tool to set priorities for project scope definition elements, based on feedback from experts.

Data Collection

Approach

The data collection was conducted through semi-structured interviews. Yin (2009) stated that semi-structured interviews can maximize the flexibility of the interview and provide the capability to shape the interview to suit individuals. The questionnaires were designed to extract information from the participants. Prior to data collection, ethics approval was sought from the Faculty of Architecture, Building and Planning Human Ethics Advisory Group at The University of Melbourne, Australia. All participants were given a plain language statement (PLS) at the start of the interview. The PLS included an invitation, research summary, what participants would be asked to do and for how long. It also addressed confidentiality issues. The participants were also asked to sign a consent form if they chose to participate.

Sampling and Profile of Respondents

A total number of 16 respondents participated in the interview. They were selected non-randomly through the use of purposeful sampling. Participants were identified through the first author's personal contacts and by snowball sampling method; they had expertise in managing public building projects in Saudi Arabia. The majority of the respondents were either project managers or construction general managers and had over 20 years of experience in the construction industry as well as over 20 years of experience in public building projects. Also most respondents had a mixed background, which was considered to be good for this research, because it required both technical experience as well as management experience.

The interrelationships among the elements examined are complex. Thus the study uses the ANP technique. ANP is not a traditional quantitative method; instead, it is a technique in which statistical sampling is not the issue in all circumstances. In fact, seeking a large number of participants is not a necessity in ANP (Lam & Zhao 1998). ANP is a technique in which an analytical manner of sampling is targeted, rather than a statistical one (Herath 2004; Sambasivan & Fei 2008). The sample size (16) is sufficient for understanding the interrelationships among the elements. The participants are experts with sufficient and relevant background knowledge and experience; and their responses can be confidently relied upon; essential criteria when using the ANP technique.

Data Analysis and Results

Step One: Selection of Early Stage Scope Definition Elements

To achieve the first objective of the study, a semi-structured interview was conducted. The PDRI for building was used as a point of departure for the data collection. Participants were presented with the list of project definition elements recommended in the PDRI for building, which includes 11 categories and 64 elements. Participants were allowed to add and delete categories and elements from the list in order to identify those that are applicable to the Saudi Arabian construction context.

The outcome gained from this step is a list of all categories and their elements that should be considered when defining a public building project in Saudi Arabia, at the pre-project planning stage. The list includes 9 categories and 42 elements, presented in Table 3. In the data analysis, an element is included if it has been selected by two or more respondents. An element is excluded either because it is relevant to a later stage, not pre-project planning stage, or it is not applicable in the Saudi Arabian construction context.

Out of eight elements under the *'Business Strategy'* category, the majority of respondents excluded one element *'Economic Analysis'*. This is because public projects in Saudi Arabia are not-for-profit social infrastructure projects owned and maintained by the government and these projects serve the community needs. Public construction organisations finance their projects from the fund provided in the annual national budget specified for each project. In the process of defining the economic analysis element, specific information should be determined such as the viability of the project, evaluation of other alternatives, length of ownership and the economic impact of early or late delivery. This kind of information is not necessary in public sector projects therefore the element was excluded from the list.

In the *'Project Requirement'* category, *'Value-Analysis'* element was excluded, because it is relevant to the design stage of a project life cycle and not at the pre-project planning stage. The value-analysis should be in place to consider the cost effectiveness of design and material alternatives. However, at the pre-project planning stage in Saudi Arabian construction projects, the project design is not available. Additionally the architectural and engineering consultant usually provides this kind of information at the design stage of the project. Thus, element *'Value-Analysis'* was excluded from list for this stage, but it still needs to be defined later by the architectural and engineering consultant at the design stage.

The *'Building Programming'* category consists of thirteen elements, but only four elements *'Program Statement'*, *'Building Summary Space List'*, *'Growth and Phased Development'* and *'Transportation Requirements'* were included because of their relevance to the pre-project planning stage. The elements are significant at the pre-project planning stage because they form the baseline for the architectural and engineering consultant design role in subsequent stages. The remaining elements in this category were excluded because they consist of technical information too early to be defined at this stage of the project. These are more relevant to the design stage, specifically in the concept development phase.

All elements under *'Building/ Project Design Parameters'* and *'Equipment'* categories were excluded from the list, as they are considered technical information and relevant to later stages. Specifically *'Building/ Project Design Parameters'* category is required at detailed design stage and often provided by the architectural and engineering consultants. Usually in Saudi Arabian construction projects, the *'Equipment'* category is developed by the contractor and approved by the owner and the project management team.

The rest of the categories *'Owner Philosophies'*, *'Site Information'*, *'Procurement Strategy'*, *'Deliverables'*, *'Project Control'* and *'Project Execution Plan'* were included with all their elements because respondents recognised their importance at the pre-project planning stage. All the elements included in the list need to be defined properly at the pre-planning stage in order to increase the possibility of having a complete project scope definition package.

Step Two: Interrelationships and Interactions among Elements

Following the identification of elements that should be included in project scope definition at the pre-project planning stage, the next stage of the data analysis addresses the second objective of the study. This is necessary because the ANP approach requires these connections in order to formulate the structure of the network among elements. To investigate the interrelationships and interactions, a contextual relationship of 'leads to' type was chosen. This means that one element leads to another elements (Singh et al. 2003). Based on this, a contextual relationship is developed. This step is called structural self-interaction matrix (SSIM), which is usually used in interpretive structural modelling (ISM) technique. The SSIM step can be used to indicate pair-wise relationships among variable of the system under consideration (Ahuja et al. 2009). The direction of the relation between any two elements (*i* and *j*) is assessed. Four symbols are used to indicate the type of the relation that exists between two elements as follows:

V: element *i* will help achieve element *j* but not in both directions;

A: element *j* will help achieve element *i* but not in both directions;

X: elements *i* and *j* will help achieve each other; and

O: elements *i* and *j* are unrelated.

The interrelationships and interactions among elements matrix was completed in a focus group discussion with three expert project managers. In order to complete this step, 6 sessions with 3 hours in each were required to determine 882 contextual relationships. Based on the responses, the SSIM has been developed and presented in Table 2.

Table 2: Interrelationships and interactions among elements matrix, SSIM

	L5	L4	L3	L2	L1	K5	K4	K3	K2	K1	J2	J1	H2	H1	E9	E5	E2	E1	D8	D7	D6	D5	D4	D3	D2	D1	C6	C5	C4	C3	C2	B4	B3	B2	B1	A8	A7	A6	A5	A3	A2	A1
A1	O	O	O	O	O	O	O	O	O	O	O	O	O	O	O	O	V	X	O	O	O	O	O	O	O	O	O	O	O	O	X	X	X	X	A	X	X	V	V	A	X	
A2	O	O	O	O	O	O	O	O	O	O	O	O	O	O	O	O	O	O	O	O	O	O	O	O	O	O	O	O	O	O	O	O	O	O	O	O	X	V	O	X		
A3	O	O	O	A	O	O	O	O	X	O	O	O	O	O	O	O	O	O	V	V	V	V	V	V	V	V	V	V	V	V	X	X	X	X	X	V	V	X	X			
A5	O	O	O	A	O	O	O	O	O	O	O	O	V	V	O	X	V	X	O	O	V	A	O	O	O	O	V	O	O	O	O	O	O	O	O	V	X	V				
A6	O	X	O	O	O	O	O	O	O	O	O	O	O	O	O	X	O	A	O	O	X	O	O	O	O	V	O	O	V	X	X	X	O	O	V	V	X					
A7	O	O	O	O	O	V	V	O	O	O	O	O	O	O	O	O	O	V	O	A	A	O	O	A	O	O	X	O	V	X	X	X	O	V	O	X						
A8	O	V	V	O	V	V	V	V	V	V	O	O	O	O	O	V	O	V	V	V	O	V	O	O	O	V	V	V	V	O	V	X	X	X	X							
B1	O	O	O	O	V	O	O	O	O	O	O	O	O	V	O	O	O	A	O	O	A	V	A	O	O	O	V	O	O	O	X	X	X	X								
B2	O	O	O	O	O	O	O	O	O	O	O	O	O	O	O	O	O	A	X	O	X	A	O	O	O	O	X	O	O	X	X	X	X									
B3	O	O	O	O	O	O	O	O	O	O	O	O	O	O	O	V	O	A	V	O	O	O	O	O	O	O	V	O	V	A	V	X										
B4	O	O	O	A	A	O	O	O	O	O	O	V	O	O	O	X	X	V	O	O	O	X	A	A	A	X	V	V	V	A	X											
C2	O	O	O	O	O	O	O	O	O	O	O	A	O	O	A	O	O	O	A	O	A	A	A	A	O	O	O	O	O	X												
C3	O	V	O	O	O	O	O	O	O	X	O	O	O	O	O	O	V	O	X	O	X	V	O	O	O	O	V	V	O													
C4	O	V	V	V	V	O	O	O	O	V	V	O	V	V	V	V	O	O	A	X	X	A	X	A	A	A	V	V														
C5	O	O	O	A	O	O	A	A	O	O	X	O	O	V	O	O	O	O	O	O	O	O	O	O	O	O	O															
C6	O	V	O	X	X	O	A	A	A	X	A	O	A	A	A	A	O	O	A	A	A	A	A	A	A	A																
D1	O	O	O	O	O	O	O	O	O	O	O	O	O	O	A	A	O	O	A	A	A	A	A	A	X																	
D2	O	O	O	O	O	O	O	O	O	O	V	O	O	O	O	O	O	O	O	O	V	A	A	O																		
D3	O	O	O	O	O	V	V	O	O	O	O	O	O	O	V	V	O	O	O	O	O	O	O																			
D4	O	O	O	O	O	O	O	O	O	V	V	V	O	O	O	V	O	O	O	O	O	X																				
D5	O	O	O	O	O	O	V	O	O	O	O	O	O	O	O	V	O	O	X	O	O																					
D6	O	O	O	O	O	O	O	O	O	O	O	O	O	O	O	V	O	O	V	O																						
D7	O	O	O	O	O	A	V	O	O	O	O	O	O	O	O	V	O	O	O																							
D8	O	O	O	O	O	O	O	O	O	O	O	O	O	O	O	V	O	O																								
E1	O	O	O	O	O	O	O	O	O	O	O	O	O	O	O	V	V																									
E2	O	O	O	A	O	O	O	O	O	O	O	O	O	O	O	O																										
E5	O	O	O	A	O	O	O	O	O	O	O	O	O	O	O																											
E9	O	O	O	O	O	O	O	O	O	O	O	O	O	O																												
H1	O	O	O	O	O	O	V	V	X	O	O	O	X																													
H2	O	O	O	A	O	O	V	O	O	A	O	O																														
J1	V	O	O	A	O	O	O	O	O	A	X																															
J2	V	A	O	X	X	O	O	X	X	X																																
K1	V	V	O	A	V	O	O	O	O																																	
K2	O	O	O	O	V	O	O	O																																		
K3	O	O	O	V	O	O	O																																			
K4	O	O	A	O	A	O																																				
K5	O	O	O	O	O																																					
L1	O	X	A	A																																						
L2	V	V	O																																							
L3	O	V																																								
L4	O																																									
L5																																										

Where V: element *i* will help achieve element *j* but not in both directions; *A:* element *j* will help achieve element *i* but not in both directions; *X:* elements *i* and *j* will help achieve each other; and *O:* elements *i* and *j* are unrelated.

The outcome from this step is the formulation of the network structure among the elements, which is used as an input for the ANP model in the next step. Figure 1 presents the network for the ANP step.

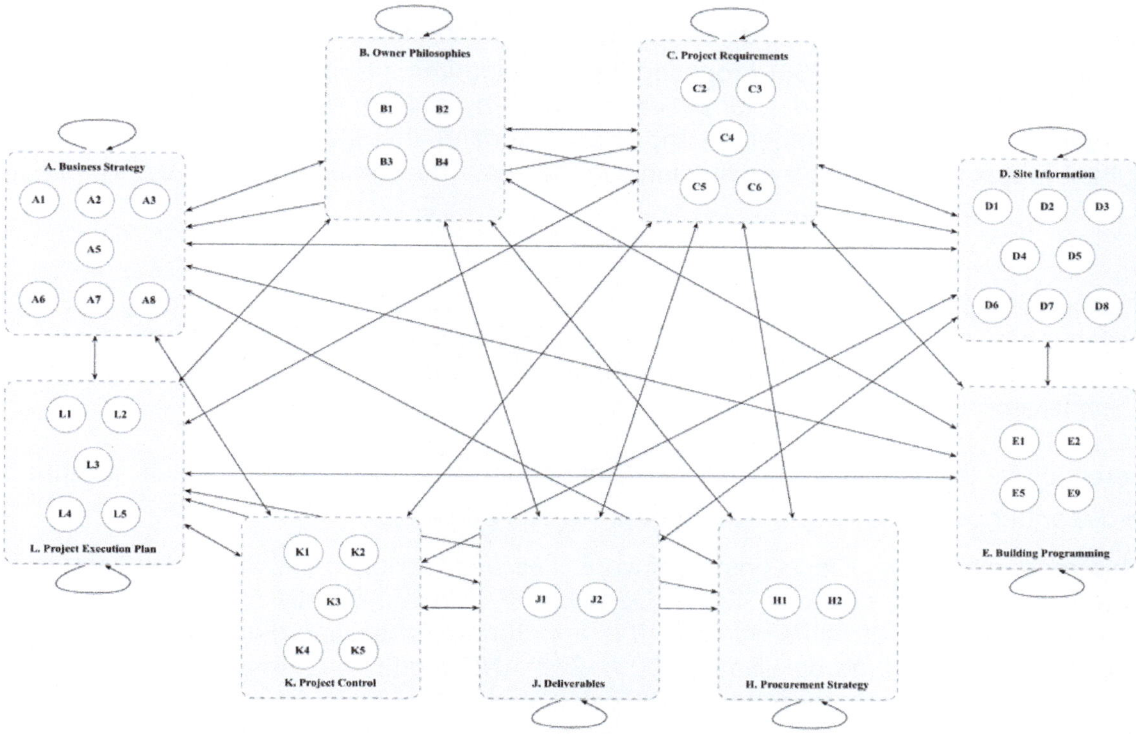

Figure 1: Interrelationships and interactions among elements, ANP model

Step Three: Contribution of Elements to Overall Project Scope Definition Completeness

The ANP was used to address the third objective of the study. Pair-wise comparisons were adopted in semi-structured interviews with the aid of structured questionnaires using the ANP model. The questionnaire was based on using the ANP model presented in Figure 1. The questionnaire included a total number of 340 pair-wise comparisons. Five project managers completed the third questionnaire based on interview sessions. They were required to answer the question that stated, *'How much importance does an element have compared to another element with respect to a preference?'* In other words, participants were asked to compare between two elements and assign a score to each element in term of its importance to each other elements on the list, using Saaty's scale of judgements. The ANP uses a fundamental scale of absolute value to carry out the comparison judgements. The relative importance values were determined using a scale of 1 to 9, where a score of 1 indicates equal importance between the two elements and 9 represents the extreme importance of the *i*th element compared to the *j*th element. Participants responded to a series of pair-wise comparisons and assigned a score for all the elements to be evaluated in term of their contribution to project scope definition completeness. For example, 'with respect to element *"Business Justification"*, the comparison was conducted in this way: please compare the relative significance between *"Building Use"* element and *"Business Plan"* element using the scale provided'. These pair-wise comparisons were conducted several times, each time with respect to an element and so on for all the relationships identified based on SSIM in the previous step. Thus, each evaluation can be represented by an eigen-vector, and the relative importance values are determined. Pair-wise comparison in ANP is performed in the framework of a matrix, and a local priority vector can be derived as an estimate of the relative importance associated with the elements being compared by solving the following equation:

$$A \times w = \lambda_{\max} \times w$$

Where A is the matrix of pair-wise comparison, w is the eigenvector, and $\lambda_{\max}$ is the largest eigen-value of A. Saaty (1980) proposed several algorithms to approximate w. The process of averaging over normalised columns can be done by dividing each element in a column by the sum of the column elements, and then summing the elements in each row of the resultant matrix and dividing by *n* elements in the row. This can be deducted by:

$$w_i = \frac{\sum_{i=1}^{I}\left(\frac{a_{ij}}{\sum_{j=1}^{J} a_{ij}}\right)}{J}$$

Where w_i is the weighted priority for component *i;* a_{ij} is a matrix value assigned to interdependence relationship of component *i* to component *j.*

In this research, Super Decisions® software was used to compute the eigen-vectors from pair-wise comparison matrices. Super Decisions®, is commercially available software that has been developed for AHP and ANP. It is appropriate for solving decision problems with a hierarchy and network model (Saaty 2003). The ANP model was entered into the software and the pair-wise comparisons were calculated. The software also determined the consistency ratio (CR), which is the degree to which the pair-wise comparisons are consistent. According to Saaty (1994), people are often not consistent in their judgements. However, the recommended level of CR should be less than 5% for pair-wise comparisons between three elements, 8% for pair-wise comparisons between four elements, while for more than four elements it should be less than 10%. A discussion about the CR was carried out with participants to inform them that their judgments in each pair-wise comparison should be consistent. Otherwise, in order to resolve the inconsistencies, some or all of their judgments for the comparisons must be repeated. The CR of each comparison in the ANP model for this research was in the acceptable range with respect to the matrix size.

Once the pair-wise comparisons were completed, the global supermatrix for project scope definition elements was generated. The overall normalised priorities were then obtained by the Super Decisions® software calculations for the supermatrix. The outcome from this ANP step, is the list of elements that are required in defining the project scope as well as their weights of importance to the scope definition completeness for a project. These weights account for the interrelationships and interactions among elements. Categories' weights were excluded from the model and the category name was the only identification for its elements, as the elements were more essential in this study. Thus, each category weight is the sum of its elements. Table 3 presents the significant weights for project scope definition elements for public building projects in Saudi Arabia.

Distribution Pattern of Project Scope Definition Elements' Weights

In the ANP analysis, the 42 elements that affect the completeness of the project scope definition package were prioritised. In this section, Pareto principle was used to investigate the distribution pattern and classify the elements into the significant few that principally contribute to project scope definition completeness, if any. Pareto analysis is a quality control tool that ranks data, in descending order from the highest frequency of occurrences to the lowest frequency of occurrences. Vilfred Pareto, an Italian economist, presented Pareto analysis in 1897. He suggested that 80% of the nation's wealth was held by 20% of its population. Juran (1962) used Pareto analysis to classify the problems of quality into vital few and trivial many. In the field of construction management, Aibinu and Odeyinka (2006) used this concept and revealed that 88% of the factors were responsible for 90% of the overall construction delays.

Table 3: Significance weights for project scope definition elements

Category		Element		Element Weight	Total Category Weight
A.	Business Strategy	A1.	Building Use	2.1173	
		A2.	Business Justification	0.3482	
		A3.	Business Plan	1.0139	
		A5.	Facility Requirements	0.7817	
		A6.	Future Expansion/Alternate Consideration	3.6373	
		A7.	Site Selection Consideration	3.3444	
		A8.	Project Objective Statement	1.5962	12.8390
B.	Owner Philosophies	B1.	Reliability Philosophy	2.1463	
		B2.	Maintenance Philosophy	3.3303	
		B3.	Operating Philosophy	1.3729	
		B4.	Design Philosophy	7.4721	14.3216
C.	Project Requirements	C2.	Project Design Criteria	3.9751	
		C3.	Evaluation of Existing Facilities	3.3611	
		C4.	Scope of Work Overview	1.6615	
		C5.	Project Schedule	4.3799	
		C6.	Project Cost Estimate	7.7240	21.1016
D.	Site Information	D1.	Site Layout	3.0512	
		D2.	Site Surveys	1.0761	
		D3.	Civil/Geotechnical Information	0.0421	
		D4.	Governing Regulatory Requirements	0.3420	
		D5.	Environmental Assessment	0.9509	
		D6.	Utility Sources with Supply Conditions	1.1394	
		D7.	Site Life Safety Consideration	0.2392	
		D8.	Special Water and Waste Treatment	1.0695	7.9104
E.	Building Programming	E1.	Program Statement	1.5111	
		E2.	Building Summary Space List	1.1051	
		E5.	Growth and Phased Development	3.4000	
		E9.	Transportation Requirements	0.0419	6.0581
H.	Procurement Strategy	H1.	Identify Long-Lead/Critical Equip. and	3.9120	
		H2.	Procurement Procedures and Plans	2.5827	6.4947
J.	Deliverables	J1.	CADD/Model Requirements	3.4492	
		J2.	Documentation/Deliverables	7.6478	11.0970
K.	Project Control	K1.	Project Quality Assurance and Control	3.6561	
		K2.	Project Cost Control	0.4982	
		K3.	Project Schedule Control	2.1833	
		K4.	Risk Management	2.7012	
		K5.	Safety Procedures	0.1472	9.1860
L.	Project Execution Plan	L1.	Project Organisation	2.9268	
		L2.	Owner Approval Requirements	1.5570	
		L3.	Project Delivery Method	0.1117	
		L4.	Design/Construction Plan & Approach	3.6221	
		L5.	Substantial Completion Requirements	2.7740	10.9916
	Total			100	100

Table 4: Pareto analysis of project scope definition elements

Rank		Elements	Cumulative number of elements %	ANP Weight %	Cumulative ANP weight %
1	C6	Project Cost Estimate	2.38	7.72	7.72
2	J2	Documentation/Deliverables	4.76	7.65	15.37
3	B4	Design Philosophy	7.14	7.47	22.84
4	C5	Project Schedule	9.52	4.38	27.22
5	C2	Project Design Criteria	11.90	3.98	31.20
6	H1	Identify Long-Lead/Critical Equip. & Materials	14.29	3.91	35.11
7	K1	Project Quality Assurance and Control	16.67	3.66	38.77
8	A6	Future Expansion/Alternate	19.05	3.64	42.40
9	L4	Design/Construction Plan & Approach	21.43	3.62	46.03
10	J1	CADD/Model Requirements	23.81	3.45	49.48
11	E5	Growth and Phased Development	26.19	3.40	52.88
12	C3	Evaluation of Existing Facilities	28.57	3.36	56.24
13	A7	Site Selection Consideration	30.95	3.34	59.58
14	B2	Maintenance Philosophy	33.33	3.33	62.91
15	D1	Site Layout	35.71	3.05	65.96
16	L1	Project Organisation	38.10	2.93	68.89
17	L5	Substantial Completion Requirements	40.47	2.77	71.66
18	K4	Risk Management	42.85	2.70	74.36
19	H2	Procurement Procedures and Plans	45.24	2.58	76.95
20	K3	Project Schedule Control	47.62	2.18	79.13
21	B1	Reliability Philosophy	50.00	2.15	81.28
22	A1	Building Use	52.38	2.12	83.39
23	C4	Scope of Work Overview	54.76	1.66	85.06
24	A8	Project Objective Statement	57.14	1.60	86.65
25	L2	Owner Approval Requirements	59.52	1.56	88.21
26	E1	Program Statement	61.90	1.51	89.72
27	B3	Operating Philosophy	64.28	1.37	91.09
28	D6	Utility Sources with Supply Conditions	66.66	1.14	92.23
29	E2	Building Summary Space List	69.04	1.11	93.34
30	D2	Site Surveys	71.42	1.08	94.41
31	D8	Special Water & Waste Treatment Requirements	73.81	1.07	95.48
32	A3	Business Plan	76.19	1.01	96.50
33	D5	Environmental Assessment	78.57	0.95	97.45
34	A5	Facility Requirements	80.95	0.78	98.23
35	K2	Project Cost Control	83.33	0.50	98.73
36	A2	Business Justification	85.71	0.35	99.08
37	D4	Governing Regulatory Requirements	88.10	0.34	99.42
38	D7	Site Life Safety Consideration	90.48	0.24	99.66
39	K5	Safety Procedures	92.86	0.15	99.80
40	L3	Project Delivery Method	95.24	0.11	99.92
41	D3	Civil/Geotechnical Information	97.62	0.04	99.96
42	E9	Transportation Requirements	100.00	0.04	100.00

Drawing on the Pareto principle, this research conducted an analysis of the distribution pattern of the 42 projects scope definition elements, in order to possibly identify the trivial and important elements. The relative average weights were computed based on the ANP results in Table 3 and equated to 1. Table 4 shows the result of Pareto analysis and Figure 2 illustrates it.

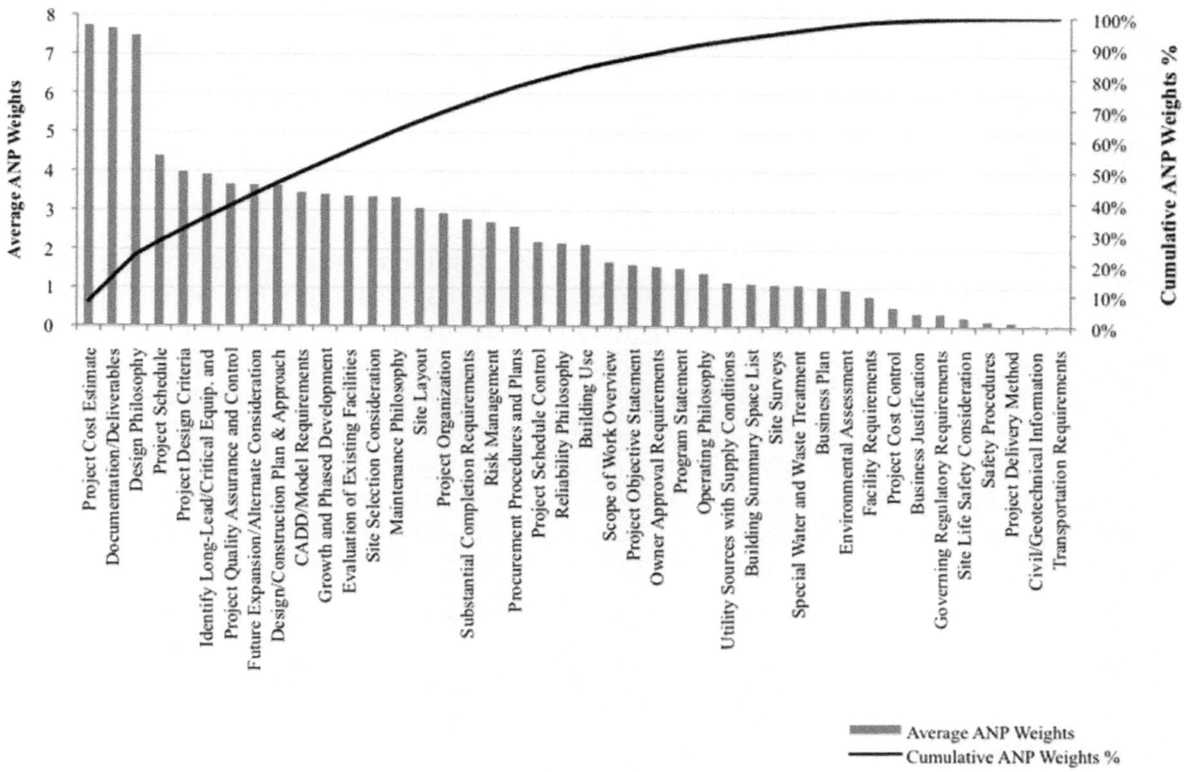

Figure 2: Pareto analysis of project scope definition elements

The result of Pareto analysis on the project scope definition elements revealed that 80.95% of the elements, that is 34 highest priority elements, contribute 98.23% of all elements' weights. More than 80% of the identified project scope definition elements are responsible for about 98% of the completeness of the project scope definition document. However, about 7% of the elements, that is the 3 highest priority elements, are the single largest contributors to the total weights. The three elements contributed 22.84% of all project scope definition elements' weights for the completeness.

Discussion of Results

Looking at the contribution of each element to the overall project scope definition completeness (indicated by the weight), as presented in Table 4, the most influential element for project scope definition completeness is the *'Project Cost Estimate'*, weighing 7.72%. This is not surprising because the reason for cost overrun challenges during a project's execution is often inaccurate estimates. Inadequate or unclear cost estimation for every single direct and indirect item in a project affects the cost performance during the implementation stage and can lead to delays, cost overruns, schedule overruns, and project failure due to lack of finance. Additionally, in order for Saudi Arabian public construction organisations to be able to finance their projects, they need to present a clear and detailed cost estimation document for the project in the annual national budget. This is because each year the government of Saudi Arabia announces the annual national budget together with the share for each ministry including construction related organisations. In defining the project cost estimate element, all costs necessary for completion of the project should be addressed. It may include information such as: construction contract estimate, architectural and engineering consultants fees, administrative costs, insurance requirements, utility costs during construction and all technical information related to the site and the project.

The second most influential element is the *'Documentation and Deliverables'* weighing 7.65%. Documentation and deliverables are required during the project execution and must be identified in pre-project planning in order to avoid uncertainty and rework during execution. They are also important during project execution and after completion for best practice benchmarking. Deliverables may include all drawings and specifications, project correspondence, permits, quality assurance documents, warranties, working drawings, maintenance and operation information.

Project *'Design Philosophy'*, weighing 12.84%, is the third most influential element. In accordance with the project functionality and environments, a listing of the design philosophy issues should be developed. This listing may include design life, aesthetic requirements, quality of life, sustainability, levels of certification, and requirements of any adopted anti-terrorism design standards. This information is necessary at the pre-project planning stage before commencing the design stage in order to avoid major changes and rework. Even though all project scope definition elements are very significant for planning the project, some of the elements have low contribution weights in relation to the completeness of the project scope definition package. This is because these elements may be defined briefly at the planning stage, and then in more detail at later stages of the projects' development cycle.

When comparing the result of project scope definition elements' weights in this study with PDRI elements' weights, some differences are evident. The list provided in this study includes 42 elements instead of the 64 elements included in the PDRI's list. *'Building Use'* element is the most influential element in the PDRI for building whereas in this study it occurs after 50% of all elements, which is the 22nd element in the prioritised list. The reason is that the PDRI is a generalised tool for all building types regardless of whether they are public or private projects. Therefore, it is vital to know the use and type of the building because this could affect other scope definition elements. This study focuses on public building projects, which usually are facility services projects, thus this could account for the differences in the importance of '*Building Use'* in the PDRI list.

Unexpectedly, this study discovered that *'Facility Requirements'* element is within the lowest 10 influential elements. In contrast, it is the second influential element in PDRI. The reason behind this deviation is that respondents consider this element as a technical issue, which is discussed briefly in pre-project planning stage and in more detail in the design stage. This practice is one of the reasons for uncertainties and can lead to project delays and incomplete projects.

Element *'Project Cost Estimate'* is significant in both PDRI and this study. It is the primary contributor element to the completeness of project scope definition in this study and ranked the 4th contributor element in the PDRI for building.

Element *'Architectural Design'* is within the highest 10 contributors to the project scope definition completeness. Conversely, it was excluded from the list in this study. This study is limited to the pre-project planning stage of the project life cycle. Similarly to all the other elements of the *'Building/ Project Design Parameters'* category, architectural design at this stage has not yet been developed. Usually the building/project design parameters are provided at the design stage by the architectural and engineering consultants.

The second most influential element in this study is *'Documentation and Deliverables'* and the 10th element is *'CADD/Model Requirements'*. However, in the PDRI, these occur as the lowest three elements contributing to scope definition completeness.

Summary of Findings

Compared to industrialised countries, Saudi Arabia is less developed and the challenges are certainly different. There is a noticeable lack of research on the knowledge required to define project scope in Saudi's construction context. Therefore, this research identifies 9 categories including 42 elements that should be identified and evaluated at the pre-project planning stage in public construction projects in Saudi Arabia. Also it identifies the elements' relative significant weights for the project scope definition document completeness with respect to the interrelationship and interactions among them.

The result from Pareto analysis on the elements revealed that there is no discernable difference between the 42 elements except the three highest elements. The analysis shows that more than 80% of the elements are responsible for about 98% of the completeness, which indicates that all the elements are important for ensuring the project scope definition completeness, thus enhancing the likelihood of achieving better project outcomes. An inspection of the elements indicates that the 34 highest priority project scope definition elements is a combination from all the nine categories. This suggests that even though the project scope definition elements can be prioritised, the contribution of all the nine categories to the overall scope definition completeness is not negligible. The result implies that all the categories are important for ensuring project scope definition completeness.

Even though construction projects may share the same procedures globally, the PDRI is a good approach for evaluating the completeness of project scope definition. The PDRI was developed in the US mainly for the US construction context. Therefore, it is necessary to identify a different set of information that respects the nature and the environment of a project's context.

Conclusion

The developed priorities list is useful as it can guide decision-makers in evaluating project scope definition completeness and deciding whether to proceed with a project or not. In other words, the project management team can measure the level of each project scope definition completeness element by using a Likert scale of satisfaction as follows: Incomplete or Poor Definition (0%), Major Deficiencies (25%), Some Deficiencies (50%), Minor Deficiencies (75%) and Complete Definition (100%). Then all scores are calculated according to each element priority weight, presented in Table 3. The final score of the level of project scope definition completeness would be the total of all the elements scores. The higher the total score the more well defined the project is. This procedure allows project management teams to take actions that can help improve the scope definition of those elements that have low scores, which indicate the area of risk to the project, and maximize the chance for a successful project.

The ANP is an innovative tool for multi-criteria decision-making. Both researchers and industry practitioners should find it useful in different ways. Most of the data obtained for the ANP were obtained from an expert panel, and potentially can be analytically generalised. Even though the technique does not require a large sample size, increasing the sample size could improve the result. In addition, although the study focuses on public building projects only in Saudi Arabia, the results are applicable to other developing countries with similar environments of delivery methods and industry practices. The study can be replicated in the context of private sector projects.

References

Abdul-Hadi, N, Al-Sudairi, A.S. & Alqahtani, S. 2005, 'Prioritizing barriers to successful business process re-engineering (BPR) efforts in Saudi Arabian construction industry', Construction Management and Economics, 23 (3), 305-315.

Ahuja, V, Yang, J. & Shankar, R. 2009, 'Benefits of collaborative ICT adoption for building project management', Construction Innovation: Information, Process, Management, 9 (3), 323-340.

Aibinu, A.A. & Odeyinka, H.A. 2006, 'Construction delays and their causative factors in Nigeria', Journal of Construction Engineering and Management, 132 (7), 667-677.

Al-Humaidan, M. 2011, Al-Eqtisadiah, Online. Available: http://www.aleqt.com/2011/12/09/article_605385.html. Accessed 09-12-2011.

Al-Khalil, M.I. & Al-Ghafly, M.A. 1999, 'Delay in public utility projects in Saudi Arabia', International Journal of Project Management, 17 (2), 101-106.

Al-Kharashi, A. & Skitmore, M. 2009. 'Causes of delays in Saudi Arabian public sector construction projects', Construction Management and Economics, 27 (1), 3-23.

Alsehaimi, A, Koskela, L. & Tzortzopoulos, P. 2013, 'Need for Alternative Research Approaches in Construction Management: Case of Delay Studies', Journal of Management in Engineering, 29 (4), 407-413.

Arain, F.M, Pheng, L.S. & Assaf, S.A. 2006, 'Contractors' views of the potential causes of inconsistencies between design and construction in Saudi Arabia', Journal of Performance of Constructed Facilities, 20 (1), 74-83.

Assaf, S.A. & Al-Hejji, S. 2006, 'Causes of delay in large construction projects', International Journal of Project Management, 24 (4), 349-357.

Atkinson, R, Crawford, L. & Ward, S. 2006, 'Fundamental uncertainties in projects and the scope of project management', International Journal of Project Management, 24 (8), 687.

Cano, J.L. & Lid6N, I. 2011, 'Guided reflection on project definition', International Journal of Project Management.

Central Department of Information and Statistics, 2013, National Accounts Indicators 2013, Online. Available: http://www.cdsi.gov.sa/dmdoc/nat-2013.pdf.

Cheng, E.W. & Li, H. 2001, 'Analytic hierarchy process: an approach to determine measures for business performance', Measuring Business Excellence, 5 (3), 30-37.

Cho, C.S. & Gibson, E. 2001. 'Building project scope definition using project definition rating index', Journal of Architectural Engineering, 7 (4), 115.

Chritamara, S, Ogunlana, S.O. & Bach, N.L. 2001, 'Investigating the effect of initial scope establishment on the performance of a project through system dynamics modelling', Engineering Construction and Architectural Management, 8 (5-6), 381.

Cii 1994, Pre-project planning: Beginning a project the right way, Construction Industry Institute, The University of Texas at Austin.

Gibson, G.E, Kaczmarowski, J. & Lore Jr, H. 1995, 'Preproject-planning process for capital facilities', Journal of Construction Engineering and Management, 121-312.

Gibson, G.E, Wang, Y.R, Cho, C.S. & Pappas, M.P. 2006, 'What Is Preproject Planning, Anyway?', Journal of Management in Engineering, 22-35.

Haponava, T. & Al-Jibouri, S. 2009, 'Identifying key performance indicators for use in control of pre-project stage process in construction', The International Journal of Productivity and Performance Management, 58 (2), 160.

Hatush, Z. & Skitmore, M. 1998, 'Contractor selection using multicriteria utility theory: an additive model', Building and Environment, 33 (2), 105-115.

Herath, G. 2004, 'Incorporating community objectives in improved wetland management: the use of the analytic hierarchy process', Journal of Environmental Management, 70 (3), 263-273.

Hwang, B.G, Thomas, S.R, Haas, C.T. & Caldas, C.H. 2009, 'Measuring the impact of rework on construction cost performance', Journal of Construction Engineering and Management, 135-187.

Juran, J.M. 1962, 'Quality control handbook', New York, McGraw-Hill.

Kähkönen, K. 1999, 'Multi-character model of the construction project definition process', Automation in Construction, 8 (6), 625.

Lam, K. & Zhao, X. 1998, 'An application of quality function deployment to improve the quality of teaching', International Journal of Quality & Reliability Management, 15 (4), 389-413.

Lordsleem Jr, A.C. & Melhado, S.B. 2014, 'Scope of design for production of wall partitions', Journal of Engineering, Design and Technology, 12 (2), 263-279.

Love, P.E.D. 2002, 'Influence of project type and procurement method on rework costs in building construction projects', Journal of Construction Engineering and Management, 128(1), 18-29.

Love, P.E.D, Irani, Z. & Edwards, D.J. 2004, 'A rework reduction model for construction projects', Engineering Management, IEEE Transactions on, 51 (4), 426-440.

Mirza, M.N, Pourzolfaghar, Z. & Shahnazari, M. 2013, 'Significance of Scope in Project Success', Procedia Technology, 9 (2013), 722-729.

Saaty, R.W. 2003, 'Decision making in complex environments', Pittsburgh, Creative Decisions Foundation.

Saaty, T.L. 1980, The analytic hierarchy process: planning, priority setting, resources allocation

Saaty, T.L. 1994, 'How to make a decision: the analytic hierarchy process', Interfaces, 19-43

Saaty, T.L. 1996, Decision Making with Dependence and Feedback: The Analytic Network Process, Pittsburgh, RWS Publications.

Sambasivan, M. & Fei, N.Y. 2008, 'Evaluation of critical success factors of implementation of ISO 14001 using analytic hierarchy process (AHP): a case study from Malaysia', Journal of Cleaner Production, 16 (13), 1424-1433.

Singh, M, Shankar, R, Narain, R. & Agarwal, A. 2003, 'An interpretive structural modeling of knowledge management in engineering industries', Journal of Advances in Management Research, 1 (1), 28-40.

Wang, Y.R. 2002, 'Applying The PDRI in Project Risk Management', Ph.D Thesis, University of Texas at Austin, Texas.

Wang, Y.R. & Gibson, E. 2010, 'A study of preproject planning and project success using ANNs and regression models', Automation in Construction, 19 (3), 341-346.

Woodhead, R.M. 2000, 'Investigation of the early stages of project formulation', Facilities, 18 (13/14), 524-534.

Yin, R.K. 2009, Case study research: Design and methods, 4th ed., Los Angeles, Sage publications, INC.

A Study of Joint Venture Formation between Construction Organizations in Tanzania

15

Steven J. Minja, and Geraldine J. Kikwasi (Ardhi University, Tanzania)
Wellington D. Thwala (University of Johannesburg, South Africa)

Abstract

Formation of joint ventures between construction companies is one of the recent efforts in combating contractors problems in Tanzania and addresses one of the key challenges facing the construction industry today in the country especially when large and complex projects are involved. The main objective of the research is to examine the formation of joint ventures in Tanzania, specifically to: study joint ventures formation procedures; identify criteria in selection of joint venture partner; and identify risks associated with and challenges facing joint venture undertakings in Tanzania. The study adopted a descriptive research design and purposive sampling. Multiple sources of evidence was used to collect information mainly literature review, questionnaires and interviews. Key findings reveal that legal and statutory requirements in the formation of joint ventures include registration by Business Registration and Licensing Agency (BRELA), Contractors Registration Board (CRB) and under Registration of Documents Act (RDA). All JV respondents were registered by BRELA and CRB which are mandatory but only 7 were registered by RDA. Most important factors considered during formation of JVs are: contract agreement, financial stability and commitment while key risks associated with JVs are cultural and social differences, delays in approvals and financial risks. In addition, main challenges facing joint ventures are: identification of possible risks and joint venture agreement interpretation. The study concludes that formation of JVs have been addressing some of emerging challenges facing local construction organizations despite the risks and challenges that exist. The study recommends that firms entering joint ventures should explore the benefits of registering with RDA and ensure risks associated with their JV are properly assessed. The study also recommends introduction of an incentive scheme through Tanzania Investment Centre to grant tax relief for foreign contractors going into joint venture with local construction firms.

Keywords: Construction organizations, joint ventures, process, risks, challenges

Introduction

Joint venture formation between construction companies is one of the recent efforts in combating contractors problems in Tanzania and addresses one of the key challenges facing the construction industry today in the country especially when large and complex projects are involved. Construction joint ventures are formed for a number of reasons including pooling of resources, sharing of risks, undertaking large and complex projects, entry to a foreign market and business diversification (Minja, 1986; Dalle and Ports, 1999; Kwok, Then and Skitmore, 2000; Adnan and Morledge, 2003).

In formation of joint venture interested parties need to set criteria that will lead to successful execution of envisaged construction project. Various studies (Kwok, Then and Skitmore, 2000; Adnan and Morledge, 2003; Rowan 2005; Adnan, Chong and Morledge,2011) have identified criteria pertinent to JV partner selection such as ability, experience and skills, financial stability, partners' commitment, understanding and commitment to JV objectives, inter-partner trust and personal knowledge of the partner organizations.

Risks are inherent in construction Joint venture is not an exception some of the risks such as the agreement of the contract, partner selection, potential financial distress in JV partner, improper project feasibility study, project delay, inadequate forecast about market demand, loss due to bureaucracy for late approvals and design changes have been identified worldwide (Kwok et al, 2000; Shen, Wu and Catherine, 2001; Adnan and Morledge, 2003; Rowan 2005; Adnan, 2008; Jamil, Mufti and Khan, 2008; Adnan et al, 2011). Likewise construction joint ventures face challenges which to be known addressed by the potential partners. The purpose of this research is to examine the formation of joint ventures in Tanzania particularly, JV formation procedures, partner selection criteria and associated risks.

Review of Literature

Overview of Joint Venture

Gaeton and Keith (1999) state important resources to successfully undertake construction projects to completion as: adequate financial resources, managerial skills, plant and equipment, appropriate technology and skilled and unskilled labour. Depending on the nature of the project, at times these resources are not readily available to construction organizations thus necessitating formation of joint ventures. A number of authors (Gaeton and Keith 1999; Wolf, 2000; Miller, 2002; Watermeyer, 2005) have defined joint venture broadly, basically as a partnership or coalition between two or more companies which can be short or long term to carry out an activity.

Globally, construction businesses are becoming more competitive due to complexity of clients requirements and technology advancement. Construction organizations are forced to form alliances at varying levels in order to stay in business. Some of the reasons dictating such alliances include: surfacing of large and complex projects; pooling of resources; sharing of risks; entry to foreign market; access to technological improvements; political or diplomatic improvements; tied aids (Minja, 1986; Dalle and Ports, 1999). By pooling resources and complementary strengths companies can increase productivity and competitive standing in ways they could not do by themselves. It is important for companies which are considering entering into a joint venture to examine strategies applied in order to uncover what makes them successful (or unsuccessful) and why they work differently in different situations (Kathryn, 1986). In general joint ventures are not the easiest forms of organisation to manage and operate. Therefore there must be compelling reasons why parties to a construction contract resort to the formation of a joint venture in contrast to the conventional contractor/subcontractor relationship (Dalle and Ports, 1999).

Adnan and Morledge (2003) identify important objectives to be achieved through joint venture works which are: new market opportunity, profit, financial objectives, increased project scale, reducing risks and organizational learning. Similarly Kwok, Then and Skitmore (2000) have listed motivators for joint ventures in Singapore as strategic market entry, business diversification, sharing of resources and risk sharing.

Joint Venture Agreements

The joint venture agreement on a sophisticated international construction project is a complex document that usually is the product of extensive negotiations between the parties. There is no single standard form of Joint Venture Agreement because there can be a range of characteristics. But general consideration of Joint Venture Agreement should bear in mind important issues as follows (Rowan, 2005):

- defines the parties to the agreement;
- defines the purposes and objectives of the joint venture;
- defines the monetary and non-monetary contributions to be made by each of the parties;

- defines the management and structure of the joint venture and the associated appointment mechanism;
- defines the (corporate) vehicle under which the joint venture will be carried on;
- defines the basis on which the participants share in the profits and losses of the joint venture;
- defines the liabilities of the joint venture partners;
- provides for a conflict resolution mechanism, and
- provides for a termination mechanisms.

Joint venture agreements depend on the form of joint venture adopted. Rowan (2005) indicates three forms of joint ventures which are: joint venture in form of company; joint venture in form of partnership; and joint venture on contractual basis (unincorporated joint venture).

Joint Venture Partner Selection

Prior to selection of a joint venture partner firms aspiring to form joint ventures should formulate or adopt criteria that will guide their undertakings. The main criteria considered when an organization is looking for possible partners to go into joint venture partnership are as follows (Rowan 2005):

Ability, experience and skills: Require organization with proven ability experience and skills

Financial stability: Financial suitability is an important requirement for joint venture partnership

Similar business philosophy and concepts: This factor constitutes a basis for building a cohesive team with common goals

Strength of partner's willingness to form joint ventures: The joint venture must be driven by both parties

Spoken language: Most contractors prefer to look for partners abroad who speak the same language to avoid communication problems

Kwok, Then and Skitmore (2000) reveal criteria used in selection of a joint venture partner in Singapore as partners' commitment, understanding and commitment to JV objectives, past JV experience, compatibility between partners and understand own and shared risks. Adnan and Morledge (2003) identify criteria for partner selection among12 critical success factors for joint venture projects. Additionally, Adnan, Chong and Morledge (2011) identify critical factors in the international JV partner selection in which criteria were related to the reputation, inter-partner trust, experience, personal knowledge of the partner organizations, commitment and human resources management

Risks in joint venture undertakings

Joint venture undertaking involves risks that need to be known to the partners such that appropriate measures can be taken. Adnan (2008) identifies risk factors in joint venture formation in Malaysia as the agreement of the contract, partner selection, control/equity, sub-contractors, renegotiations and training. Other most important risk factors in Singapore were revealed by Kwok et al (2000) and these are disagreement in accounting profit and loss, potential financial distress in JV partner, potential financial distress in JV, partner's lack of management competence and resourcefulness, over-interference by parent companies of either parties, and disagreement in allocation of work.

A number of studies have identified risks in construction joint venture undertakings in various parts of the world. Jamil et al (2008) identify 10 most important risks in international joint venture as financial risks due to delays in payments, foreign currency fluctuation, incompatibility of local policies with international practice, pollution and safety rules, changes in laws and regulations, design problems, inflation and cost overrun, delays in approvals and permits from government, inadequate technical specifications in tender documents and site access/right of way. Shen et al (2001) identify top ten risks in construction joint venture in China and these are: cost increase due to changes of policy, improper project feasibility study, project delay, inadequate forecast about market demand, improper selection of project location, improper selection of project type, increase of resettlement costs, inadequate choice of project partner, loss due to bureaucracy for late approvals and design changes.

Challenges Facing Joint Ventures

Challenges are part of any project undertakings including joint ventures. According to Watermeyer (2005) the following are the challenges facing construction companies going into joint ventures:

- identification of possible risks likely to be in a joint venture undertaking and finding ways which will enable losses to be minimized and opportunities to be maximized;
- to determine forms and reasons for joint venture formation most appropriate in Tanzania;
- how to grade a joint venture capability and capacity;
- how to encourage joint venture formation, assuming that it is a desirable practice;
- decision on how two or more contractors with different capacities can enter into joint venture agreement; and
- how does one prevent misinterpretation of the facts in order to influence the tender process or the award of a contract arising from a tender offer or collusive practices intended to subvert developmental objectives during the period of performance of the contract.

Methodology

Research design

The research design of this study is descriptive which is attempting to put to light issues on construction joint venture formation in Tanzania.

Population and Sample size

The population of the study is the number of projects registered by Contractors Registration (CRB) undertaken under joint venture arrangement between 2007 and 2010. Statistics from CRB indicates that the number of Joint ventures registered during the period were 102. The sample size was 45 comprising 30 of joint ventures, 10 consultants and 5 statutory bodies.

Sampling procedure

Some criteria were set to obtain the sample from joint ventures and these are: at least one of joint venture partner has the office located in Dar es Salaam for easy communication and has execute a project with a value of not less than TAS 500 million. Using the criteria, only 30 joint venture companies were selected. Consulting firms involved in the study were selected basing on the fact that they were established before 2001 and therefore have been in practice for over ten years, and have been involved in joint venture projects for between two to six years. Purposive sampling was used to select consultants from quantity surveying, architectural firms and statutory and consultative bodies.

Data Collection

Multiple sources of evidence were used to collect data. Literature was reviewed to establish what has been written on formation of joint ventures between construction firms. Questionnaires were sent to selected joint ventures, consulting firms, statutory and consultative bodies. Statutory and consultative bodies are: Business and Licensing Agency (BRELA), Contractors Registration Board (CRB), Architects and Quantity Surveyors Registration Board (AQRB), Public Procurement Regulatory Authority (PPRA) and National Construction Council (NCC). Responses from the respondents are indicated in Table 1 below.

Table 1: Responses from Consultants, Statutory Bodies and Joint Ventures

S/N	Respondent	Questionnaires sent	Questionnaires received
1	Statutory and consultative Bodies (AQRB, CRB, BRELA, PPRA and NCC)	5	5
2	Consultants	10	6
3	Joint Venture Companies	30	24
	Total	**45**	**35**

Analysis and Discussion

The process of data process was done by arranging the collected data in frequency tables and then computing mean, standard deviation and relative importance index. Relative Importance Indices (RIIs) and cross comparison tables are then used to rank the results obtained.

- Calculation of Mean: This is the average of all values in a set of data. The mean is calculated by adding all the values in then group and then dividing by the number of values.
- Standard deviation: This is the measure of the degree to which data is spread around the mean. It is a positive square root of the variance. Standard Deviation is calculated as follows:

$$\sqrt{\frac{\Sigma (x - \bar{x})^2}{N}}$$

Where

X = Value, $\bar{X}$ = Mean value, N = total number of respondents

- Relative Importance Index (RII): This is calculated as follows:

RII = ΣW/AxN

Where;

W = weight given to each factor by respondents

A = highest weight, N = total number of respondents.

Relative Importance Indices (RIIs) comparison tables can be used to rank the results by taking into account the average scores and the RII as follows:

Table 2: Relative Importance Indices (RIIs) comparison tables

Average Score	RII	Rankings
4.0 to 5.0	0.80 to 1.00	High (H)
3.0 to 4.0	0.60 to 0.80	Medium (M)
1.0 to 3.0	0.20 to 0.60	Low (L)

Source: Chileshe, Haupt and Fester (2007)

Respondents' Profile

The joint ventures involved in the study comprised of 12 local to local, 10 local to foreign and 2 foreign to foreign joint ventures. Statutory bodies selected have the following roles: BRELA which is responsible for registration of companies; CRB which is responsible for registration of contractors; AQRB is responsible for registration Architects and Quantity Surveyors; PPRA is responsible for regulating procurement of public services and goods; and NCC is an institution that gives directives on construction related matters. Consulting firms comprised of architectural and quantity surveying firms.

Process of Formation of Joint Ventures

The formation of joint venture between two companies in Tanzania like in many other countries involves registration with various entities namely: BRELA, CRB and under Registration of Documents Act (RDA). Registration with BRELA and CRB is mandatory; however, registration under RDA is optional. Assessment of 24 joint ventures registration status is indicated in Table 3 below.

Table 3: Registration of JVs by BRELA, CRB and under RDA

S/NO	ENTITY	JVs
1	Registration by Business Registrations (BRELA)	24
2	Contractors Registration Board (CRB)	24
2	Registration of Documents Act (RDA)	7

Results in Table 3 above indicate that all joint ventures were registered by BRELA and CRB and only 7 were registered by RDA. The fact that registration by BRELA and CRB is mandatory and no company in Tanzania can undertake any construction work without fulfilling it, all JVs were registered and they did not find any importance of registering with RDA. The joint venture companies obviously overlook the advantage of this registration which gives more weight to the document as the registration gives more confidence and trust to the partners themselves and even to the outsiders who wish to join the joint ventures.

Forms of Joint Ventures in Construction Projects in Tanzania

There are three forms of joint ventures classified as integrated, non-integrated and combination joint ventures. Responses on the type of joint ventures commonly applicable in Tanzania are as shown in Table 4 below.

Table 4: Forms of Joint Ventures

S/N	Type	Total	Ranking
1	Integrated JV	19	1
2	Non -Integrated JV	8	2
3	Combination JV	2	3

Basing on Table 4 above the most preferred form of joint venture is integrated followed by non-integrated. These results suggest that integrated joint venture is preferred due to the fact that it involves sharing profits and losses by 50% - 50% and therefore reducing investment risks.

Factors guiding selection of a JV partner

Selection of joint venture partners is a crucial factor which determines the success of the joint venture undertaking. Some aspects must be carefully considered before the selection of appropriate partner is done. A list of factors was compiled for joint ventures to assign weightings as follows: 4=Very Important, 3=Important, 2= Moderately Important, 1=Less Important and 0=Not Important at all. The results are summarized in Table 5 below.

Table 5: Key factors guiding selection of a JV partner- JVs

S/N	Factor	Total	Mean	Std Dev.	RII	Rank
1	Willingness	90	3.75	0.44	0.94	1
2	Financial stability	83	3.46	0.51	0.86	2
3	Competetiveness of required task	80	3.33	0.70	0.83	3
4	Similar business philosophy	71	2.96	1.12	0.74	4
5	Firm's ability	68	2.83	0.96	0.71	5
6	Firm's experience	48	2.00	1.14	0.50	6
7	Political stability	44	1.83	1.05	0.46	7
8	Track record in JV's	40	1.67	1.17	0.42	8
9	Spoken language	36	1.50	0.98	0.38	9

Results indicate that highly ranked key factors for success in joint ventures are willingness, financial stability and competitive of required task while similar business philosophy and firm's ability are considered to be medium. Similarly the factors were examined using responses from JVs and Consultants and the results are shown in Table 6 below.

Table 6: Key Factors guiding selection of a Joint Venture partner- Consultants and JVs

S/N	Factor	Total	Mean	Std Dev.	RII	Rank
1	Contract agreement	109	3.63	0.56	0.91	1
2	Financial stability	100	3.33	0.61	0.83	2
3	Commitment	97	3.23	0.77	0.81	3
4	Mutual understanding	88	2.93	0.87	0.73	4
5	Coordination	84	2.80	0.96	0.70	5
6	Cooperation	59	1.97	1.00	0.49	6
7	Partner selection	57	1.90	0.96	0.48	7
8	Local conditions	50	1.67	1.15	0.42	9
9	Communication	51	1.70	0.95	0.43	8
10	Past experience	13	0.43	1.04	0.11	10

Results of combined responses indicate that highly ranked key factors for success in joint ventures are contract agreement, financial stability and commitment while mutual understanding and coordination are considered to be medium.

Risks associated with Joint Venture undertakings in Tanzania

It is desirable for construction firms to identify the risks and find probable impact of these risks as early as possible, so that suitable strategies are made before actual execution of project commences. Risks in international construction projects are more critical as compared to domestic projects because they play a vital role for successful project delivery and increase efficiency and profitability of construction projects (Jamil, Mufti and Khan, 2008). Assessment of risks associated with joint venture formation in Tanzania is summarized in Table 7 below.

Basing on Table 7, the highly perceived risks are cultural and social differences, delays in approvals and financial risks while foreign currency fluctuation, loss of trust, incompatibility of local policies with international practice, termination of joint venture agreements and cost escalation were perceived to be medium. In international joint ventures, incompatibility of local policies with international practice affects mainly the foreign partner who is not familiar with the policies of the host country. Therefore in planning for joint ventures there is a need to minimize or eliminating such risks through identifying, assessing, analyzing, ranking risks and finally developing handling options. Fluctuations in rates directly impact on projects which have a significant component of imports but all remaining types of projects also get affected to some extent.

Table 7: Risks in JV undertakings in Tanzania

S/N	Risk	Total	Mean	Std Dev.	RII	Rank
1	Cultural and social differences	103	3.43	0.68	0.86	1
2	Delays in approvals	98	3.27	0.78	0.82	2
3	Financial risks	97	3.23	0.73	0.81	3
4	Foreign currency fluctuation	94	3.13	0.73	0.78	4
5	Loss of trust	92	3.07	0.91	0.77	5
6	Incompatibility with local policy	90	3.00	0.87	0.75	6
7	Termination of JV agreements	89	2.97	0.67	0.74	7
8	Cost Escallation	80	2.67	0.92	0.67	8
9	Poliyical stability	66	2.20	0.81	0.55	9

Challenges facing JVs Undertakings in Tanzania

Joint ventures may encounter challenges during the course of execution of joint venture projects or even at the close out stage. These challenges may have repercussions on the joint venture and therefore they should be well planned ahead of contract implementation to avoid frustration or even termination of joint venture agreement. Table 8 below summarizes challenges facing JVs.

Table 8: Challenges facing JV Undertakings in Tanzania

S/N	Challenge	Total	Mean	Std Dev.	RII	Rank
1	Identification of Possible Risks	103	3.43	0.77	0.86	1
2	JV Agreement Interpretaion	100	3.33	0.84	0.83	2
3	Operations by Different Contractors	94	3.13	0.82	0.78	3
4	Alignment of Partner Strategies	90	3.00	0.69	0.75	4
5	Management Control over Local JV	86	2.87	1.07	0.72	5
6	Grading JV ability and capacity	78	2.60	1.04	0.65	6
7	Competitive cost Structure	59	1.97	0.81	0.49	7
8	Joint Venture Formation	54	1.80	1.24	0.45	8
9	Establishing Joint Goals	47	1.57	1.19	0.39	9
10	Tender document Pricing	43	1.43	1.04	0.36	10

Results in Table 8 above reveal that the highly perceived challenges in JVs undertakings are identification of possible risks and joint venture agreement interpretation followed by operations by different contractors, alignment of partners strategies, management control over local joint venture and grading joint venture ability and capacity which are perceived as medium. This suggests that prior to entering into joint venture agreements, joint venture partners must satisfy themselves that the undertaking has overcome all possible risks and set procedures to settle such problems should they occur in the course of executing the contract. However due to the urgency of drawing up the contract and joint venture formation some of these aspects may be overlooked resulting into problems during the course of executing the contract.

Conclusions and Recommendations

Conclusion

Formation of joint ventures between construction organizations has been an important endeavour in overcoming problems facing local contractors such as inadequate work opportunities, weak financial capability, lack of plant / equipment and lack of management skills. These problems can be addressed by forming joint ventures between foreign and local, local and local or foreign and foreign companies. The usual purpose of joint venture is to spread a risk inherent in large projects and to pool resources in a way that permits to execute a project.

The formation of joint venture companies in Tanzania need to follow all legal and statutory requirements including registration with registration by Business Registrations and Licensing Agency (BRELA), registration by Contractors Registration Board (CRB) and registration under Registration of Documents Act (RDA). In planning for a joint venture arrangement, it is important to consider all important factors in order to have a project delivered successfully. Factors that guide joint venture formation in Tanzania are contract agreement, financial stability and commitment. The most important risks associated with joint venture undertakings are cultural and social differences, delays in approvals and financial risks. Similarly, joint ventures as any other undertaking do encounter challenges. Challenges facing joint ventures in Tanzania include: identification of possible risks and joint venture agreement interpretation.

Recommendations

Based on the findings and conclusion of this study, it is recommended that firms entering joint ventures should explore the benefits of registering with RDA and ensure risks associated with their JV are properly assessed. The study also recommends introduction of an incentive scheme through Tanzania Investment Centre to grant tax relief for foreign contractors going into joint venture with local construction firms.

References

Adnan, H. 2008. An assessment of risk management in joint venture projects in Malasyia, *Asian Social Sciences*, 4 (6) pp 99-106

Adnan, H and Morledge, R. 2003. *Joint Venture Projects in Malaysian Construction Industry: Factors Critical to Success*, A Paper presented at the 19th Annual Association of Researchers in Construction Management Conference, University of Brighton, Malaysia.

Adnan, H. Chong, H.Y and Morledge, R. 2011. Success criteria for international joint ventures: The experience of Malaysian contractors in the Middle East. *African Journal of Business Management* 5(13), pp. 5254-5260

Aimin, Y. and Yadong, L. 2001. *International Joint Ventures: Theory and Practice,* 2nd Edition, M. E. Sharpe, Inc, 80 Business Park Drive, Armonk, New York 10504, United States of America.

Bairi, M. and Njaa, S. 2005. *Registration of Joint Ventures Contracting Firms*, A Paper presented in a Workshop on Joint Ventures Partnership for Contractors in Tanzanian, 14th and 15th April, Dar-Es-Salaam, Tanzania.

Chileshe, N, Haupt, T., and Fester, F. 2007. Assessing the Readiness of Building Diplomates for the South African Construction Industry, *Journal for Education in the Built Environment*, 2 (2)

Dalle, G .and Potts, K. 1999. *Joint ventures in the construction industry*, A Paper presented at the Annual Workshop, School of Engineering and the Built Environment, University of Wolverhampton, UK

Gaeton, D. and Keith, P. 1999. *Joint Ventures in the Construction Industry*, School of Engineering and the Built Environment, University of Wolverhampton, UK

Jamil, M., Mufti, N.A.and Khan, A.H. 2008. Risk Identification for International Joint Venture Construction Projects, *First International Conference on Construction In Developing Countries (ICCIDC–I)*, Karachi, Pakistan

Kathryn, R. H. 1985. *Joint Ventures, Alliances, and Corporate Strategy*, Beard Books, New York.

Kathryn, R. H. 1986. *Managing for Joint Venture Success*, Lexington Books, New York.

Kwok, H.C.A. and Then, D. and Skitmore, M. 2000. Risk management in Singapore construction joint ventures. *Journal of Construction Research 1*(2):pp. 139-149.

Minja, S. J. 1986. *Joint Ventures between Local and Foreign Construction Firms*, Problems and Prospects, ARI, Dar Es Salaam.

Rockart, J.F. 1994. The changing role of the information systems executive: a critical success factors perspective. *Sloan Management Review*. 241(1), 3-13.

Rowan, V. 2005. *How joint ventures are organized, operated on international construction projects*. An outline used in the presentation to the Overseas Construction Association of Japan, Inc. (OCAJI)

Shen, L. Y., Wu, G. W. C., and Catherine S. K. N. 2001. Risk assessment for construction joint ventures in china. *Journal of Construction Engineering and Management*, 127 (1), pp 76-81 source: http//jingpin.szu.edu.cn accessed on 31/5/2012

Watermeyer, R. 2005. *Fundamentals of a Joint Venture*, A Paper presented in a

Wolf, R. C. 2000. *Effective International Joint Venture Management*, M .E.Sharpe, Inc 80 Business Park Drive, Armonk, New York 10504.

Overcoming Inadvertent Barriers to Entry in Large Infrastructure Projects

16

Peter Livesey, (University of Technology Sydney, Australia)

Justin Bold, (Bold Consulting Pty Ltd, Australia)

Abstract

The history behind the award of Brisbane City Council's Legacy Way project is discussed. The discussion focus on the possible impact of cognitive bias in the Expression of Interest (EOI) process together with the steps that were taken during the EOI development and evaluation phase to reduce the impact that this may have had on the selection of Contractors from their EOI submissions. The paper concludes that Cognitive Bias may have created a greater barrier to entry to Contractors attempting to enter the Australian PPP market than has been previously realised and makes suggestions as to how this effect could be minimised in the future.

Keywords: PPP, Infrastructure Projects, Barriers to Entry, Cognitive Bias

Introduction

Australian Government agencies and their state and local counterparts have been courting international construction companies for several years. The government has stated its desire to improve the efficiency of major infrastructure procurement (particularly PPPs) for the sake of securing the best possible value for money. To international contractors seeking to deploy highly skilled, highly experienced resources in new jurisdictions the government's intent is attractive, and with such capable resources on the prowl it's no wonder the Australian Government continues to entice.

Despite continued effort of both the Australian government and international contractors, very few projects in the civil infrastructure domain have been awarded to (or won by) international contractors. There have been numerous discussions and reports conducted with key stakeholders in an attempt to address this issue and unlock the potential of foreign involvement in major infrastructure projects. Interestingly, the conclusions from much of the discussion and debate have been remarkably similar.

A recent KPMG report (KPMG 2010) on behalf of Infrastructure Australia (IA) provides a good summary and concludes that most Participants (both private and public sector) felt that there was good competition within the Australian market, however factors identified as deterrent for both Australian and international competitors were:

- The unknown pipeline of projects and their sporadic nature
- A perceived lack of commitment to PPPs across all jurisdictions in Australia
- The size of the bid costs

An interesting point of difference was that several international participants considered that the requirement to demonstrate local experience was also a barrier. However, this view was not shared by the Federal Government.

These points are highly representative of the conclusions by many on both sides of the relationship and both sides of the debate. So, why do so few international contractors proceed beyond the Expressions of Interest (EOI) Stage? The answer may not lie in the consistency of views proposed by various stakeholders but in the inconsistency of views between government and internationals regarding barriers to entry as outlined above.

One project to successfully attract international contractors, see them progress through EOI and deliver compelling value for government was Legacy Way in Brisbane, Australia. However, the companies within the Joint Venture responsible for Legacy Way have been comparatively unsuccessful in proceeding beyond the EOI phase of subsequent projects. Given the significant similarities in size and complexity of Legacy Way and the projects being targeted by these contractors, this presents a confounding concern for government and international contractors. Has the courtship been a brief fling or is there potential for a fulfilling marriage?

This paper uses the authors' experience gained during the Legacy Way project as a case study and hypothesises why Legacy Way was effective in securing international contractors and perhaps why other contemporary projects may not be as successful - as this is clearly not the intent of the various public bodies involved in the letting of PPPs - the phrase 'inadvertent barriers to entry' is coined. In so doing, the authors make the assessment that significant barriers to entry do exist in the Australian construction industry but are cautious to highlight that they may be the consequence of cognitive biases. In the case of Legacy Way particular changes were made to the Expressions of Interest evaluation process and criteria to reduce the effect of any bias and it is the authors' belief that if these changes had not been made the eventual winner would not have qualified past the EOI stage.

The authors' intent is to highlight the frame of mind adopted on Legacy Way, and the resultant actions undertaken in shaping the procurement process (particularly the EOI phase) and contrast them with operating philosophies and processes of conventional projects. The inference is that by repeating some of the behaviour adopted by the client on Legacy Way, it is possible that the procurement results may be replicated in part i.e. inclusion of international project teams beyond the EOI phase in a fair and transparent process.

We pose practical solutions that government agencies may consider adopting when engaging international companies. The initiatives span the full procurement cycle from initial market sounding through to interactive RFP processes. The aim of the paper is not to comment on the postulate that international competition is desirable and in the interests of the Australian public and taxpayer.

Background

Project Description

Brisbane City Council has undertaken several major infrastructure projects since 2005. Two of these projects were constructed by 'new' international contractors and second tier contractors. The $300m Go Between Bridge was delivered under an Alliance between Brisbane City Council, Bouygues Traveaus Publics, Seymour Whyte and MacMahon and the $1.7bn Legacy Way is under construction by the Transcity Joint Venture (Acciona, Ghella, BMD).

Brisbane City Council's initiatives appear, on the surface, to be consistent with, and remarkably similar to, other government's seeking international interest. However, their results provide compelling evidence to investigate a little deeper.

Legacy Way is the fourth of five major projects that form Brisbane City Council's TransApex programme. TransApex comprises:

- CLEM7 (formerly North South Bypass Tunnel) – currently in operation with RiverCity Motorway SPV as concessionaire
- Airport Link – currently under construction with BrisConnections SPV as concessionaire
- Go Between Bridge – currently in operation with Brisbane City Council as owner/operator
- Legacy way (formerly Northern Link) – currently under construction with Brisbane City Council as owner/operator
- East-West Link – currently forecast for delivery by 2026

The Procurement Cycle

The procurement of large infrastructure projects in Australia usually follows a six step process involving:

1. Market Sounding
2. Issue of Request for Expression of Interest (EOI)
3. Short listing from the EOI's received
4. Issue of a Request for Proposal (RFP) to the shortlisted EOI respondents
5. Evaluation of RFP responses
6. Negotiation leading to the award of contract

This paper deals largely with steps 1-3 and is based on the experience developed by the Transaction Team during the procurement of Legacy Way (formally known as the Northern Link Project).

Development of the Procurement Process

The procurement phase for Legacy Way was conducted over 24 months from September 2008 to September 2010. It followed a typical project lifecycle with prefeasibility in 2004, a preliminary assessment in 2007 and detailed feasibility in 2007-8. The procurement phase included a first phase of market sounding and EOI, a second phase of market sounding and EOI and the Request for Proposals phase. During the procurement phase the operating environment was particularly dynamic. An understanding of the macro environment, the industry dynamics and resources required/available was highly powerful for guiding the project through its development.

The project was initially conceived as a PPP, but due to the Global Financial Crisis and the poor comparison of forecast traffic with actual traffic experienced on other similar projects it quickly became apparent that there was not sufficient interest in the market to enable a competitive PPP process to proceed. Council therefore decided to undertake international market sounding to reignite market appetite in the project.

The feedback from the international companies canvassed in this campaign brought Council's attention to the following:

1. Large international companies were interested in the Australian market
2. These companies were discouraged from proceeding with EOI's due to a continued list of failures
3. They regarded the Australian Market as "a closed shop" dominated by the Major Australian Construction Companies. They were particularly concerned about members of the same group being allowed to submit competing offers
4. There appeared to be a lack of transparency in the evaluation process thus making the reason for failure hard to determine

5. These companies were successful both in their own countries and in a number of overseas countries and other than for point 3 were unable to explain their lack of success in the Australian market
6. None of the companies visited had the appetite for taking greenfield traffic risk

It soon became apparent that overseas companies whilst interested in the Project were not ready to commit the resources required to submit an EOI unless some or all of points 3, 4 and 6 were adequately addressed.

Demonstrating Intent

Based on the above feedback Council was determined to demonstrate that Legacy Way would be different; principally that the Council project team had listened to concerns of the Internationals and that positive steps were taken to present Legacy Way as a new project and minimise the risk of another failed EOI. This point is critical. Council was in a do or die situation. Not all projects experience the heightened sensitivity to failure that focuses efforts on finding new solutions to old problems.

This intent was predominantly demonstrated through visible changes to the EOI process and evaluation criteria that were difficult for many project team members and advisors to accept.

Changes to Council EOI Evaluation Criteria

The review resulted in the modifications to evaluation criteria as shown in table 1 below:

Criteria Prior to Review	Criteria Post Review	Risks In New Criteria	Mitigation Strategy
Experience of the Brisbane Market Required	Proven experience in working in an overseas market with a similar environment to Brisbane required and demonstration of capability to successfully complete projects in new jurisdictions	Overseas proponents would not have experience of specific Australian requirement in terms of I.R., safety and environmental requirements	Strongly encourage any overseas partner to obtain an Australian partner with good local knowledge
Full Teams needed to be formed	Full teams need not be formed at the EOI stage. Other members could be added during the RFP stage.	Shortlisted proponents may not be able to form full bid teams in the required time frame and some key members of the team may not be acceptable to council.	Identify key consortium members and require their identification at the EOI stage. Require unsuccessful EOI Respondents to release team members thus making them available for the RFP phase
Evaluation criteria somewhat vague	Dictionary type definitions used to define evaluation criteria requirements	Difficulty in ensuring definitions were accurate	Continue review by senior project staff
Evaluation weighting not published	Evaluation weighting published in the EOI request	Resistance from team members used to subjective format	Carry out in-house reviews and ensure weightings truly represented Councils concerns
Lack of Understanding of EOI requirements	Interactive EOI phase proposed to enable clarification from bidders	Probity Concerns	Plan briefings with probity auditor and of course have probity auditor present during briefings

Table 1 Modifications to the EOI Criteria

Recognising Cognitive Biases of the Client

Cognitive biases were first popularised by Kahneman and Tversky in the 70s (Kahneman and Tversky 1972, Tversky and Kahneman 1973, Kahneman and Tversky 1977). Since then, a growing list of cognitive biases has been developed by numerous researchers of decision making and theory in fields such as cognitive science, social psychology and behavioural economics (Gilovich, Griffin et al. 2002, Kahneman and Frederick 2002). Of particular interest in this case is the rational and irrational routes to status quo bias and the interplay between status quo bias, loss aversion and omission bias.

Of critical importance in this discussion is the objective of the shortlisting process. At EOI stage, clients intend to decide, using incomplete information, which entities are most likely to provide the most compelling binding offer following the RFP phase. This establishes a decision making process that is complex with essentially incomplete information (committed bids are unknown) and thus introduces the need for 'intuitive' responses by advisors reporting on the submissions of each EOI respondent.

It could be argued that a rational path to status quo bias exists – that is, it is 'safer' to select known choices (in this case contractors with reputable records in Australia with whom advisors are familiar) because local contractors generally deliver acceptable results. However, this line of interrogation leads to maintaining the status quo and in the case of Legacy Way would have risked a failed process. More broadly, adopting this view fails to deliver on the outcomes sought by Infrastructure Australia – i.e. removing barriers "that might reduce the competitiveness of the PPP procurement process such as to prevent Governments from achieving **best** value for money (even if they still achieve value for money in an absolute sense)" (KPMG 2010)

Instead, irrational paths to status quo bias include loss aversion, and omission bias. We propose that when advisers are faced with making assessments about potential bidders they subconsciously emphasise the prospect of a loss (i.e. arriving at a decision that is unfamiliar, not the norm and results in a worse outcome than traditionally experienced). This results in assessments of potential bidders that would follow the function presented by Kahneman and Tversky's prospect theory (Kahneman and Tversky 1979). Australian and/or familiar Contractors, sub-contractors and personnel would sit in the right hand upper quadrant and unfamiliar Contractors in the lower left. These weightings are unintentional and occur prior to any formal scoring of participants and are therefore unlikely to be observed in the majority of projects.

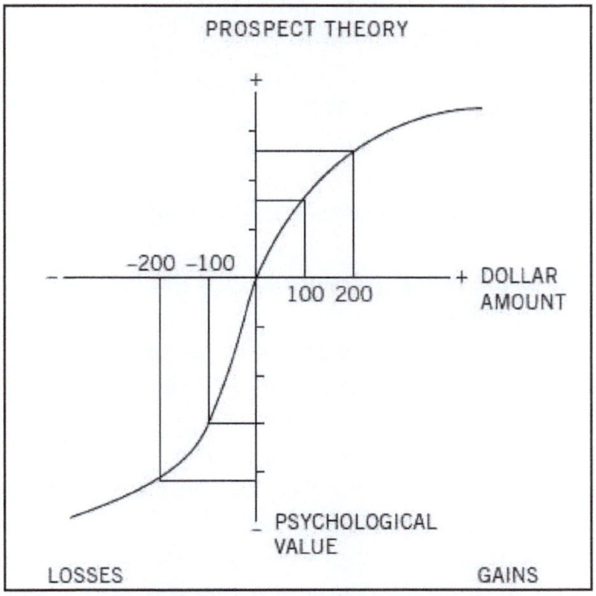

Figure 1 Prospect Theory

Legacy Way was somewhat unique due to the environmental conditions of the time in that the prospect of a failed process was acting against the status quo (i.e. the prospect of a loss through inaction meant changing behaviour was positively valued). Further, the project team was made up of individuals with less personal familiarity with local Contractors and through numerous market soundings became arguably equally familiar with international contractors thereby minimising (not eradicating) possible cognitive biases.

The Evaluation Process

A two stage evaluation process was developed:

Stage 1 Compliance: Requiring that the required information had been adequately submitted:
Stage 2 Comparative Assessment: In which those EOI's passing stage 1 were scored against the evaluation criteria.

Request for expression of interest were sent out in December 2009 and six responses, four of which were from overseers groups, were received in March 2010.

The Result

The RFP Process and Project Award

RFP's were issued to:

- Transcity (a consortium of Aciona, Ghella and BMD)
- Northern Direct (Bouygues and Laing O'Rourke)
- LBRJV (a consortium consisting of Leightons, Baulderstone and Razel)

After a detailed evaluation process the project was awarded to the TransCity joint venture for a bid price well below Brisbane City Council's budget and substantially below alternatives proposed by Contractors more familiar with the Australian environment. Needless to say, speculation was rife in industry regarding the likely performance of TransCity and Council. The final outcome for the project financially and technically is of interest to many in the industry.

Current status

Two years into the construction of Legacy Way, publicly available information on financial and technical progress is relatively limited. Anecdotal evidence points to a sound relationship between government and Contractor with positive comments regarding stakeholder management. Technically, recent tunnelling rates have resulted in setting a world record for 11m TBMs with 48m of advance in one week.

Conclusion

Using Legacy Way as a case study provides several insights:

- Inadvertent barriers to entry may help to explain why international contractors systematically fail to pass EOIs
- Demonstrably changing the EOI process to herald the intent of the client may assist in increasing international contractor appetite (eschewing status quo for the sake of building trust)
- Maintaining internal focus on the objectives of EOI assessments, the relative incompleteness of information and hence complexity of the decision may highlight potential cognitive biases
- Gaining equal exposure to potential international and local bidders and structuring a project team with broad international experience may balance the cognitive biases

- Environmental conditions may serve to exacerbate or counteract loss aversion of selecting unfamiliar contractors
- The RFP stage is where a more complete picture of each bidder's offer is presented but cognitive biases may still exist in complex decisions requiring subjective assessments

Being aware of and taking steps to mitigate potential cognitive bias may hold the key to nuptial bliss between International Contractors and the Australian Government. In our ongoing commitment to achieving best value for money on major infrastructure projects, can we afford to maintain the status quo?

References

Gilovich, T., et al. (2002) *Heuristics and biases: The psychology of intuitive judgment*, Cambridge University Press

Kahneman, D. and S. Frederick (2002) 'Representativeness revisited: Attribute substitution in intuitive judgment', *Heuristics and biases: The psychology of intuitive judgment,* 49-81

Kahneman, D. and A. Tversky (1972) 'Subjective probability: A judgment of representativeness', *Cognitive psychology,* **3** (3), 430-454

Kahneman, D. and A. Tversky (1977) Intuitive prediction: Biases and corrective procedures, DTIC Document

Kahneman, D. and A. Tversky (1979) 'Prospect theory: An analysis of decision under risk', *Econometrica: Journal of the Econometric Society*, 263-291

KPMG (2010) Advisory report for Infrastructure Australia

Tversky, A. and D. Kahneman (1973) 'Availability: A heuristic for judging frequency and probability', *Cognitive psychology* **5** (2), 207-232

Assessing Occupant Comfort in an Iconic Sustainable Education Building

17

Rick Best and Brian Purdey, (Bond University, Australia)

Abstract

The building that houses the Mirvac School of Sustainable Development at Bond University is the first educational building to achieve a six Green Star rating from the Green Building Council of Australia. It has won numerous awards since opening in August 2008 including being judged the RICS (Royal Institution of Chartered Surveyors) Sustainable Building of 2009. After more than two years in use a post-occupancy evaluation study was carried out to assess the performance of the building from the viewpoint of the users; both resident staff and transient students. Results for factors such as lighting, thermal comfort, noise and air quality were compared to benchmarks established by the Usable Buildings Trust. The evaluation also assessed the occupants' perceptions of the building's impact on their own productivity. Users generally find the building provides a comfortable work environment although a number of areas of performance were noted as posing some concerns. These included intrusive noise in some parts of the building and some issues with glare in daylit teaching spaces. Such concerns were found to be in accord with the results of previous studies and they highlight some recurrent problems in "green" buildings designed to maximise the use of natural ventilation and natural light. These design challenges and how occupant satisfaction is to be measured and benchmarked are also discussed in the context of this comparative building study.

Keywords: Green building, Post occupancy evaluation, Education facility, Usable Buildings

Introduction

It seems obvious that any manufactured product, whether a car, a television or a building, would be tested in use to see if it works well. Buildings, although very expensive products, tend to be one-off creations so unlike cars and televisions they are seldom tested, re-designed and refined in the way that mass produced items are. It could be argued, however, that as each building is individually designed that feedback from completed buildings, particularly those of similar function located in comparable environments (geographical, social, financial and so on) would routinely be sought in order to inform the design of subsequent buildings and thus to improve these designs. In reality this is not the case and good buildings often go unnoticed while poor building performance is often not publicised (Leaman *et al.*, 2010). The result is that the same mistakes are repeated and the process of improvement that occurs in manufacturing is seldom duplicated in the design and construction of buildings.

The advent of green buildings in recent years has seen some major changes in the way base buildings are designed with much greater emphasis on energy efficiency, material choice and water usage. These new designs have also provided opportunities to create different built environments – workspaces which emphasise quality as well as cost performance criteria (Vischer, 2008).

Understanding how satisfied occupants are with the workspaces they occupy has underpinned post occupancy evaluation of buildings since the 1980's (Marans and Spreckelmeyer, 1982; Oldham and Rotchford, 1983; Ornstein, 1999). But with the advent of green buildings there has been a renewed interest in the satisfaction of occupants of these buildings. There is some evidence that greener buildings enhance satisfaction, reduce absenteeism or sick leave and promote improved productivity as a result of the improved internal environment conditions (e.g. Romm and Browning, 1994; Paul and Taylor, 2008; Armitage *et al.*, 2011).

However, based on user assessments, not all green buildings perform well, while others are rated better than conventional ("non-green") buildings (Edwards, 2006). Leaman *et al.* (2007) in their Australian comparison of 22 "green design intent" buildings and 23 conventional buildings found that the most comfortable green buildings in terms of noise, temperature, air and lighting, were rated better than any of the conventional buildings. Conversely the buildings that were most poorly rated were also green buildings. Leaman and Bordass (2007) found similar results in their review of 177 UK buildings. These results point to the challenges in translating building design intent into occupant satisfaction given the likelihood that there will be other influencing factors. When green buildings work well they work very well but when, for whatever reason, predicted building performance is not realised in use then the occupants express dissatisfaction.

Murugan and Kato (2010) carried out a limited comparison of two conventional buildings and the one green building on the Bond University campus (the Mirvac building) and found that the green building scored better in regard to nearly all the factors included in their survey which included temperature, lighting, incidence of fatigue, headaches and colds and general satisfaction with the workplace.

The Mirvac School of Sustainable Development

The Mirvac School of Sustainable Development at Bond University is located on the Gold Coast in Queensland, Australia (Bond, 2011). The climate in the region is described as "humid sub-tropical" (DERM, 2010) with hot, humid summers and mild to cool winters. The Mirvac School comprises two buildings opened in August 2008 that are generally considered as a single unit having been designed and built together. The building received the RICS sustainability design award in 2009 and achieved a six Green Star (Design) rating (GBCA, 2011), the first education building in Australia to receive this top rating (Best, 2010).

The main building has three floors with teaching spaces at ground level and offices, meeting and utility rooms on the upper floors. The smaller building, known as The Living Lab, is a single storey building that is used for meetings, seminars and occasional teaching.

A sophisticated building management system (BMS) controls various building functions including switching to natural ventilation mode when external conditions (temperature and humidity) are such that 100% outside air can be utilised. Most spaces have manually operated windows and individually switched ceiling fans that can be operated independent of the air conditioning system which is available in enclosed offices via individual occupant controlled units and other spaces such as teaching studios and the Living Lab on an as-needed basis. Most of the artificial lighting in the building is controlled by movement and lighting level sensors.

Research Method

The concept of green building design and the techniques used are still relatively new, every green building is, to some extent, a prototype as different design concepts are tested in use. In order to increase our understanding of what works best in practice and to improve future designs it is important that newly constructed buildings are formally scrutinised. To this end a detailed post occupancy evaluation (POE) was carried out in mid-2011 to gather information from the users' perspective on the performance of the Mirvac School building.

The evaluation separated building users into two groups: resident users (staff and researchers who occupy office space in the building) and transient users (students) and applied the paper based *BUS Methodology 2 page Occupant Questionnaire 2011 Workplace with Response* instrument (UBT, 2011) developed by the Usable Buildings Trust (UBT) in the UK. The BUS methodology has been in use for over 20 years having been developed and refined over time and is supported by an extensive database with benchmarks, possibly useful for comparison purposes.

The survey questionnaire covers a variety of factors related to occupant satisfaction. It is not limited to indoor environment factors such as thermal comfort and noise intrusion but includes general questions regarding the occupants' perception of the building (e.g. image, safety, cleaning, availability of meeting rooms) and operational issues such as speed and effectiveness of responses to problems related to building use (adjustments to temperature and lighting for example). To achieve higher response rates UBT recommended using a paper-based survey questionnaire which was hand delivered to staff and students over a period of two weeks in June 2011. Staff who were away from the university during the survey period were given the option of completing an online version to maximise responses.

Respondents rated their level of satisfaction with the various building attributes, generally on a seven point Likert type scale. Some questions provided the opportunity for respondents to add written comments.

Key Findings

A total of 85 survey responses were collected; 50 were transient users and 35 resident users. This represents an overall response rate of around 20% of the total number of regular building users which is at an acceptable level. Survey results were processed by the Usable Buildings Trust as part of the licence agreement. The consolidated results for the whole sample and those for the two distinct user groups are summarised in Table 1.

Assessment of occupant comfort using the BUS methodology relates to functional comfort (Vischer, 2007) and the overlapping concepts of occupant satisfaction with the built environment and its possible impact on work performance. Occupant comfort is addressed in the survey through questions relating to four basic parameters (air, lighting, noise and temperature) and an overall comfort question.

Each comfort parameter is rated using three types of seven point scale with 4 being the neutral mid-point. A type scales have best results at the high point (i.e. a score of 7), B type scales are centred with the best result at the midpoint (i.e. a score of 4) while C scales have the best result at the lower end (i.e. a score of 1).

Occupants also rate their perception of self-assessed productivity impact on a nine point Likert type scale (from 1 = productivity decreased by 40% or more, to 9 = productivity increased by 40% or more), in 10% increments.

To simplify interpretation, the BUS methodology also colour categorises results for each parameter as "red" signifying a problem area, "amber" as a potential problem and "green" to indicate a satisfactory result. The colour code is determined statistically by comparing the Mirvac building's mean score for each variable (e.g. *Air in summer*) against the average score for the last 60 buildings surveyed by BUS (Leaman, 2011). In this case the average scores are for other Australian buildings.

Using this colour categorisation the Mirvac building study, in relation to occupant comfort, returned 15 satisfactory results (green) including *Overall comfort*, nine potential problem areas (amber) and three clear problem areas (red).

Indoor Air Quality

Air quality in the building is generally well regarded by occupants with mean scores for overall perceptions of air quality in both winter and summer greater than 5. Air in summer was the only variable to be rated "red", while Air in winter was the only other variable in this section not to be rated "green". Occupants perceived the air to be a little humid in summer and a little dry in winter.

Parameter	Residents	Transients	All users	Scale type and range	Rating
Air in summer					
dry/humid	4.88	4.28	4.54	B 1: dry - 7: humid	Red
fresh/stuffy	3.24	3.54	3.45	C 1: fresh - 7: stuffy	Green
odourless/smelly	2.81	2.85	2.85	C 1: odourless - 7: stuffy	Green
still/draughty	4.13	3.89	3.99	B 1: still - 7: draughty	Green
overall	4.94	5.13	5.06	A 1: unsatisfactory – 7: satisfactory	Green
Air in winter					
dry/humid	3.62	3.68	3.66	B 1: dry - 7: humid	Amber
fresh/stuffy	2.62	3.28	3.01	C 1: fresh - 7: stuffy	Green
odourless/smelly	2.68	2.72	2.72	C 1: odourless - 7: stuffy	Green
still/draughty	4.12	3.72	3.89	B 1: still - 7: draughty	Green
overall	4.82	5.27	5.11	A 1: unsatisfactory – 7: satisfactory	Green
Lighting					
artificial light	3.79	4.02	3.93	B 1: too little - 7: too much	Green
glare from lights	2.97	3.77	3.45	C 1: none - 7: too much	Amber
natural light	4.24	4.40	4.33	B 1: too little - 7: too much	Amber
glare from sun and sky	3.76	3.54	3.64	C 1: none - 7: too much	Amber
overall	5.09	5.29	5.23	A 1: unsatisfactory – 7: satisfactory	Amber
Noise					
noise from colleagues	4.35	3.96	4.12	B 1: too little - 7: too much	Green
other noise from inside	3.85	3.77	3.81	B 1: too little - 7: too much	Red
unwanted interruptions	4.18	3.66	3.85	C 1: not at all - 7: very frequently	Amber
noise from outside	4.85	4.33	4.53	B 1: too little - 7: too much	Red
overall	4.97	5.06	5.04	A 1: unsatisfactory – 7: satisfactory	Green
Temperature in summer					
hot/cold	3.31	3.80	3.60	B 1: too hot - 7: too cold	Amber
stable/varies	4.45	3.46	3.88	B 1: stable - 7: varies during day	Amber
overall	4.88	4.72	4.77	A 1: uncomfortable – 7: comfortable	Green
Temperature in winter					
hot/cold	4.68	4.33	4.47	B 1: too hot - 7: too cold	Amber
stable/varies	4.59	3.62	4.02	B 1: stable - 7: varies during day	Green
overall	4.97	5.63	5.38	A 1: uncomfortable – 7: comfortable	Green
Comfort overall	5.77	5.60	5.69	A 1: unsatisfactory – 7: satisfactory	Green

Table 1 Mean scores for key comfort parameters

Lighting

Lighting does not appear to satisfy occupants with the exception of natural lighting. Glare from both external and internal sources is a problem for some, with *Glare from lights* being one of the few variables to show a substantial difference in perceptions between residents and transients. *Lighting overall* was perceived by both groups to be acceptable but not completely satisfactory (mean scores: 5.09 for residents and 5.29 for transients).

Occupant comments identified two specific issues of concern. For residents, office lights that are controlled by sensors dim in the late afternoon, or switch off when occupants are at their desks and there is insufficient movement to keep the lights on. For transients a common complaint was the difficulty reading whiteboards, particularly when used as projector screens in the teaching studios.

Noise

Noise is an area of some concern with two red, one amber and three green ratings. The problem areas identified were *Noise from outside* and *Other noise from inside.* The transient users' scores were more favourable across all of the five noise parameters than those of the sedentary users.

Temperature

Overall temperature in winter and summer returned satisfactory scores (green). The variables *Temperature in summer: hot/cold* and *Temperature in winter: hot/cold* were both identified as potential problems, but no temperature parameters were rated in the red category.

Overall Comfort

The *Overall comfort* score was a healthy 5.69; this suggests that although there are some concerns in regard to particular aspects of comfort, both resident and transient users are generally comfortable in the building.

The BUS methodology includes the calculation of a *Comfort Index* and a *Forgiveness Index*. The *Comfort Index* is derived from seven key variables: temperature in summer and winter, air in summer and winter, noise, lighting and overall comfort. It is calculated by averaging the Z scores for the seven "overall" comfort variables (Baird *et al.*, 2011). Figure 1 shows the *Comfort Index* for the Mirvac building (the filled circle) in the context of the Australian comparison buildings; the result shows the building to be in the top 10% of buildings in the dataset.

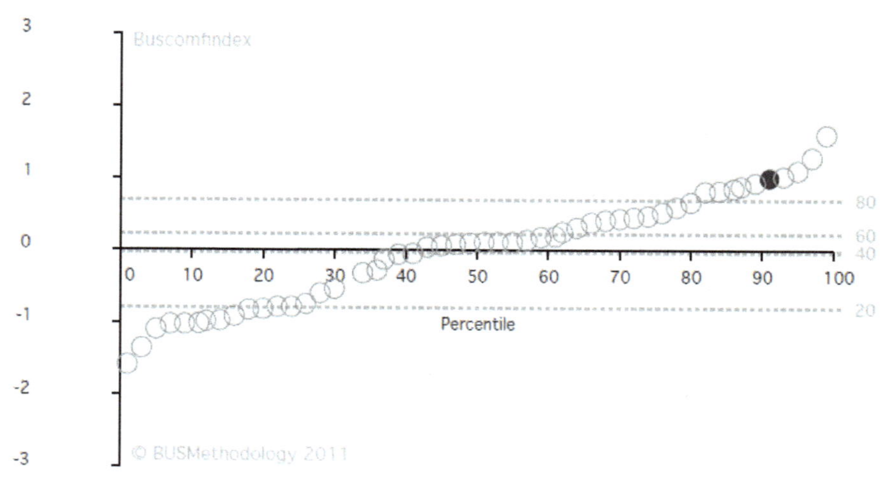

Figure 1 Comfort Index scores for the Mirvac building and the benchmark buildings

The *Forgiveness Index* is an indicator of how tolerant occupants are of the conditions in a building (Leaman and Bordass, 2007). It is calculated by dividing the mean score for the variable *Comfort overall* by the average of the mean scores for the variables *Temperature in summer overall, Temperature in winter overall, Air in summer overall, Air in winter overall, Lighting overall* and *Noise overall.* Higher values suggest that occupants are more inclined to "forgive" the building if comfort conditions are not exactly as they would prefer. Figure 2 shows the *Forgiveness Index* for the Mirvac building (the filled circle) in the context of the Australian comparison buildings; the result shows the building to be within the top 15% of buildings in the dataset and only marginally below those buildings that scored higher for this indicator. While encouraging this result is not particularly significant as the range of scores is quite narrow and it appears that most buildings in the UBT benchmark dataset score similarly for this index.

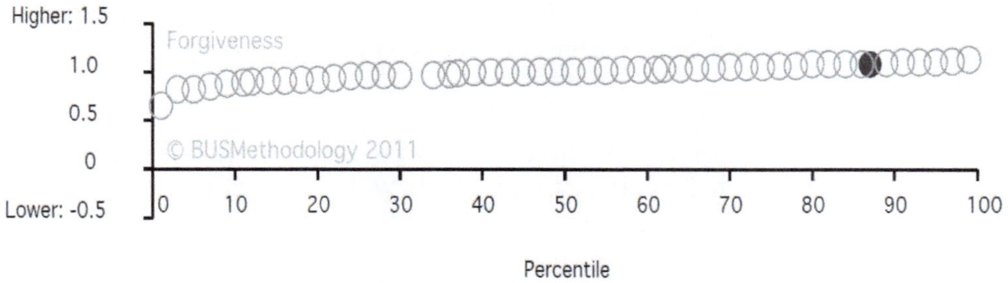

Figure 2 Forgiveness Index scores for the Mirvac building and the benchmark buildings

Benchmarks

The UBT benchmarks utilised in this study of the Mirvac building are based on the last 60 buildings studied using the BUS methodology. The benchmarks are calculated using a combination of the mean scores and scale midpoints so that a "green" score for any variable indicates performance well above the benchmark (Leaman, 2011). As the benchmarks are based on a rolling data set it is to be expected that as building design improves and more recent buildings are studied and added to the dataset that the benchmarks will rise. Leaman (2010) analysed samples of building data from the periods 1990-1999 and 2000-2008 and concluded that this does appear to be happening.

Discussion

One of the difficulties in studies such as this is finding suitable benchmarks to measure perceived performance against. Wilkinson *et al.* (2011) tried to establish benchmarks based on the expectations of respondents who completed their questionnaire. They then compared the same users' perceptions of performance in regard to the various parameters against their prior expectations. Two interesting results emerged. One was that in most cases the average expectation of the users was around 6 on a 7 point scale (where 7 = "very satisfactory"), suggesting that few users actually expect their workplace conditions to be better than just satisfactory. The second was that with almost every parameter the two target buildings, both described in the paper as "relatively new sustainable office buildings", scored well below the users' fairly modest expectations.

In terms of occupant comfort the Mirvac building was found to perform very well when compared to the UBT benchmarks.

While there are some specific concerns with noise and lighting, the overall scores for these variables were both above 5, indicating satisfactory performance The Mirvac building is located next to a moderately busy road and on a roundabout that includes a bus stop, these factors contributing to problems associated with noise intrusion from outside the building for both resident and transient users. In their written comments a number of respondents noted that traffic noise does intrude on the side of the building that faces the road and that having windows open on that side can interfere with activities such as meetings. The scores for transient users were more favourable across all of the five noise parameters compared with resident users.

At first glance this may appear to be counter-intuitive given that most of the resident users occupy individual offices and thus have the option of closing doors and windows to isolate themselves (acoustically at least) from the rest of the building and the outside world. In fact those who choose this option must rely on their air conditioning unit to supply fresh air and many prefer to have doors and windows open; the trade-off is that they may be exposed for more noise from both inside and outside the building.

Transient users spend less time in the building and so may be less concerned by noise simply because they are not subject to interference on a daily basis. Residents who trade off acoustic comfort for thermal comfort conditions are likely to be more dissatisfied with noise given the evidence that people do not adapt to noise disturbance over time (Witterseh *et al.*, 2004).

Baird (2010) used the UBT survey to gather data from eight buildings in warm-temperate climates in Australia and New Zealand, all of which were considered to be "cutting edge, sustainable buildings". Of the 32 mean scores for the four main noise variables (*Noise from colleagues, Noise from other people, Other noise from inside* and *Noise from outside*) 26 were below the mid-point (neutral) score. In Baird's study the mean scores for *Noise overall* included two buildings below the mid-point, two above the mid-point and four that were neutral, indicating generally satisfactory overall performance. In two instances while the mean scores for the four main variables were below the mid-point the *Noise overall* mean scores were neutral.

By comparison, all respondents to the Mirvac building survey rated satisfaction with *Noise overall* above the mid point at 5.04, with only *Noise from outside* and *Other noise from inside* being identified as problem areas.

Noise intrusion from outside is a function of the choices individual occupants make regarding a trade off between thermal and acoustic comfort, which is related to green building design.

Noise intrusion becomes a problem because of the ventilation strategy with openable windows and the intent to reduce use of air conditioning in all spaces. This is a particular problem in the Living Lab where it is largely impossible to open the large glass doors on the side facing the roundabout – the trade-off is always to keep the building closed and use the air conditioning or struggle to be heard over the traffic noise during meetings/classes. This problem is, of course, not restricted to the study building but is a major problem for most sustainable buildings, particularly those located in city areas where city noise can be intrusive and prohibits or severely limits the adoption of natural ventilation strategies.

Dissatisfaction with other noise inside the Mirvac building may be related to green design. Residents leaving office doors open to allow natural ventilation expose themselves to noise disturbance from inside the building. One respondent mentioned noise from one of the print (photocopier) rooms, another mentioned noise from work being done by students in the internal courtyard (this is a temporary measure) and one complaint related specifically to the noisy closing of office doors due to the closers on those doors. Several residents, however, commented that simply closing their office doors was sufficient to control noise from inside the building.

Materials choice aimed at reducing embodied energy in partitions may contribute to noise disturbance in classrooms for residents and transients alike. Two transient users commented that the operable walls between the teaching spaces on the ground floor did not block intrusive sound sufficiently while another suggested that the bells located in an adjacent tower (which chime at 9am, noon and 3pm each day) are "not needed".

These results suggest that while users do note specific noise concerns in green buildings noise problems are not a serious issue overall.

While the *Noise overall* result was satisfactory, scores for the four specific variables under the lighting heading were adjudged as amber, or problematic These apparent contradictions in noise and lighting are not uncommon. Previous studies (Leaman and Bordass, 2007; Baird *et al.*, 2011) have shown similar results with concerns identified in regard to particular variables but overall scores indicating a reasonably high level of satisfaction.

Leaman and Bordass (2007) suggest that this is due to occupants of green buildings being more tolerant of their buildings' shortcomings; the *Forgiveness Index* result shown above might support this view. Baird *et al.*(2011) link this sort of result to the nature of the client organisation; in their study they found that with more or less every building they studied, and they were all buildings that had won awards for sustainable design and/or achieved high ratings using tools such a LEED, BREEAM or Green Star, the client had "a particular commitment to sustainability" and "planned to be the long-term owner-occupier of the building". This is certainly true of the Mirvac building, as it houses the School of Sustainable Development and, being a university building, it is owned, occupied and operated by the university and is expected to be in use for many years.

The perceived problems with variables relating to noise and lighting comfort cannot be ignored even though overall results were good. Leaman and Bordass (2007) note that "[o]ccupants are complaining much more about noise than they used to" and suggest that noise is the major cause of problems in some buildings with natural ventilation or advanced natural ventilation.

Lighting is a recurrent problem in green buildings with designs often intended to maximise use of daylight and provide connection to the outside world for occupants. Armitage *et al.* (2011) found that "abundant natural light" and "having a view" were two factors that employees liked most in the workplace. Attempts to provide these attributes can often lead to problems with glare from outside and a number of respondents from both groups commented on the problem of intrusive daylight when projectors are in use in teaching spaces. This illustrates another basic problem: the lights in the teaching studios are controlled by sensors so when blinds are down to reduce the incoming daylight the sensors automatically increase artificial light levels. There is no manual switching in these spaces so slide shows and movie projections are difficult to see. Leaman and Bordass (2007) describe situations where blinds (often retrofitted after occupants complain about workspace receiving too much daylight) are often drawn and as a result artificial lights have to be on most of the day. In the Mirvac building where lights are sensor-controlled there is little or no scope for occupants to control the balance between artificial and natural light and this does lead to frustration for users particularly in the teaching spaces.

TIME	DEC	JAN	FEB	JUN	JUL	AUG
9am	67	70	72	69	65	61
3pm	69	70	70	58	55	58

Table 2 Monthly average humidity (%): Gold Coast (BOM, 2011)

The only other problem area identified was for *Air in summer: hot/humid*, however, all other results in this section were green (including the *Air in summer:* overall). Humidity is rated higher in summer, and this is to be expected given that in the December-February period average humidity in the area is generally higher than in the June-August period (see Table 2 for average Gold Coast humidity levels) and most of the building is not routinely sealed and air conditioned.

It is not unexpected that staff and students in a School of Sustainable Development would be more inclined to make allowances for factors such as humid conditions in summer and thus rate comfort overall to be good in spite of having to tolerate some periods with humidity above comfortable levels. The relatively high *Forgiveness Index* result reinforces this view.

TEMP	DEC	JAN	FEB	JUN	JUL	AUG
Max	27.7	28.6	28.5	21.3	21.1	21.8
Min	20.5	21.8	21.8	13.1	12.0	12.5

Table 3 Average monthly maximum and minimum temperatures: Gold Coast (BOM, 2011)

Summer temperatures can be moderately high (see Table 3 for monthly average temperatures) and while winters are mild (see Table 3) in the month before the survey was conducted (in June and thus during winter) the BMS did have a malfunction with the comfortable temperature somehow resetting to 15^{0}C; this resulted in the BMS putting the building into natural ventilation mode. This meant that windows in corridors and in the top of the atrium were automatically opened (and could not be manually closed) as well as the air conditioning units being disabled so that occupants had no heating option. For several days the occupants were not able to control the indoor temperature and as this coincided with a period of cool daytime temperatures it is very likely that occupants' perceptions of winter temperature would have been negatively influenced at the time of the survey. In spite of that event the temperature scores were good with no parameters in the red category.

Conclusion

The study confirms the continuing challenges associated with evaluating and benchmarking occupant satisfaction with comfort conditions in green buildings. The historical evidence is that not all green buildings perform well, while some have been rated better than conventional ("non-green") buildings the buildings most poorly rated by occupants are also green buildings. The results from the Mirvac building survey can only be compared with a rolling data set of buildings with variable green design intent – an "apples versus oranges" comparison with perhaps limited value. This points to the need for further research to develop better evaluation and benchmarking methodologies.

However, in terms of occupant comfort the Mirvac building performs well in its own right and also when compared to the Usable Buildings Trust benchmarks for Australian buildings and those studied in warm-temperate climates. The survey results do identify a handful of specific areas where occupants perceive that there are problems - variables relating to noise, lighting and thermal comfort which cannot be ignored. The study has produced occupant ratings that run generally true to form for green buildings. Scores for summary variables such as *Overall comfort* were slightly better than specific variables under the same heading, while noise intrusion, lighting quality and sensor control were the problematic areas. Given occupants are complaining much more about work disturbance than they used to, this suggest noise is the major cause of problems in most modern workplaces with naturally ventilated buildings perhaps making further contributions. The results for lighting also point to the complex task of balancing natural light, artificial light and glare in design in order to deliver optimum lighting conditions to a majority of building occupants.

Improved occupant satisfaction and performance may be less about green design intent but occupant knowledge of how to manage the situational constraints presented by this new type of built environment in order to deliver the outcomes required.

Acknowledgements

The study was funded by an internal grant for Early Career Researchers offered by the Institute of Sustainable Development and Architecture at Bond University. Judy Doherty assisted with data collection.

References

Armitage, L., Murugan, A. and Kato, H. (2011) 'Green offices in Australia: a user perception study', *Journal of Corporate Real Estate,* **13** (3) 169-180

Baird, G. (2010) *Sustainable Buildings in Practice: What the Users Think,* Routledge

Baird, G., Christie, L., Ferris, J., Goguel, C. and Oosterhoff, H. (2011) 'User perceptions and feedback from the 'best' sustainable buildings in the world', *Ecolibrium*, June, 30-37

Best, R. (2010) 'Specifying and sourcing materials for a best practice education building', In: Zachar, J., Claisse, P., Naik, T. and Ganjian, E. (eds) *Proceedings of Second International Conference on Sustainable Construction Materials and Technologies*, June 28-30, Ancona, Italy

BOM (2011) *Climate statistics for Australian locations.* Gold Coast Seaway. Bureau of Meteorology. http://www.bom.gov.au/climate/averages/tables/cw_040764.shtml

Bond (2011) *Facilities: Mirvac School of Sustainable Development*
www.bond.edu.au/faculties-colleges/institute-of-sustainable-development-architecture/about-the-institute/facilities/index.htm

DERM (2010) *Queensland Köppen Climate Zones.* Queensland Department of Environment and Resource Managemnt
www.derm.qld.gov.au/waterwise/plantselector/documents/koppen.pdf

Edwards, B. (2006) 'Benefits of Green Offices in the UK: Analysis from Examples Built in the 1990s', *Sustainable Development,* ***14***, 190-2004

GBCA (2011) *What is Green Star?* Green Building Council of Australia
http://www.gbca.org.au/green-star/green-star-overview/ Accessed 19 October 2011.

Leaman, A., Thomas, L. and Vandenberg, M. (2007) 'Green' buildings: What Australian users are saying', *EcoLibrium*, **6 (**10) (November) 22-30, Australian Institute of Refrigeration, Air Conditioning and Heating (AIRAH)

Leaman, A. and Bordass, B. (2007) 'Are users more tolerant of 'green' buildings?', *Building Research and Information*, **35** (6) 662-673

Leaman, A. (2010) 'Are buildings getting better?', *Arup Journal,* 1/2010, 4-8

Leaman, A., Stevenson, F. and Bordass, B. (2010) 'Building evaluation: practice and principles', *Building Research and Information*, **38** (5) 564-577

Leaman, A. (2011) Personal e-mail communication: 18 October.

Marans, R. W. and Spreckelmeyer, K. F. (1982) 'Measuring Overall Architectural Quality - A component of building evaluation', *Environment and Behaviour,* **14** (6), 652-670

Murugan, A. and Kato, H. (2010) *Is it worthwhile going green?* Unpublished internal report, Bond University

Oldham, G. R. and Rotchford, N. L. (1983). Relations between office characteristics and employee relations; A study of the physical environment. *Administrative Science Quarterly,* **28**, 542-556.

Ornstein, S. W. (1999). 'A postoccupancy evaluation of workplaces in Sao Paulo, Brazil', *Environment and Behavior,* ***31*** (4), 435-462

Paul, W. and Taylor, P. (2008) 'A comparison of occupant comfort and satisfaction between a green building and a conventional building', *Building and Environment,* **43** (11) 1858-1870

Romm, J. and Browning, W. (1994) *Greening the building and the bottom line,* Rocky Mountain Institute, Snowmass, Colorado

UBT (2011) Usable Buildings Trust – Portfolio http://www.usablebuildings.co.uk/fp/index.html

Vischer, J. C. (2007) 'The concept of environmental comfort in workplace performance', *Ambiente Construido,* **7** (1), 21-34

Vischer, J. C. (2008) 'Towards an Environmental Psychology of Workspace: How People are Affected by Environments for Work', *Architectural Science Review,* **51** (2), 97-108

Weatherzone (2011) *Gold Coast Climate* http://www.weatherzone.com.au/climate/station.jsp?lt=site&lc=40764 Viewed: 17 October 2011

Wilkinson, S., Reed, R. and Jailani, J. (2011) 'User Satisfaction in Sustainable Office Buildings: A Preliminary Study', In: *Proceedings: 27th PRESS Conference,* Gold Coast, 16-19 January

Witterseh, T., Wyon, D. and Clausen, G. (2004) 'The effects of moderate heat distress and open-plan office noise on SBS symptoms and on the performance of office work', *Indoor Air*, **14**, 30-40

Zhang, Y. and Altan, H. (2011) 'A comparison of the occupant comfort in a conventional high rise office block and a contemporary environmentally-concerned building', *Building and Environment*, **46** (2), 535-545

Valuing the Contribution of Knowledge-oriented Workers to Projects: A Merit Based Approach in the Construction Industry

18

Mehrdad Arashpour, (RMIT University, Australia)

Majeed Shabanikia, (Isfahan University of Technology, Iran)

Mohammadreza Arashpour, (Azad University, Iran)

Abstract

Evidence points to the fact that frequent resignation of project engineers from construction companies is primarily the result of dissatisfaction with the factors that shape the salary scale. This research aims to identify the major influencing factors in merit based salary calculation systems for knowledge-oriented engineers so as to more accurately reflect their contribution to construction projects. Results from a questionnaire sent to managers, engineers and HR professionals throughout the Iranian construction industry revealed that while there was overall agreement on principles for a merit-based approach, engineers in particular stressed 'professional skills', 'experience' and 'creativity'. Management-oriented parties should take into account engineer perspectives in order to more accurately reflect the self-assessed contribution of these workers to construction projects. This research provides a basis for understanding the key factors in the merit based salary scale formulation through the construction industry.

Keywords: Construction industry, knowledge-oriented workers, salary scale formulation

Introduction

As a rule of thumb the objective of any payment system is to attract, hold, and motivate employees. Consequently, such a system should be designed to not only help organisations to meet their goals, but also to improve employee performance and company productivity. Because construction projects are very employee intensive and need to attract a knowledge-oriented and professional workforce (Shabanikia 2009), the design of the payment system is crucial to employee retention.

The aim of this study is to explore how managers, engineers and HR professionals in the Iranian construction industry perceive the value of the contribution of engineers to construction projects. It will then discuss a merit based wage system that includes project process and outcome in the salary scale formulation for "knowledge-oriented" engineers.

This research has a direct application to construction industry supply chain companies that are seeking to reform their salary scales so as to retain their valuable assets by rewarding the contribution of a knowledge-oriented workforce.

Literature Review

Engineering companies work in a knowledge-oriented atmosphere. They are also production-oriented provided that drawings and document are considered as products. From this perspective, the construction industry needs engineers as knowledge-oriented workers. Knowledge workers are active in analyzing information, solving problems by means of their expertise, proposing opinions, and creating new items and services. In other words, engineers are idea and information processors. They are individuals who deploy their analytical and mental strengths more than hands in adding value to workplaces and are likely to be autonomous, active and proud (Abdul-Aziz & Law 2012; Arif et al. 2012; Horibe 1999).

Compensation systems can produce competitive advantages for construction firms as salary, working conditions, opportunities for promotion and lifelong learning are reported by both male and female employees as the most important factors (Chileshe & Haupt 2010). Valuing the contribution of knowledge-oriented HR to projects can increase productivity and skills of the personnel (Zakeri et al. 1996). One of the main methods of achieving job satisfaction in companies is to improve the salary structure and the social security system especially in developing countries (Fan & Yong 2011; Yisa, Holt & Zakeri 2000).

Compensation structure acts as a tool for achieving companies' strategic goals. Human resource performance appraisal is the core of project human resources management and the objective basis for implementing the scientific and rational salary management in project organisation (Cheng & Fan 2010). In fact, compensation strategy is the major component of the companies' strategy, creating competitive advantages and success as well (Ferguson and Parsons 1992). Valuing the contribution of knowledge-oriented HR has two functions of attracting and retaining skilled workforce and also directing their efforts towards organisational goal (Chen & Edgington 2005). Therefore, the most important outcome of a merit-based compensation system is to create a fair dealing atmosphere within organisations (Konopaske & Werner 2002; Milkovich et al. 1999).

In terms of employees' demonstrable qualities, knowledge-oriented workers usually have four characteristics: performing cross-disciplinary tasks, acting with flexibility, attempting towards organisational objectives and collaborating with team members (Arashpour & Arashpour 2012; Flannery, Hofrichter & Platten 2003). Motivational pay systems aim to arouse the initiatives of staff by means of perfecting the structure of pay management, expanding the proportion of unfixed pay and defining the bonuses based on measures of performance (Anantatmula & Shrivastav 2012; Li 2011).

In the field of human resource studies, little if any research has analysed interdependent factors associated with valuing the contribution of knowledge-oriented workers in the construction industry. The present research tries to bridge this gap.

Research Method

The method for the present study comprises a comprehensive literature review and a survey of the Iranian construction industry. Finally factors influencing the salary scale formulation in the construction industry are ranked based on respondents' perspectives and significant differences in perspectives are discussed comprehensively.

The questionnaire was designed to elicit perspectives that respondents had with regard to the significant factors for salary scale formulation. As such, the main feature of the survey asked respondents to classify influencing factors on salary calculation by their level of importance in the construction industry. A bipolar scaling system, which helps the respondents to express their ideas clearly, was employed (Sekaran 2006). Respondents identified each of the factors as "very insignificant = 1", "insignificant = 3", "neither significant nor insignificant = 5", "significant = 7" or "very significant = 9".

A total of 185 survey questionnaires were sent out in April-August 2011 to selected construction firms. The respondents were an array of experts from the construction supply chain. Managers, engineers and HR professional were targeted as responding groups due to the aim of the research. The majority of participants came from firms employing more than 200 people (46.5 per cent), whilst 33.7 per cent were employed by firms with 50- 200 personnel. Only 19.8 per cent of respondents worked for firms employing less than 50 people. Care was taken to select experienced individuals. No respondent had less than 10 years of experience in the construction industry.

Of 185 sent questionnaires, 150 responses were deemed valid. The survey listed 20 factors associated with job and personal characteristics, which were sourced from a number of references presented in Table 1.

Personal and Job related factors	Who and when
1. Professional skills	(Clemens 2009)
2. Leadership skills	(Morrison 1994)
3. Compliance with organisations norms and values	(Becker, Huselid & Ulrich 2001; Weaver & Trevino 1999)
4. Academic records	(Chen & Edgington 2005; Robinson & Sexton 1994)
5. Creativity	(Tian & Belk 2005)
6. Duration of employment in the current company: long term commitment	(Cowherd & Levine 1992; O'Malley 2000)
7. Education relevancy	(Arashpour & Farzanehfar 2011; Coelho 2010)
8. Experience quality: Local experience	(Monks 1996)
9. Graduating university	(Dowling & Fisher 1997)
10. Level of responsibility	(McAuliffe et al. 2009)
11. Number of employees under management	(Arashpour & Arashpour 2010; Delaney & Huselid 1996)
12. Overtime amount	(Wilson 2008)
13. Performance speed and efficiency	(Flannery, Hofrichter & Platten 2003)
14. Relevant past experience	(Medoff & Abraham 1981)
15. Tasks complexity	(Allan 2000; Verburg, Den Hartog & Koopman 2007)
16. Working condition: Physical and psychological	(Milost 2007)
17. Years of experience	(Arashpour, Wakefield & Blismas 2012)
18. Working hours	(Johnson & Lipscomb 2006)
19. Workload	(Bradley et al. 2010; Hollands & Wickens 1999)
20. Performance quality: outcomes	(Paauwe & Boselie 2005)

Table 1 Summary of personal and job related factors influencing salary calculations for knowledge-oriented workers

The final question asked respondents to evaluate their own organisational attitudes towards knowledge-oriented workers. They were asked to select 'satisfactory' or 'unsatisfactory' in response. They had opportunity to make suggestions for improving salary scales through the construction industry.

Data Analysis

The analysis included the frequencies of the various respondents' answers and the relative importance or rank of answers compared to others based on analysis of variance (ANOVA)

test. The results demonstrated significant differences among the means of the various responding groups. The F statistic demonstrated if two given variances differed from one another or were from the same population (Sekaran 2006).
A post-hoc test complemented the ANOVA to identify the groups with opposing views. Tukey's HSD (honestly significant difference) test was selected as the post-hoc to compare all possible pairs of means based on a studentized range distribution (q) which is very similar to t-test distribution.

First and foremost, to evaluate the reliability (providing same results in different situations by measurement instrument), and validity (exact measurement of the variables by the instrument), the following results were extracted by SPSS. Software evaluation illustrates the reliability of survey outcomes. Since Cronbach's Alpha is 0.771 and it is greater than the benchmark of 0.7, the reliability of questionnaire results analysis is approved (See table 2). This table also illustrates the validity of results analysis for the two groups of variables. For the set of variables the *Kaiser-Mayer-Oklin Measures of Sampling Adequacy* is 0.575. As it is greater than the benchmark of 0.5, the validity of results is approved.

Reliability of results	**Validity of results**
Cronbach's alpha= 0.771	KMO measure of sampling adequacy= 0.575

Table 2 Data reliability and validity test

Results and Discussions

A key measure for the validity of this research was the distribution of survey responses across knowledge-oriented workers in the construction industry. In the current research, engineers (41.3%), and managers (40.8%) made up the majority of respondents and the remainders were HR professionals (17.9%). All respondent were involved in the construction industry projects. Of the respondents surveyed, 40.1% had 10-20 years of experience in the construction industry and 54.7% had more than 20 years. The wealth of experience is evenly dispersed throughout the industry and thus the high percentage of experienced respondents also validates the results of the survey. Furthermore, none of the HR professionals who responded to the survey had less than 10 years of experience.

Influencing factors on salary calculations in construction industry

Table 3 presents factors influencing the salary scale of knowledge-oriented personnel identified by the respondents. It compares and contrasts opinions of managers, engineers and human resource professionals. The most important factor identified was 'performance quality' with a mean of 8.34 and standard deviation (σ) = 0.98. It was followed by 'compliance with organisation norms and values' with a mean of 7.81 and σ = 0.90. Other important factors averaging above significance (scale 7) were 'professional skills', 'relevant past experience', 'creativity', 'educational relevancy', 'leadership skills', and 'level of responsibility'. The factors considered to be least important were 'academic records', 'graduating university' and 'duration of employment in the current company'.

Some significant factors such as 'compliance with organisations norms and values' and 'professional skills' identify that personal characteristics are important factors in the salary calculation of knowledge-oriented workers. This finding is consistent with the literature where Horibe (1999) identified similar results, indicating this as a characteristic for unique knowledge workers. A great evidence to validate the survey design and factor selection is that none of the 20 investigated factors were identified 'insignificant' (i.e. factors with average means below 3) by the survey respondents.

Key factors to calculate salary			Total perspective		Managers' perspective		Engineers' perspective		HR professionals' perspective	
			Mean	SD	Mean	SD	Mean	SD	Mean	SD
1	Performance Quality (outcomes)	Most important factors	8.34	0.98	8.40	1.20	8.27	0.84	8.35	1.20
2	Compliance with organisation norms and values		8.29	0.90	8.01	0.64	7.98	0.79	8.87	0.80
3	Professional skills		7.69	0.88	7.47	0.70	8.53	0.70	7.07	1.10
4	Relevant past experience		7.63	0.88	7.00	0.89	8.43	1.00	7.47	1.07
5	Creativity		7.43	0.85	7.43	0.68	8.41	0.89	6.45	0.85
6	Educational relevancy		7.23	0.83	7.02	0.83	7.00	1.01	7.67	1.40
7	Leadership skills		7.19	1.01	7.76	0.71	7.08	0.81	6.73	1.25
8	Level of responsibility		7.01	1.00	7.02	1.23	7.74	0.70	6.27	0.65
9	Years of experience		6.64	0.93	6.45	0.65	7.09	0.85	6.38	1.00
10	Work load		6.46	1.05	6.02	1.06	7.43	0.99	5.93	1.10
11	Working hours		6.45	0.90	6.79	0.63	5.35	0.95	7.21	1.00
12	Performance speed		6.24	1.03	6.31	0.99	6.70	0.93	5.71	1.05
13	Experience quality (Local experience)		6.21	0.97	6.29	0.76	5.93	0.85	6.41	1.25
14	No. of employees under management		6.05	0.89	6.18	0.61	5.81	0.79	6.16	1.20
15	Complexity of tasks		5.88	1.16	6.01	0.84	7.00	0.80	4.63	0.70
16	Working condition	Least important factors	5.54	0.80	5.01	0.89	6.61	0.91	5.00	0.52
17	Overtime amount		5.43	1.06	4.63	0.98	6.76	0.70	4.90	1.00
18	Duration of employment in the current company		5.16	1.05	6.09	1.00	4.10	1.00	5.29	0.40
19	Graduating university		5.06	1.13	4.94	1.00	5.26	0.70	4.98	1.00
20	Academic records		5.03	1.11	4.83	1.07	5.23	0.90	5.03	1.40

Table 3 Influencing factors on salary calculation: Diverse perspectives across construction industry

Interestingly, engineers considered 'professional skills' as the most important factor. This is perhaps due to the high emphasis that knowledge-oriented workers place on the professional promotion. In the case of managers, 'performance quality' ranked as the most important factor in the salary calculation within the construction industry. This finding implies that there are continuous concerns at the top managerial levels about actual quality performance of highly skilled recruited employees. This result is alarming considering that compensation practices should contribute to competitive advantage by promoting more productive workers. High-ranking factors reflected by HR professionals were 'Compliance with organisation norms and values', 'Performance Quality (Best effort)' and 'Relevant professional experience'. As the frontline agent in recruiting, HR professionals consider the organisation norms and values as very important and employees' compliance is regarded as highly significant factor in the salary scale formation. Table 4 summarizes significant factors (i.e. factors averaging above scale 7) identified in valuing HR contribution by the knowledge-oriented workers in the construction industry.

Respondent	Factors (In Order of Importance)	Mean
Manager Perspective (8 factors)	Performance Quality	8.40
	Compliance with organisation norms and values	8.01
	Leadership skills	7.76
	Professional skills	7.47
	Creativity	7.43
	Level of responsibility	7.02
	Educational relevancy	7.02
	Relevant past experience	7.00
Engineer Perspective (11 factors)	Professional skills	8.53
	Relevant past experience	8.43
	Creativity	8.41
	Performance Quality	8.27
	Compliance with organisation norms and values	7.98
	Level of responsibility	7.74
	Workload	7.43
	Years of experience	7.09
	Leadership skills	7.08
	Educational relevancy	7.00
	Complexity of tasks	7.00
HR professional Perspective (6 factors)	Compliance with organisation norms and values	8.87
	Performance Quality	8.35
	Educational relevancy	7.67
	Relevant past experience	7.47
	Professional skills	7.07
	Working hours	7.00

Table 4 Most significant factors in valuing knowledge-oriented workers' contribution to construction projects

Surprisingly, whereas engineers identified 11 significant factors that should contribute to their salary calculation, managers and HR professionals only suggested eight and six factors, respectively. It shows that engineers expect the salary scales to include more contributing factors.

To find the controversial factors that attracted different perspectives, analysis of variance (ANOVA) was used for further statistical analysis of the survey results. As there were 20 variables and three groups of respondents, namely managers, engineers and HR professionals, ANOVA is the best mean to find out significant differences among survey participants' opinions. "Whereas the (independent samples) t-test indicates whether or not there is a significant mean difference in a dependent variable between two groups, an analysis of variance (ANOVA) helps to examine the significant mean differences among more than two groups" (Sekaran 2006, p. 347). Table 5 summarizes the ANOVA results, highlighting six factors with significance difference (less than 0.05).

No.	Factor	Sum of Squares	Mean Square	F statistic	Significance of F
1	Performance Quality	.430	.215	526.750	0.043
2	Compliance with organisation norms and values	25.543	12.772	31290.583	0.065
3	Professional skills	56.920	28.460	69727.000	0.034
4	Relevant past experience	53.123	26.562	65076.083	0.057
5	Creativity	96.040	48.020	117649.000	0.048
6	Educational relevancy	14.530	7.265	17799.250	0.066
7	Leadership skills	27.430	13.715	33601.750	0.045
8	Level of responsibility	54.030	27.015	66186.750	0.059
9	Years of experience	15.310	7.655	18754.750	0.061
10	Workload	70.770	35.385	86693.250	0.000
11	Working hour	90.750	45.375	111168.750	0.066
12	Performance speed	24.870	12.435	30465.750	0.052
13	Experience quality (Local experience)	6.240	3.120	7644.000	0.062
14	No. of employees under management	4.330	2.165	5304.250	0.067
15	Complexity of tasks	96.330	48.165	118004.250	0.043
16	Working condition	85.870	42.935	105190.750	0.062
17	Overtime amount	134.490	67.245	164750.250	0.056
18	Duration of employment in the current company	1.270	.635	1555.750	0.054
19	Graduating university	3.040	1.520	3724.000	0.071
20	Academic records	4.003	2.002	6538.778	0.066

Table 5 ANOVA analysis results: significantly different factors by participating groups (Sig<0.05 highlighted)

As it is evident from the results, there are six significantly different factors by responding groups: compliance with organisation norms and values, professional skills, creativity, leadership skills and complexity of tasks, applying a confidence level of 95%.

In the next step of the data analysis, Tukey's HSD (honestly significant difference) test was used to compare all possible pairs of means based on a studentized range distribution. Table 6 illustrates the results of post-hoc Tukey test.

Dependent Variable	(I) Person	(J) Person	Mean Difference (I-J)	Std. Error	Significance
Performance quality	engineer	manager	-.13000	.00407	.045
		HR	-.08000	.00516	.011
Professional skills	engineer	manager	1.06000	.00362	.037
		HR	1.46000	.00601	.008
Creativity	engineer	manager	.98000	.00404	.030
		HR	1.96000	.00516	.016
Leadership skills	engineer	manager	-1.06000	.00209	. 063
		HR	.40000	.00103	.038
Workload	engineer	manager	1.41000	.00415	.039
		HR	1.50000	.00328	.022
Complexity of tasks	engineer	manager	.99000	.00439	.047
		HR	2.37	.00401	.003

Table 6 Tukey's HSD (honestly significant difference) test: significantly different perspectives among groups

The Post-hoc test reveals that engineers have significantly different perspectives about performance quality than have managers and HR professionals. Engineers rank this factor below other factors such as professional skills and creativity. This could be due to the fact that engineers consider the process of doing tasks as more important than outcomes; however, managers and HR professionals focus more on outcomes to measure performance quality as evidenced by their high ranking of this factor in salary scale formulations. In addition, the similar perspectives of management-oriented parties unveils the deep influence of managers on HR practices in engineering companies (Paauwe & Boselie 2005).

Tukey's HSD results for the Post-hoc test reveal significantly different perspectives about professional skills and complexity of tasks. Both managers and HR professionals underestimated these factors compared to engineers, which is evident from the negative mean difference illustrated in table six. It is probable that engineers regard professional skills as a kind of toolkit to get the work done. However, the other two management-oriented groups paid less attention to these factors focusing more heavily on the best results from their knowledge-oriented workers.

Creativity is also a source of controversy. While engineers consider creativity as a very significant factor, HR professionals did not believe so. This might be because of the fact that HR professionals tend to quantify outcomes of employees' jobs; however, creativity is very hard to quantify. It is not surprising that managers ranked creativity below compliance to organisational norms and values. In fact, some managers consider this quality as a potential threat to established rules and work routines in their organisation (Arashpour & Arashpour 2011).

Interestingly, HR professionals did not rate leadership skills as a significant factor but managers put more stress on this factor. It is likely that managers prefer a more autonomous workforce without the need for continuous instructions. Despite the managers' desire to maintain hierarchy in their organisations, this finding shows their reluctance to be involved in administrative tasks.

Results illustrate that 'workload' is another significant factor identified by engineers. It is likely that engineers, who work at the frontline of construction projects, are more concerned about balancing professional and family lives. However, management-oriented respondents did not consider workload as important as the performance quality.

In response to the question about the degree of satisfaction about organisational attitudes towards knowledge-oriented workers, the majority of respondents (62.5%) felt that their organisation's attitude was unsatisfactory, and this perception was stronger downstream in the industry (i.e. engineers). Most of the 27.5% of respondents considering their organisation's performance as satisfactory were managers. Less dispersed in the results was the typical consensus that respondents' organisations could implement the merit-based approach to salary calculation, paying special attention to both job outcomes and the processes involved to do jobs (83% of survey participants).

Conclusion

Prior work has documented the measures assessing performance of employees in the construction industry; Becker, Huselid and Ulrich (2001) for example, report how to link people, strategy and performance in firms. However, few researchers have analysed interdependent factors associated with valuing the contribution of knowledge-oriented workers in the construction industry This paper has presented the results of a survey on factors influencing merit based salary scale formulation in order to optimize the contribution of knowledge-oriented engineers in construction projects. It was assumed that process-oriented (engineers) and management-oriented (managers and HR professionals) parties would be able to provide valuable perspectives on the issue. The engineers perceive their work as complex and the salary scale should reflect that complexity by the inclusion of more process-oriented factors. This view conflicts with that of the two management-oriented parties, which focus more on outcome-related factors.

While all parties agree in principle to a merit-based approach to salary calculation, each identified different key criteria for measuring merit. The survey illustrated that 'performance quality' was the most important factor identified by all parties. Other important shared perspectives included 'Compliance with organisation norms and values' and 'Professional skills'. However, variations were identified in the ranking of these factors. From the HR professionals' perspective, 'Compliance with organisation norms and values' was identified as the most important factor followed by 'Performance Quality', factors which emphasize job outcomes over processes. This is in direct contrast with factors identified by engineers as most significant, such as professional skills, experience and creativity.

The findings suggest that if companies are to value the work of these knowledge-oriented engineers, not only the job outcomes but also the processes involved in doing the job should be considered in the formulation of a merit based salary scale. After all, the construction industry is only as strong as its knowledge-oriented workers.

Replication of this study in construction industries in other regions of the world is recommended. Similar results might be achieved because of the significant difference in focus between management-oriented and process-oriented individuals, which is a characteristic of the different work aims and intentions.

References

Abdul-Aziz, A.R. and Law, Y.H. (2012) 'Civil Engineering Consultants from Developing Countries: An Exploration of Malaysian Firms', *International Journal of Construction Management*, **12** (2) 1- 21

Allan, C. (2000) 'The hidden organisational costs of using non-standard employment', *Personnel Review*, **29** (2) 188-206

Anantatmula, V.S. and Shrivastav, B. (2012) 'Evolution of project teams for Generation Y workforce', *International Journal of Managing Projects in Business*, **5** (1) 9-26

Arashpour, M. and Arashpour, M. (2010) *Gaining the Best Value from HR in Construction Companies, Proceedings of the 6th European Conference on Management Leadership and Governance*

---- 2011, 'Important factors influencing personnel performance of construction companies', paper presented to *International Conference on Economics, Business and Management*, Manila, Philippines

Arashpour, M. and Arashpour, M. (2012) 'A collaborative perspective in green construction risk management', paper presented to *37th Annual Conference Of The Australasian Universities Building Educators Association (AUBEA)*

Arashpour, M. and Farzanehfar, P. (2011) *Project management and control*, 1 edn, vol. 1, ISBN: 978-600-5442-74-8, Tehran

Arashpour, M, Wakefield, R & Blismas, N 2012, 'Role of simulation in construction processes- Harmony in capturing resources- DOI: 10.3850/978-981-08-7920-4', paper presented to Research, Development, and Practice in Structural Engineering and Construction, Curtin University, Perth.

Arif, M., Khalfan, M.M.A., Barnard, J.H. and Heller, N.A. (2012) 'Assessing knowledge retention in construction organisations: Cases from the UAE', *Australasian Journal of Construction Economics and Building*, **12** (2) 55-71

Becker, B.E., Huselid, M.A. and Ulrich, D. (2001) *The HR scorecard: Linking people, strategy, and performance*, Harvard Business Press

Bradley, L., Brown, K., Lingard, H., Townsend, K. and Bailey, C. (2010) 'Talking the talk and walking the walk: How managers can influence the quality of work-life balance in a construction project', *International Journal of Managing Projects in Business*, **3** (4) 589-603

Chen, A.N.K. and Edgington, T.M. (2005) 'Assessing value in organizational knowledge creation: Considerations for knowledge workers', *Management Information Systems*, **29** (2) 279-309

Cheng, Y-J amd Fan, X-M (2010) 'The study of DEA method in performance evaluation of project human resource management', paper presented to 17th *International Conference on Management Science and Engineering, ICMSE 2010,* November 24, 2010 - November 26, 2010, Melbourne, VIC, Australia

Chileshe, N. and Haupt, T.C. (2010) 'An empirical analysis of factors impacting career decisions in South African construction industry: Male and female high school students' perspectives', *Journal of Engineering, Design and Technology*, **8** (2) 221-39.

Clemens, M. (2009) 'Skill flow: A fundamental reconsideration of skilled-worker mobility and development', Working paper 180. Centre for Global Development.

Coelho, JVP 2010, 'Competence certification as a driver for professional development: A IT-related case-study', paper presented to Education Engineering (EDUCON), Lisbon, Portugal

Cowherd, D.M. and Levine, D.I. (1992) 'Product quality and pay equity between lower-level employees and top management: An investigation of distributive justice theory', *Administrative Science Quarterly*, **37** (2) 302-20.

Delaney, J.T. and Huselid, M.A. (1996) 'The impact of human resource management practices on perceptions of organizational performance', *Academy of management Journal*, **39** (4) 949-69.

Dowling, P.J. and Fisher, C. (1997) 'The Australian HR professional: A 1995 profile', *Asia Pacific Journal of Human Resources*, **35** (1) 1-20

Fan, G. and Yong, S. (2011) 'A demonstration research on job satisfaction of the employee from the private enterprise - Based on LH Group Co., LTD. Tang Shan', paper presented to *2011 International Conference on Business Management and Electronic Information, BMEI 2011*, May 13, 2011 - May 15, 2011, Guangzhou, China

Flannery, T.P., Hofrichter, D.A. and Platten, P.E. (2003) *People, Performance, & Pay: Dynamic Compensation for Changing Organizations*, Free Press.

Hollands, J.G. and Wickens, C.D. (1999) *Engineering psychology and human performance*, Prentice Hall New Jersey

Horibe, F. (1999) *Managing knowledge workers: New skills and attitudes to unlock the intellectual capital in your organization*, Wiley

Johnson, J.V. and Lipscomb, J. (2006) 'Long working hours, occupational health and the changing nature of work organization', *American Journal of Industrial Medicine*, **49** (11) 921-929

Konopaske, R. and Werner, S. (2002) 'Equity in non-North American contexts: Adapting equity theory to the new global business environment', *Human Resource Management Review*, **12** (3) 405-41.

Li, Y. (2011) 'Exploring and analysis on strategies of SMEs' human resource management in the period of post financial crisis', paper presented to *2nd International Conference on E-Business and E-Government, ICEE 2011*, May 6, 2011 - May 8, 2011, Shanghai, China

McAuliffe, E., Bowie, C., Manafa, O., Maseko, F., MacLachlan, M., Hevey, D., Normand, C. and Chirwa, M. (2009) 'Measuring and managing the work environment of the mid-level provider-the neglected human resource', *Human Resources for Health*, **7** (1) p. 13

Medoff, J.L. and Abraham, K.G. (1981) *Experience, performance, and earnings*, National Bureau of Economic Research Cambridge, Mass., USA

Milkovich, G.T., Newman, J.M., Milkovich, C. and Mirror, T. (1999) *Compensation*, Irwin/McGraw-Hill

Milost, F. (2007) 'A dynamic monetary model for evaluating employees', *Journal of Intellectual Capital*, **8** (1), 124-38

Monks, K. (1996) 'Global or local HRM in the multinational company: the Irish experience', *International Journal of Human Resource Management*, **7** (3), 721-735

Morrison, E.K. (1994) *Leadership skills: Developing volunteers for organizational success*, Da Capo Press

O'Malley, M. (2000) *Creating commitment: How to attract and retain talented employees by building relationships that last*, John Wiley & Sons

Paauwe, J. and Boselie, P. (2005) 'HRM and performance: what nexta?', *Human Resource Management Journal*, **15** (4), 68-83

Robinson, P.B. and Sexton, E.A. (1994) 'The effect of education and experience on self-employment success', *Journal of Business Venturing*, **9** (2), 141-156

Sekaran, U. (2006) *Research methods for business: A skill building approach*, Wiley-India

Shabanikia, M. (2009) 'An investigation and recommendation for a system of salary and fringe benefit in consultant companies' Unpublished MS dissertation, Grenoble University.

Tian, K. and Belk, R.W. (2005) 'Extended self and possessions in the workplace', *Journal of Consumer Research*, **32** (2) 297-310

Verburg, R.M., Den Hartog, D.N. and Koopman, P.L. (2007) 'Configurations of human resource management practices: a model and test of internal fit', *The International Journal of Human Resource Management*, **18** (2), 184-208

Weaver, G.R. and Trevino, L.K. (1999) 'Compliance and values oriented ethics programs: Influences on employees' attitudes and behavior', *Business Ethics Quarterly*, **9** (2) 315-35.

Wilson, A.L. (2008) *Financial Management for Health-System Pharmacists*, ASHP publications.

Yisa, S.B., Holt, G.D. and Zakeri, M. (2000) 'Factors affecting management motivation in the Iranian construction industry: a survey of site managers', paper presented to *Proceedings of the 16th annual conference, ARCOM Association of Researchers in Construction Management*, 6–8 September.

Zakeri, M., Olomolaiye, P.O., Holt, G.D. and Harris, F.C. (1996) 'A survey of constraints on Iranian construction operatives' productivity', *Construction Management & Economics*, **14** (5), 417-426

Permissions

All chapters in this book were first published in AJCEB, by UTS ePress; hereby published with permission under the Creative Commons Attribution License or equivalent. Every chapter published in this book has been scrutinized by our experts. Their significance has been extensively debated. The topics covered herein carry significant findings which will fuel the growth of the discipline. They may even be implemented as practical applications or may be referred to as a beginning point for another development.

The contributors of this book come from diverse backgrounds, making this book a truly international effort. This book will bring forth new frontiers with its revolutionizing research information and detailed analysis of the nascent developments around the world.

We would like to thank all the contributing authors for lending their expertise to make the book truly unique. They have played a crucial role in the development of this book. Without their invaluable contributions this book wouldn't have been possible. They have made vital efforts to compile up to date information on the varied aspects of this subject to make this book a valuable addition to the collection of many professionals and students.

This book was conceptualized with the vision of imparting up-to-date information and advanced data in this field. To ensure the same, a matchless editorial board was set up. Every individual on the board went through rigorous rounds of assessment to prove their worth. After which they invested a large part of their time researching and compiling the most relevant data for our readers.

The editorial board has been involved in producing this book since its inception. They have spent rigorous hours researching and exploring the diverse topics which have resulted in the successful publishing of this book. They have passed on their knowledge of decades through this book. To expedite this challenging task, the publisher supported the team at every step. A small team of assistant editors was also appointed to further simplify the editing procedure and attain best results for the readers.

Apart from the editorial board, the designing team has also invested a significant amount of their time in understanding the subject and creating the most relevant covers. They scrutinized every image to scout for the most suitable representation of the subject and create an appropriate cover for the book.

The publishing team has been an ardent support to the editorial, designing and production team. Their endless efforts to recruit the best for this project, has resulted in the accomplishment of this book. They are a veteran in the field of academics and their pool of knowledge is as vast as their experience in printing. Their expertise and guidance has proved useful at every step. Their uncompromising quality standards have made this book an exceptional effort. Their encouragement from time to time has been an inspiration for everyone.

The publisher and the editorial board hope that this book will prove to be a valuable piece of knowledge for researchers, students, practitioners and scholars across the globe.

List of Contributors

Mehrdad Arashpour
RMIT University, Australia

Ron Wakefield
RMIT University, Australia

Nick Blismas
RMIT University, Australia

Eric Wai Ming Lee
City University, Hong Kong

Grace Ding
University of Technology Sydney, Australia

Malcolm Abbott
Swinburn University of Technology, Australia

Chris Carson
Unitech Institute of Technology, New Zealand

Suhair Zaid Alkilani
University of Technology Sydney, Australia

Julie Jupp
University of Technology Sydney, Australia

Anil Sawhney
Indian Institute of Technology, India

Michael Johnston
School of Design, Queensland University of Technology, Australia

Mirko Guaralda
School of Design, Queensland University of Technology, Australia

Sukanlaya Sawang
QUT Business School, Queensland University of Technology, Australia

Kevin McLain
Missouri Department of Transportation, USA

Douglas Gransberg
Iowa State University, USA

Micheal Loulakis
Capital Project Strategies, LLC, USA

Claus Nesensohn
Liverpool John Moores University, UK

David Bryde
Liverpool John Moores University, UK

Edward Ochieng
Liverpool John Moores University, UK

Damian Fearon
Liverpool John Moores University, UK

Graham Brewer
University of Newcastle, Australia

Thayaparan Gajendran
University of Newcastle, Australia

Goran Runeson
University of Western Sydney, Australia

Imriyas Kamardeen
University of New South Wales, Australia

Rosemarie Rusch
Bond University, Australia

Rick Best
Bond University, Australia

Joseph Ssegawa
University of Botswana, Botswan

De-Graft Owusu-Manu
Kwame Nkrumah University of Science and Technology, Ghana

Edward Badu
Kwame Nkrumah University of Science and Technology, Ghana

Gary D. Holt
University of Central Lancashire, UK

David Edwards
Birmingham City University, UK

Che Maznah Mat Isa
Universiti Teknologi MARA, Malaysia

Hamidah Mohd Saman
Universiti Teknologi MARA, Malaysia

Christopher Nigel Preece
Universiti Teknologi Malaysia, Malaysia

Mohammed K. Fageha
University of Melbourne, Australia

Ajibade A. Aibinu
University of Melbourne, Australia

Xosé Luís Fernández-López
University of Cantabria, Spain

Pablo Coto-Millán
University of Cantabria, Spain

Peter Livesey
University of Technology Sydney, Australia

Justin Bold
Bold Consulting Pty Ltd, Australia

Mostafa Babaeian Jelodar
Department of Civil and Environmental, Faculty of Engineering, University of Auckland, New Zealand

Tak Wing Yiu
Department of Civil and Environmental, Faculty of Engineering, University of Auckland, New Zealand

Suzanne Wilkinson
Department of Civil and Environmental, Faculty of Engineering, University of Auckland, New Zealand